AF342134

The AVIATION
Fact Book

Compiled and Edited by Daryl Murphy

- ◆ **General Aviation manufacturing and use statistics**
- ◆ **Technical data--conversions, glossary of terms**
- ◆ **Histories of the world's prominent manufacturers**
- ◆ **Specifications and Performance of current world aircraft**
- ◆ **Pilot and airport data, FAA facilities**
- ◆ **All registered U.S. aircraft by make and model**
- ◆ **Directory of clubs and organizations, Air Force bases**
- ◆ **Halls of Fame--awards and trophies**
- ◆ **Aviation's historical timeline**
- ◆ **A directory of the world's airlines**
- ◆ **History--speed, altitude and distance**
- ◆ **Air Racing from 1908 to 1996**
- ◆ **American military aircraft 1908 to 1996**
- ◆ **The aircraft of World War II**
- ◆ **Aviation anniversaries throughout the year**

McGraw-Hill

New York   San Francisco   Washington, D.C.   Auckland   Bogotá   Caracas   Lisbon
London   Madrid   Mexico City   Milan   Montreal   New Delhi   San Juan
Singapore   Sydney   Tokyo   Toronto

The AVIATION
Fact Book

Compiled and Edited by Daryl Murphy

Introduction

Aviation is generally considered as a tree with three main branches: civil or commercial, military and general aviation. Civil or commercial is concerned solely with flight for profit, whether it's passengers, cargo, or sundry other aerial specialities. Military aviation does the same duty as commercial aviation, but in the service of government. General Aviation is a catch-all title for an industry which is easier to describe by what it is not. It is not military aviation, nor is it commercial aviation--but is nearly everything else. It is private and corporate flying, cargo hauling, aerial farming, homebuilts, antiques and flying for sport, and nearly all facets of it can be for pleasure as well as for business.

The collection of diverse information contained in *The Aviation Fact Book* had its beginnings some years ago when I was Editor of *General Aviation News*. In addition to regularly receiving a great deal of industry and governmental statistical information and data, the publication (which was founded in 1948) had its roots in postwar civil aviation, and was often called upon as a source by related businesses, historians, and the media.

In order to answer these queries efficiently, I created a personal data base containing the most popular information such as individual model specifications, performance, numbers of various aircraft in use, safety statistics, and industry manufacturing and sales figures. That collection of information has grown into this 300-page Fact Book.

There is something on these pages for everyone with an interest in aviation, from the businessman to the casual hobbyist, and the information has been researched and compiled from many public and private sources, governmental and commercial, and to the best of our knowledge is accurate as of the date of publication. In order to provide the most current statistics on aviation, the Fact Book is published as soon as possible after annual data is available. In the case of government reports, annual figures are not available until late the following year. Therefore, this edition reflects statistics for the year ending December 31, 1996.

Daryl Murphy
Editor

McGraw-Hill

A Division of The McGraw·Hill Companies

The sponsoring editor for this book was Shelley Ingram Carr.

Printed and bound by R. R. Donnelley & Sons Company.

 This book is printed on recycled, acid-free paper containing a minimum of 50% recycled, de-inked fiber.

McGraw-Hill books are available at special quantity discounts to use as premiums and sales promotions, or for use in corporate training programs. For more information, please write to the Director of Special Sales, McGraw-Hill, Inc., 11 West 19th Street, New York, NY 10011. Or contact your local bookstore.

FIFTY YEARS AGO, aviation in 1948
* X-1 sets world speed and altitude marks * Massive Airlift is lifeline for Berlin
* Transatlantic records * Orville Wright dies; Wright Flyer returns to US
* Low-cost 4-place Cessna debuts * Jet engine inventor finally gets his due
* New auto lengine utilizes aero engine * New York Idlewild Airport opens
* Johnson and P-51 top Thompson Trophy * Point-contact transistor may replace radio tubes
* Boeing's L-15 Scout slated for Army use * Texas Senator uses helicopter in campaign
* Air safety questioned in wake of crashes * Convair testing improved B-36
* Airline offers "coach" fare * US military services set up joint transport unit
* Quad-jet fighter may be Curtiss' last airplane * Auto and aircraft price list
* Two jet prototypes fly * Engine builders crank out new postwar models
* Mach 1 achieved in fighter aircraft * Air Force changes designations of aircraft
* British break sound barrier * Wee Bee world's smallest airplane
* When they say "burn off" fog, they really mean it
* Viscount is first purpose-designed turboprop airliner * In the news

About the Author

Daryl Murphy has been a writer for nearly 40 years, and involved in General Aviation since 1965, when he began a seven-year career at Cessna Aircraft Co. in Wichita, Kansas, followed by a decade in the advertising business. In 1985, he was appointed editor of *General Aviation News*. When that publication was merged with *Western Flyer* in 1991, Murphy continued to write his popular "Gadding with GAN" column and serve on the staff of *EAA Today*, the daily newspaper published during the annual Oshkosh convention. In addition, he was Southwest Correspondent for *Aviation International News* and staff writer for *HAI Convention News* and *NBAA Convention News*.

In 1991, Murphy edited and updated the 3rd edition of Paul Garrison's popular book, *Flying VFR in Marginal Weather*, for TAB Publishing,. Two years later, his first automotive work, *Carrera Panamericana: the history of the Mexican Road Race*, was published by Motorbooks International, and later a collection of his GAN columns, *Generalities, Truths, & Assorted Fables*, was released by Jones Publishing. Murphy was retained by Microsoft Corp. in 1995 as senior writer for the "World of Flight" CD-ROM program.

While his license is limited to single-engine land operation, much of the 1,000 hours Murphy has accumulated in his 30 years of flying has been for the purpose of writing pilot reports, and he has flown (with properly rated safety pilots) 38 types of civil and military aircraft, including helicopters, turbines, seaplanes, and the Goodyear blimp.

Daryl Murphy lives in Irving, Texas, with his wife, Kathleen, who is a computer teacher in the Dallas schools. During the summer months (except during the annual Oshkosh fly-in), they can usually be found in central Kansas, working on the ongoing restoration of the 120-year-old farm home his great-grandfather built.

I. GENERAL AVIATION AIRCRAFT

ANNUAL SHIPMENTS OF NEW U.S. GENERAL AVIATION AIRCRAFT

Source: GAMA

Year shipped	Units	Single-Engine	Multi-Engine	Turbo-prop	Turbo-jet	No. co.	Net $ (million)
1946	35,000 (est.)						110.0
1947	15,594					15	57.9
1948	7,037					12	32.4
1949	3,405					11	17.7
1950	3,386					13	19.1
1951	2,302					12	16.8
1952	3,058					8	26.8
1953	3,788					7	34.4
1954	3,071					7	43.4
1955	4,434					7	68.2
1956	6,738					8	103.7
1957	6,118					9	99.6
1958	6,414					10	101.9
1959	7,689					9	129.8
1960	7,588					8	151.2
1961	6,778					8	124.3
1962	6,697	5,690	1,007			7	136.8
1963	7,569	6,248	1,321			7	153.4
1964	9,336	7,718	1,606	9	3	8	198.8
1965	11,852	9,873	1,780	87	112	8	318.2
1966	15,768	13,250	2,192	165	161	10	444.9
1967	13,577	11,557	1,773	149	98	14	359.6
1968	13,698	11,398	1,959	248	93	14	425.7
1969	12,457	10,054	2,078	214	111	14	584.5
1970	7,292	5,942	1,159	135	56	13	337.0
1971	7,466	6,287	1,043	89	47	11	321.5
1972	9,774	7,913	1,548	179	134	12	557.6
1973	13,646	10,788	2,413	247	198	12	828.1
1974	14,166	11,579	2,135	250	202	12	909.4
1975	14,056	11,441	2,116	305	194	12	1,032.9
1976	15,451	12,785	2,120	359	187	12	1,225.5
1977	16,904	14,054	2,195	428	227	12	1,488.1
1978	17,811	14,398	2,634	548	231	12	1,781.2
1979	17,048	13,286	2,843	639	282	12	2,165.0
1980	11,877	8,640	2,116	778	326	12	1,486.2
1981	9,457	6,608	1,542	918	289	12	2,919.9
1982	4,266	2,871	678	458	259	11	1,999.5
1983	2,691	1,811	417	321	142	10	1,469.5
1984	2,431	1,620	371	271	169	9	1,680.7
1985	2,029	1,370	193	321	145	9	1,430.6
1986	1,495	985	138	250	122	9	1,261.9
1987	1,085	613	87	263	122	9	1,363.5
1988	1,143	628	67	291	157	9	2,918.4
1989	1,535	1,023	87	268	157	11	2,803.9
1990	1,144	608	87	281	168	14	2,007.5
1991	1,021	564	49	222	186	14	1,968.0
1992	941	552	41	177	171	16	1.839.6
1993	964	516	39	211	198	16	2,143.8
1994	928	444	55	207	222	13	2,357.1
1995	1,077	515	61	255	246	13	2,841.9
1996	1,130	530	70	289	241	13	3,142.2
1997*	237	215	7	39	63	9	885.6

*First Quarter

Annual new U.S. manufactured General Aviation unit shipments

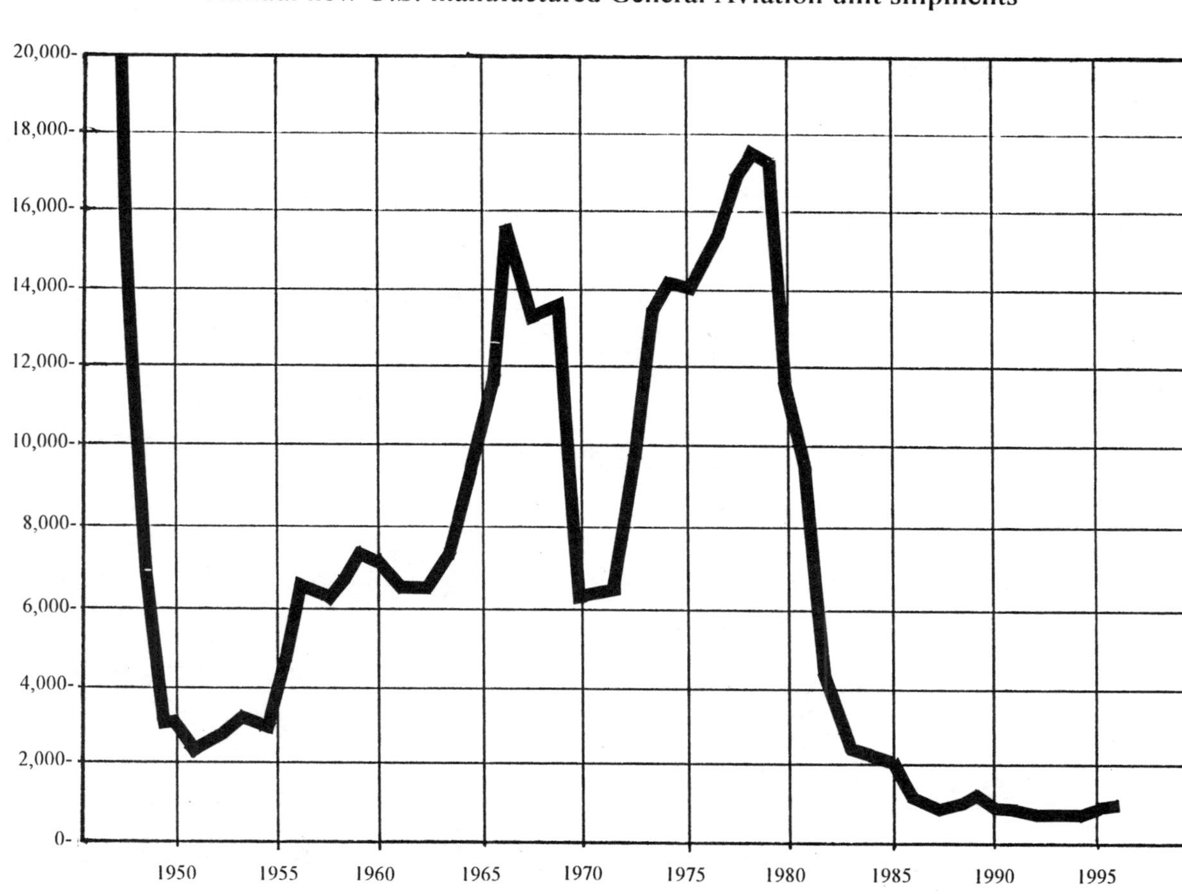

ESTIMATED VALUE OF
U.S. GENERAL AVIATION AIRCRAFT DELIVERED
($ million)

Source: GAMA

Year	Single-engine	Multi-engine	Total piston	Turbo-prop	Turbo-jet	Total turbine	Total billing
1975	290	280	570	180	281	461	1,031
1976	376	320	696	238	293	531	1,227
1977	473	390	863	296	329	625	1,488
1978	516	493	1,009	394	378	772	1,781
1979	523	555	1,078	548	540	1,088	2,166
1980	391	403	794	875	816	1,691	2,485
1981	327	348	675	1,120	1,125	2,245	2,920
1982	200	220	420	590	990	1,580	2,000
1983	145	115	260	460	750	1,210	1,470
1984	147	133	280	436	966	1,402	1,682
1985	126	68	194	524	713	1,237	1,431
1986	80	43	123	430	709	1,139	1,262
1987	80	18	98	477	789	1,266	1,364
1988	66	12	78	596	1,242	1,838	1,916
1989	104	24	128	524	1,149	1,673	1,801
1990	68	24	92	644	1,272	1,916	2,008
1991	--	--	93	527	1,348	1,875	1,968
1992	--	--	96	460	1,284	2,068	1,839
1993	--	--	76	595	1,473	2,068	2,143
1994	--	--	94	595	1,681	2,276	2,357
1995	--	--	123	653	2,066	2,719	2,842
1996	--	--	146	734	2,247	2,981	3,127

Estimated value of U.S. General Aviation aircraft delivered by type

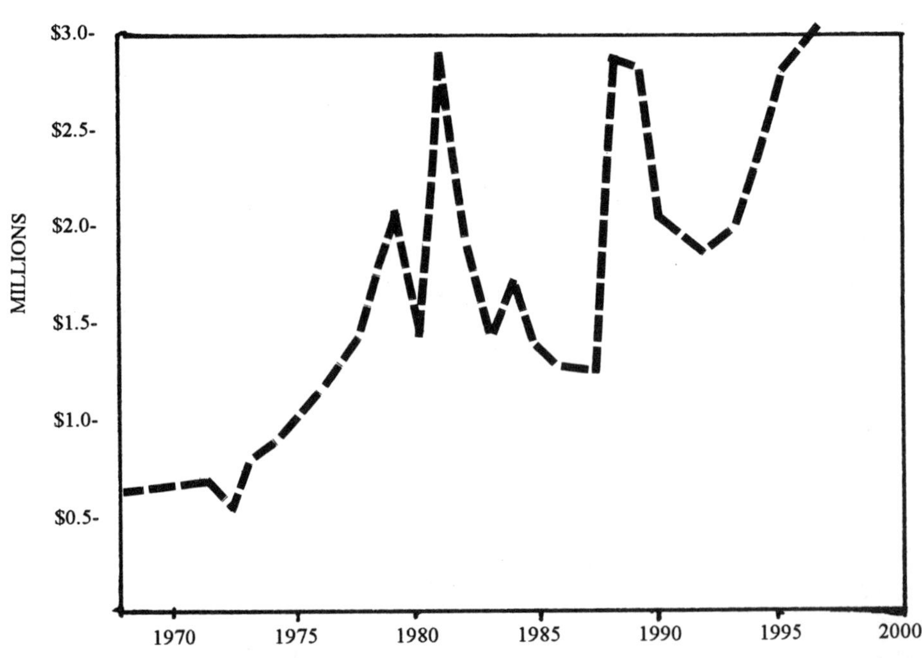

GAMA GENERAL AVIATION AIRCRAFT EXPORTS

Source: GAMA

Year	Units exported	% of total production	Factory net billings	% of total dollars
1965	2325	19.6	$61,200,000	19.2
1966	2903	18.4	$75,400,000	16.0
1967	3035	22.4	$76,500,000	21.3
1968	2803	20.5	$91,500,000	21.5
1969	2623	21.1	$107,100,000	18.3
1970	2170	29.8	$98,900,000	29.3
1971	1854	24.8	$95,600,000	29.7
1972	2254	23.1	$137,900,000	24.7
1973	3530	25.9	$230,200,000	27.8
1974	4248	30.0	$287,500,000	31.6
1975	3512	25.0	$308,100,000	29.8
1976	3539	22.9	$331,200,000	27.0
1977	3611	21.4	$354,500,000	23.8
1978	3612	20.3	$486,700,000	27.3
1979	3995	23.4	$600,900,000	27.8
1980	3555	29.9	$756,400,000	30.4
1981	2270	24.0	$749,000,000	25.7
1982	1162	27.2	$650,200,000	32.5
1983	513	19.1	$316,500,000	21.5
1984	334	13.7	$260,700,000	15.5
1985	354	17.4	$230,000,000	16.1
1986	441	29.5	$343,600,000	27.2
1987	439	40.5	$469,300,000	34.3
1988	425	37.2	$626,800,000	32.7
1989	566	36.9	$587,000,000	32.5
1990	442	38.6	$843,800,000	42.0
1991	385	37.7	$815,300,000	41.4
1992	363	40.4	$626,200,000	33.0
1993	349	36.2	$856,800,000	40.0
1994	277	29.8	$684,200,000	29.0
1995	310	28.8	$816,000,000	28.7
1996	345	30.5	$903,000,000	28.7

GENERAL AVIATION AIRCRAFT

Source: GAMA

GENERAL AVIATION AIRCRAFT EXPORTS BY TYPE

Year	Single-engine	Multi-engine	Turboprop	Jet
1972	1715	455	55	29
1973	2674	732	58	66
1974	3371	732	75	70
1975	2680	644	122	66
1976	2704	669	114	52
1977	2835	594	126	56
1978	2712	652	166	82
1979	2942	774	181	98
1980	2565	635	245	110
1981	1546	363	259	102
1982	718	227	135	82
1983	298	119	66	30
1984	199	79	25	31
1985	208	69	49	28
1986	272	69	68	32
1987	252	60	78	49
1988	220	52	91	62
1989	385	46	78	57
1990	224	57	86	91
1991	204	25	74	79
1992	196	16	90	51
1993	149	23	109	68
1994	84	42	84	67
1995	130	30	85	70
1996				

WORLDWIDE SHIPMENTS OF NEW CIVIL AIRCRAFT BY MANUFACTURER

	1987	1988	1989	1990	1991	1992	1994	1995	1996
Aerospatiale	81	97	121	204	203	163			
American Champ.						22	46		53
ATR						67			
Aviat			75	68	71		47	42	56
Beech	314	372	371	432	402	348	317	363	382
Bellanca							2	1	2
British Aerospace	100	102	118	108	103	70			
Britten-Norman	29	15	9	19	12	17			
Canadair	13	22	20	28	18	25			
Cessna	187	161	183	171	176	140	172	200	229
Classic							4	7	6
Commander							22	25	15
Dassault	49	43	32	33	26	19			
De Havilland	30	46	54	66	58	33			
Embraer	146	84	142	129	87	54			
Fairchild	39	29	13	14	10	14	16	7	7
Gulfstream	30	51	40	34	29	25	22	26	27
Israel Aircraft	5	8	11	9	11	6			
Lake	23	28	23	17	11	9			
Learjet	16	23	25	25	25	23	36	43	36
Maule	54	55	35	28	66	33	65	68	63
Mooney	143	142	143	147	88	69	71	84	73
Piaggio				1	6	7			
Piper	282	282	621	178	41	85	132	165	183
Short			NA	7	15				
TBM			7	7	23	30			
Total*	1556	1567	2036	1722	1548	1288	928	1077	1130

1996 SHIPMENT BY MODEL

Source: GAMA, individual manufacturers

Air Tractor
AT-401B..............................16
AT-402,502.........................91

American Champion
Adventurer 7GCAA.................1
Aurora 7ECA.........................2
Scout 8GCBC.......................7
Super Decathalon................26
Citabria Explorer.................17

Aviat
S-2B Pitts...........................10
A-1 Husky..........................46

Beech (Raytheon) Aircraft
BE-F33A Bonanza...............8
BE-A36 Bonanza................83
BE-36TC Bonanza.............14
BE-58 Baron......................44
BE-C90B King Air............38
BE-200 Super King Air.....33
BE-350 Super King Air.....27
BE-1900D Airliner......... .. 69
Starship 2000A....................8
Beechjet 400A....................29
Hawker 800XP..................26
Hawker 1000.......................3
King Air RC-12P/C012......16
400T Jayhawk....................35

Bellanca Aircraft
Super Viking 17-30A...........2

Cessna Aircraft
CE-208 Caravan I...............13
CE-208B Caravan IB..........94
CE-525 CitationJet.............44
CE-560 Citation V..............52
CE-650 Citation VII...........19
CE-750 Citation X..............7
OT-47B................................4
UC-35A................................1

Classic Aircraft
YMF-5 Waco Super...........6

Commander Aircraft
Commander 114B...............7
Commander 114TC............5
Commander 114AT............3

Fairchild Aircraft
Metro 23.............................7

Gulfstream Aerospace
Gulfstream IV....................24
Gulfstream V.......................3

Learjet Corp.
LR-31A..............................12
LR-60................................22

Maule Air, Inc.
MX-7-160...........................3
MX-7-180A.........................5
MX-7180B...........................8
MX-7-180C..........................1
MXT-7-180........................18
MXT-7-180A........................6
M-7-235B...........................11
MT-7-235............................4
M-7-235C............................7

Mooney Aircraft Corp.
M-20J MSE........................25
M20R Ovation...................33
M-20M TLS.......................15

New Piper Aircraft Corp.
PA-28-161 Warrior III..........5
PA-28-181 Archer III..........45
PA-28R-201 Arrow..............7
PA-32R-301 Saratoga II 43
PA-46-350P Mirage.............57
PA-44-180 Seminole............8
PA-34-220T Seneca IV.......18

General Aviation Manufacturers Association (GAMA)

General Aviation Manufacturers Assn.
1400 K St. NW, Suite 801, Washington, DC 20005-2485
(202) 393-1500; fax (202) 842-4063; Internet: http://www.generalaviation.org

GAMA EXECUTIVE COMMITTEE CHAIRMEN
CHAIRMAN OF THE BOARD: **Fred A. Breidenbach**, President and COO, Gulfstream Aerospace Corp.
VICE CHAIRMAN OF THE BOARD: **Arthur E. Wegner**, Chairman and CEO, Raytheon Aircraft Co.
INTERNATIONAL AFFAIRS COMMITTEE: **Peter J. Beucher**, General Manager, Aircraft Products, PPG
Industires, Inc.
FLIGHT OPERATIONS POLICY COMMITTEE: **Michael A. Smith**, Vice President and General
Manager, Honeywell, Inc.,
PRODUCT LIABILITY COMMITTEE: **Arthur R. Disbrow**, President, Hartzell Propeller, Inc.
PUBLIC AFFAIRS COMMITTEE: **Caroline Daniels**, CEO, Aircraft Technical Publishers
SAFETY AFFAIRS COMMITTEE: **Charles M. Suma**, President and CEO, The New Piper Aircraft, Inc.
TECHNICAL POLICY COMMITTEE--**Greg Summe**, President, AlliedSignal Aerospace, Engines
GA TEAM 2000, **Edward W. Stimpson**, Vice Chairman, GAMA

GAMA STAFF
PRESIDENT: **Edward M. Bolen**
VICE PRESIDENT, ENGINEERING AND MAINTENANCE: **William H. Schultz**
VICE PRESIDENT, OPERATIONS: **Ronald L. Swanda**
MANAGER OF TECHNICAL AFFAIRS AND OPERATIONS: **Molly Martin Pearce**
DIRECTOR OF ADMINISTRATION: **Darawan M.J. Gideos**
DIRECTOR OF GOVERNMENT AND INTERNATIONAL AFFAIRS: **Darby M.R. Becker**
DIRECTOR OF COMMUNICATIONS: **Shelly R. Snyder**
MANAGER, AVIATION EDUCATION/COMMUNICATIONS: **Bridgette A. Mikula**
TECHNICAL ASSISTANT: **Adam V. Welch**
ACCOUNTANT: **Jahan Ahmad**

MEMBER COMPANIES
GAMA is a national trade association representing 53 US manufacturers of general aviation aircraft, engines, avionics and related equipment. GAMA's members also operate fleets of aircraft, fixed base operations at many airports and pilot traning facilities across the United States.

Advanced Industries, Inc.
4550 Southeast Blvd.
Wichita, KS 67210

Aircraft Modular Products
4000 NW 36th Ave.
Miami, FL 33142

Aircraft Technical Publishers
101 South Hill Drive
Brisbane, CA 94005-1203

Airtechnics, Inc.
230 Ida
Wichita, KS 67211

AlliedSignal Aerospace
2525 W. 190th St.
Torrance, CA 90504

Allison Engine Co.
P.O. Box 420
Indianapolis, IN 46206-0420

Ametek, Inc. Aerospace Products
Clymer Ave.
Sellersville, PA 19860

B&D Instruments
and Avionics, Inc.
209 W. Main
Valley Center, KS 67147

Century Flight Systems, Inc.
P.O. Box 610
Mineral Wells, TX 76067

Cessna Aircraft Co.
P.O. Box 7706
Wichita, KS 67277-7706

Collins Commercial Avionics
Rockwell International
400 Collins Rd., NE
Cedar Rapids, IA 52498

Commander Aircraft Co.
7200 NW 63rd St.
Bethany, OK 73008

Cooper Industries
Champion Aviation Products
P.O. Box 910
Toledo, OH 43661

Crane Company
Hydro-Aire Division
3000 Winona Ave.
Burbank, CA 91504

The Dee Howard Co.
P.O. Box 469001
San Antonio, TX 78246

Dowty Aerospace Yakima
P.O. Box 9007
Yakima, WA 98909-9907

Dukes, Inc.
9060 Winnetka Ave.
Northridge, CA 91324

Electrosystems, Inc.
P.O. Box 273, Airport Complex
Ft. Deposit, AL 36032

ERDA, Inc.
701 Maple St.
Peshitgo, WI 54157

FlightSafety International, Inc.
Marine Air Terminal
LaGuardia Airport
New York, NY 11371-1061

GEC-Marconi Aerospace, Inc.
110 Algonquin Parkway
Whippany, NJ 07981

BFGoodrich Aerospace
250 N. Cleveland-Massillon Rd.
Akron, OH 44333-2465

Grimes Aerospace
P.O. Box 247
Dublin, OH 43078

Gulfstream Aerospace Corp.
P.O. Box 2206
Savannah, GA 31402

Hartzell Propeller, Inc.
One Propeller Place
Piqua, OH 45356-2634

Honeywell, Inc.
Space and Aviation Control
P.O. Box 29000
Phoenix, AZ 85038-9000

Jeppesen
55 Inverness Drive East
Englewood, CO 80112-5498

Kollsman
220 Daniel Webster Hwy.
Merrimack, NH 03054

Learjet, Inc.
P.O. Box 7707
Wichita, KS 67277-7707

Lucas Aerospace Power Equipment Corp.
777 Lena Drive
Aurora, OH 44202

Marathon Power Technologies
P.O. Box 8233
Waco, TX 76714-8233

Mooney Aircraft Corp.
Louis Schreiner Field
Kerrville, TX 78028

The New Piper Aircraft Corp.
2626 Piper Drive
Vero Beach, FL 32960

The Nordam Group
P.O. Box 3365
Tulsa, OK 74101

Northrop Grumman Corp.
Commercial Aircraft Div.
P.O. Box 655907
Dallas, TX 75255-5907

Parker Hannifin Corp.
Parker Bertea Aerospace
18321 Jamboree Blvd.
Irvine, CA 92715

PPG Indsustries, Inc.
P.O. Box 2200
Huntsville, AL 35804

Precision Aerospace Corp.
3220 100th St. SW #E
Everett, WA 98204

Raytheon Aircraft Corp.
P.O. Box 85
Wichita, KS 67201-0085

Sabreliner Corp.
7733 Forstth Blvd., Suite 1500
St. Louis, MO 63105-1821

SimuFlite Training International
P.O. Box 619119
DFW Airport, TX 75261

Sundstrand Corp.
Sundstrand Aerospace
P.O. Box 7002
Rockford, IL 61125

Teledyne Continental Motors
P.O. Box 90
Mobile, AL 36601

Textron Lycoming
652 Oliver St.
Williamsport, PA 17701

Trimble Navigation, Ltd.
2105 Donley Drive
Austin, TX 78746

Unison Industries, Inc.
7575 Baymeadows Way
Jacksonville, FL 32256

United Technologies Corp.
Pratt & Whitney Canada
1000 Marie-Victoria Blvd.
Longueuil, Quebec Can. J4G 1A1

Universal Avionics Systems Corp.
3260 E. Lerdo Road
Tucson, AZ 85706

Whittaker Controls, Inc.
Whittaker Aerospace
12838 Saticoy St.
North Hollywood, CA 91605

Williams International
P.O. Box 200
Walled Lake, MI 48390-0200

Woodward Governor Co,
P.O. Box 7001
Rockford, IL 61125-7001

ACTIVE GENERAL AVIATION AIRCRAFT BY TYPE AND PRIMARY USE, 1996

Source: FAA

	Corp-orate	Busi-ness	Perso-nal	Inst.	Appli-cation	Obs.	Sight-seeing	Other work	Air taxi	Other	Total active
Piston											
Single engine	997	18775	85346	12139	3724	2864	340	794	844	2975	128804
Multi-engine	2014	5493	5054	923	251	317	15	26	1772	725	16594
Turboprop	2345	560	266	27	297	26	0	59	709	238	4530
Turbojet	3444	230	119	13	0	2	3	0	326	436	4577
Rotorcraft											
Piston	10	136	345	356	313	197	42	15	0	15	1474
Turbine	882	119	257	172	289	898	52	21	458	191	3643
Gliders	0	33	1481	210	0	1	120	6	0	51	1905
Balloons	0	29	2284	219	1	30	379	208	55	165	3374
Experimental											
Homebuilt	0	270	9531	210	0	100	0	0	0	850	10964
Exhibition	0	41	1340	47	0	14	0	10	0	386	1841
Other	250	305	2464	69	42	79	35	31	102	194	3576
Total	9944	25996	108492	14389	4924	4536	990	1360	4273	6430	181341

GENERAL AVIATION HOURS FLOWN BY AIRCRAFT TYPE AND PRIMARY USE, 1996

(thousands of hours)

Source: FAA

	Corp-orate	Busi-ness	Pers-onal	Inst.	Appl-ication	Obs.	Sight-seeing	Other work	Air Taxi	Other	Total
Piston											
Single engine	200	2384	7622	3253	958	753	98	214	308	457	18886
Multi-engine	410	704	483	230	35	72	3	5	564	130	16246
Turboprop	664	76	40	9	138	6	0	16	271	136	1356
Turbojet	1134	26	23	7	0	0	1	0	118	84	1392
Rotorcraft											
Piston	1	10	21	126	75	86	10	8	1	4	341
Turbine	464	16	38	84	117	522	39	323	380	171	1992
Gliders	0	3	98	56	0	0	24	1	0	3	185
Balloons	0	1	73	6	0	1	22	9	2	20	134
Experimental											
Homebuilt	0	17	510	10	0	12	0	0	0	30	578
Exhibition	0	3	71	2	0	1	0	1	0	47	125
Other	133	43	150	6	25	13	24	4	48	13	459
Total	3007	3283	9129	3788	1349	1467	219	417	1691	1097	25447

U.S. REGISTERED CIVIL HOMEBUILT AIRCRAFT BY TYPE, 1984-94

Source: FAA

	1984	1985	1986	1987	1988	1989	1990	1991	1992	1993	1994
Fixed wing piston	13,767	13,344	14,046	14,619	15,123	16,449	17,564	18,175	19,137	22,363	22749
Fixed wing turbine	85	134	89	131	141	404	566	556	459	858	832
Rotorcraft	1,548	1,612	1.655	1,723	1,801	1,891	1,982	1,951	2,010	2,299	2283
Gliders	865	930	954	951	919	966	985	993	1,027	1,759	1792
Balloons/Dirigibles	297	318	339	365	349	451	496	496	526	501	496
Total	16,477	16,204	16,994	17,658	18,192	19,757	21,027	21,615	22,700	26,922	28152

Note: prior to 1994, Experimental and amateur aircraft included those built without a production certificate. These include research and development, amateur built, exhibition, racing, crew training, and market survey aircraft and aircraft used to show compliance with the Federal Air Regulations.

AGE OF THE GENERAL AVIATION FLEET, 1990-96
U.S. REGISTERED AIRCRAFT

Source: FAA

Aircraft type	Engine type	Seats	Average age, yr.				
			1990	1993	1994	1995	1996
Single engine	Piston	1-3	27.7	28.3	29	29	30
		4	23.4	26.3	27	27	20
		5-7	19.2	22.2	22	22	24
		8+	34.9	38.4	36	36	38
		All	24.8	28.3	27	27	29
Single engine	Turboprop	All	10.6	11.8	8	8+	8
Single engine	Turbojet	All	32.4	34.3	31	31	32
Multi-engine	Piston	1-3	28.5	32.3	32	32	32
		4	23.0	27.3	26	26	29
		5-7	21.7	26.3	25	25	27
		8+	21.8	26.3	25	25	27
		All	21.8	26.3	25	25	27
Multi-engine	Turboprop	All	14.0	17.2	16	16	17
Multi-engine	Turbojet	All	14.4	17.2	16	16	16
All			**23.8**	**27.3**	**26**	**26**	**28**

AVERAGE AGE OF THE BUSINESS AIRCRAFT FLEET (1996)

Source: NBAA/AvData, Inc.

Type	No. Aircraft	Age., yr.
Heavy jets	1881	13.96
Medium jets	2667	16.78
Light jets	4282	15.44
Heavy turboprops	220	28.07
Medium turboprops	8659	17.59
Light turboprops	114	8.08

U.S. ACTIVE GENERAL AVIATION AIRCRAFT
BY REGION & STATE 1984-1995

Source: FAA

	1984	1986	1988	1989	1992	1994	1995
Alaskan							
Alaska	7684	7557	6309	6520	6083	5479	5338
	7684	7557	6309	6520	6083	5479	5338
Central							
Iowa	3416	2683	2615	2622	2489	2028	2056
Kansas	3713	4033	3500	3466	2973	2738	2992
Missouri	4396	4192	4068	4183	3068	2822	3881
Nebraska	1805	2177	1967	1659	1721	1787	1392
	13331	13085	12150	11390	10251	9374	10321
Eastern							
Delaware	533	932	1112	15371	1047	1324	1735
DC	31	21	35	63	12	37	4
Maryland	2870	2709	3032	2897	2489	2210	2373
New Jersey	4041	4291	3755	3994	3547	2929	3384
New York	6599	6772	5804	6601	5615	5237	5209
Pennsylvania	6205	6405	5808	5956	5398	5067	5145
Virginia	3137	3225	3291	3323	2609	2795	2670
West Virginia	880	1132	1091	865	966	824	995
	24294	25487	23298	25236	21683	20419	21514
Great Lakes							
Illinois	9087	7603	7276	7581	6373	6002	6652
Indiana	3797	4212	4229	3782	3411	3560	3145
Michigan	7066	7060	6771	7875	6248	5737	6237
Minnesota	5139	4507	5040	4750	4517	3904	4252
North Dakota	1572	1615	1487	1637	1323	1143	1214
Ohio	7553	7283	7388	7896	6101	5583	6367
South Dakota	1393	1378	1192	1290	1005	919	1096
Wisconsin	4180	4180	4053	5672	3965	4037	4205
	39788	37838	37436	39383	32943	30884	33169
New England							
Connecticut	1863	1992	2231	2157	1589	1339	1561
Maine	1055	1320	1307	1473	982	835	1073
Massachusetts	3316	3248	3687	3265	2483	2605	2267
New Hampshire	1298	1443	1270	1458	1385	1026	1161
Rhode Island	396	560	486	456	382	324	334
Vermont	466	388	619	587	487	452	338
	8393	8952	9600	9396	7248	6581	6724
NW Mountain							
Colorado	5180	4273	3782	4539	3676	3604	4174
Idaho	2328	2274	1836	1998	1804	1580	2054
Montana	2472	2368	1808	2004	1923	1770	1723
Oregon	5032	4543	4126	4353	3935	4057	4562
Utah	1337	1284	1246	1388	1264	959	1218
Washington	6665	6160	6281	6439	5874	4729	5910
Wyoming	1474	1101	835	913	677	717	699
	24502	22004	19914	21634	19153	17415	20338

GENERAL AVIATION AIRCRAFT FLEET

Aircraft by region & state cont.

	1984	1986	1988	1989	1992	1994	1995
Southern							
Alabama	3234	2673	2539	3347	2873	2362	2656
Florida	12720	12882	13831	13599	11753	10574	11723
Georgia	4450	4517	4974	5340	4326	3983	3893
Kentucky	1802	1738	1740	1606	1472	1411	1398
Mississippi	2082	2060	1897	1954	1776	1661	1573
North Carolina	4412	4386	4688	5208	3729	4014	4081
Puerto Rico	422	352	72	333	370	217	166
South Carolina	1661	1787	1988	2085	1669	1566	1543
Tennessee	2884	3145	2900	3306	2763	2862	2967
	34007	33642	34629	36778	30371	28731	30007
Southwest							
Arkansas	2920	2514	2354	2879	2469	2078	2238
Louisiana	4627	3746	3250	3270	2823	2437	2454
New Mexico	2300	2302	2224	2401	2043	2180	1843
Oklahoma	5345	4147	4254	4127	2762	2858	3140
Texas	19941	19961	17424	18321	14787	13976	14705
	35341	32669	29506	30998	24884	23529	24380
Western-Pacific							
Arizona	5177	5787	5118	5234	4355	4332	4709
California	30494	30387	28910	29757	24909	21728	22357
Hawaii	463	366	600	519	372	391	460
Nevada	1823	2219	2074	2239	1684	1729	2013
	38414	38764	36702	37749	31320	28185	29547
Other U.S. Terr.	76	107	92	97	136		
Total	220943	220043	210266	219737	184434	170600	181341

ACTIVE GENERAL AVIATION AIRCRAFT

Historical and forecast units (thousands)   Source: FAA (* forecast)

	Fixed wing					Helicopter				
	S.E.	M.E.	TP	TJ	Pist.	Turb.	Exp.		Other	Total
1979	168.4	25.1	3.5	2.7	3.1	2.7			4.8	210.3
1980	168.4	24.6	4.1	3.0	2.8	3.2			4.9	211.0
1981	167.9	25.5	4.7	3.2	3.3	3.7			5.0	213.3
1982	164.2	25.0	5.2	4.0	2.4	3.7			5.2	209.7
1983	166.4	25.1	5.5	3.9	2.5	4.0			5.9	213.3
1984	171.9	25.5	5.8	4.3	2.9	4.2			6.3	220.9
1985	164.4	23.8	5.4	4.4	2.9	3.5			6.3	210.7
1986	171.8	23.9	6.0	4.5	2.9	4.0			7.0	220.0
1987	171.0	23.4	5.3	4.4	2.8	3.5			6.8	217.2
1988	164.8	22.8	5.3	4.2	2.6	3.8			6.9	210.3
1989	166.2	22.8	5.5	4.3	2.5	4.5			7.1	212.9
1990	165.0	22.6	5.6	4.4	2.6	4.7			7.0	212.3
1991	154.1	21.1	4.9	4.3	2.3	4.0			7.6	198.4
1992	143.6	18.6	4.7	4.0	2.2	3.5			7.8	184.4
1993	130.7	16.4	4.4	3.9	1.6	2.9	11.0		5.2	176.0
1994	123.3	15.6	4.2	4.1	1.4	3.0	12.9		6.2.	170.6
1995*	120.0	15.2	4.2	4.1	1.3	3.0	13.1		6.4	167.3
1996*	117.8	14.9	4.3	4.2	`.3	3.0	13.3		6.6	165.4
1997*	119.0	15.1	4.4	4.3	1.3	3.0	13.5		6.7	167.3
1998*	120.1	15.2	4.4	4.3	1.2	3.0	13.7		6.8	168.7
1999*	121.4	15.4	4.5	4.4	1.2	3.0	13.9		6.9	170.7
2000*	122.6	15.5	4.6	4.5	1.2	3.0	14.1		7.0	172.5
2001*	123.3	15.5	4.6	4.5	1.2	3.0	14.3		7.1	173.5

U.S. General Aviation Domestic Aviation Operations (x000) 1985-1995

Source: FAA

	1985	1986	1987	1988	1989	1990	1991	1992	1993	1994	1995
IFR aircraft handled at ARTCCs	8,330	8,055	8,103	8,053	8,198	7,931	7,390	7,393	7,433	7,685	7,824
Total instrument operations at FAA facilities**	16,429	16,877*	18,040*	18,377*	19,924*	19,234*	18,251*	18,335*	17,894	18,049	18,127
Total aircraft contacts at FSSs	5,826	5,383	5,209	5,008	4,873	4,795	4,376	4,130	3,703	3,509	3,204
Total airport operations at FAA control towers	37,191	37,694*	38,522*	38,252*	38,832*	40,449*	38,911*	38,355*	36,601*	34,691	
Itinerant	22,373	21,942	22,079	22,096	22,079	22,480	21,538	21,281	20,377	20,209	20,860
Local	14,818	15,158	14,752	15,407	15,674	16,690	16,040	15,664	15,851	14,484	15,066

* Includes Federal and contract operated towers and facilities
** Facilities=control towers, TRACONs, SERAPs and RAPCONs

Number of International Aircraft 1982-94

Source: ICAO

	1982	1983	1984	1985	1986	1987	1988	1989	1990	1991	1992	1993	1994
Europe	30500	30700	30200	30800	31200	31500	32000	33100	32000	31300	31100	36200	36100
Africa	4000	3700	4750	4600	4650	4600	4500	4970	4950	6200	5500	6200	6050
Middle East	500	550	550	520	540	550	600	690	670	610	580	590	580
Asia/Pacific	9500	8600	8700	8400	8500	9200	9800	10300	10200	10240	10250	11100	11500
Latin Amer.	19500	12000	14000	13700	13900	13800	13500	15200	15200	18900	18600	18800	18600
N.America	229000	231400	233000	236000	224300	224150	229320	223030	232080	224750	219000	188300	185890
TOTAL	293000	286950	291200	294020	282990	283800	289720	287290	296300	292000	285030	261190	258720

Number of International General Aviation hours flown (x000)

Source: ICAO

	1982	1983	1984	1985	1986	1987	1988	1989	1990	1991	1992	1993	1994
Europe	5850	6000	5950	6080	6400	6500	6600	6720	6870	6730	6700	7260	7240
Africa	780	750	820	790	820	800	800	820	820	700	700	800	770
Middle East	270	270	280	260	240	260	260	270	310	300	180	300	290
Asia/Pacific	2400	2400	2380	2420	2740	3060	3250	3380	3470	3500	3770	4180	4250
Latin Amer.	3890	3940	3800	3850	3380	3550	3570	3400	3300	3150	3150	3340	3280
N. America	39700	34650	33100	33920	32100	31070	31110	31610	31950	32100	26200	24220	23120
TOTAL	52940	47880	46360	46830	45870	45260	45430	46140	46710	46380	40700	40100	38950

FLIGHT PROSPECTS
By Frederick/Schneiders, Inc., October 1995
Sponosred by AEA, AOPA, GAMA, AGATE and Sporty's Academy

Demographics (%)

	Total	Male	Female
Age			
25-29	20	21	16
30-39	37	38	33
40-49	27	25	31
50-60	16	15	18
Income			
<$50,000	39	37	44
$50-75,000	26	27	23
$75-100,000	13	15	10
$100,000+	9	10	11
Marital Status			
Married	62	63	58
Divorced/widowed	16	14	21
Single	22	23	21
Have children	52	50	55
Self-employed	27	29	22

Do you have a romantic or a practical reason to learn to fly? (%)

	Rom.	Prac.
Male	49	37
Female	47	49
Age		
25-29	50	32
30-39	49	44
40-49	45	41
50-60	51	38
Income		
<$50,000	51	35
$50-75,000	58	38
$75-100,000	37	52
$100,000+	51	38
Married	47	42
Unmarried	52	37
Self-employed	51	43

Are you more likely to buy or rent? (%)

	Buy	Rent
Male	29	66
Female	19	71
Age		
20-29	33	63
30-39	26	65
40-49	26	68
50-60	22	76
Income		
<$50,000	24	69
$50-75,000	25	72
$75-100,000	26	66
$100,000+	45	50
Self-employed	37	57
Likely student	30	65

Proximity of nearest flight training (%)

<5 miles	20
5-10 miles	31
10-20 miles	27
20-50 miles	16
50+ miles	4
Don't know	2

Time and Cost Estimates (%)

	Total	Male	Female
Training time			
<6 months	29	30	27
6-12 months	35	36	31
1-2 years	23	22	25
2+ years	3	6	12
Training cost			
<$1,000	10	10	9
$1-3,000	28	28	27
$3-5,000	28	28	28
$5,000+	28	29	28
Hourly rental			
<$50	21	24	12
$50-100	30	30	28
$100-200	23	24	20
$200+	17	12	28
Used airplane			
<$10,000	13	13	14
$10-30,000	41	44	33
$30-60,000	16	15	16
$60,000+	18	19	13

Factors that would improve your interest in learning to fly (%)

	Total	Male	Female
Fly twice the speed of driving	44	43	45
Weekend adventures	43	41	50
License in 6 months	42	41	44
Mentors available	42	41	44
Entertain family/friends	42	41	45
Rent for $50/hr.	41	37	52
Recreation license for less	34	34	34
Make friends	32	30	36
Safer than biking	32	31	35
Airplane clubs	32	33	31
High self-esteem	29	28	32
Career	23	24	21
Learning cost $3,500	21	20	26
Used plane cost $30,000	18	18	16

Chances of beginning flight training in next 5 years (%)

Region	Likely	Very likely
Northeast	34	13
Midwest	43	20
South	42	13
West	62	19
Male	46	15
Female	42	18
Age		
25-29	48	17
30-39	48	17
40-49	45	16
50-10	343	12
Income		
<$50,000	38	11
$50-75,000	45	18
$75-100,000	55	17
$100,000+	49	19
Married	42	16
Unmarried	52	17
Self-employed	53	18

First interest in flying (%)

	Total	Male	Female
Child	40	43	40
Teenager	34	35	32
Adult	25	22	34

Who first interested you in flying? (%)

	Total	Male	Female
Friend	26	26	28
Relative	20	18	27
Other	37	39	30

Reason for wanting to learn to fly (%)

	Total	Male	Female
Fun	35	33	40
Always wanted to	25	26	22
Adventure	12	11	15
Transportation	5	4	6
Career	3	3	2

Characteristics which flight training prospects use to describe themselves (%)

	Total	Male	Female
Independent	71	71	71
Busy	68	70	63
Adventurous	67	69	63
Take-charge	66	69	60
Competent	64	66	59
Family-oriented	58	58	59
Happy	57	57	56
Secure	56	59	45
Technical	55	60	41
Risk-taker	53	54	52
Successful	50	54	40
Bored	18	15	17

Participation in other activities (%)

	Total	Male	Female
Bicycling	64	62	70
Fishing	59	64	46
Boating	55	56	51
Golf	40	45	26
Backpacking	39	39	37
Water skiing	35	38	27
Tennis	31	31	32
Motorcycling	29	32	19
Scuba diving	29	23	18
Mountain climbing	21	21	20
RV touring	29	18	19
Auto racing	18	19	14
Skydiving	9	8	9
Ballooning	5	4	8
Hang gliding	4	4	4

WORLD AIRCRAFT MANUFACTURERS

Advanced Aerodynamics & Structures
10703 Vanowen St.
N. Hollywood, CA 91605
(818) 753-1888

Aermacchi S.p.A.
C.P. 246 via Sanvito 80
I-21100 Varese, Italy
0332 254111

Aerospatiale
37, Blvd. de Montmorency
F-75781
Paris Cendex 16, France
1 422 42424

Agusta S.p.A.
21 Via Caldera
I-20153 Milan, Italy
0332 254111

3050 Red Lion Road
Philadelphia, PA 19114
(215) 281-1400

Air Tractor, Inc.
P.O. Box 485
Olney, TX 76374
(817) 564-5616

Allison Engine Company
P.O. Box 420
Indianapolis, IN 46206
(317) 230-2000

Alturdyne
8050 Armour St.
San Diego, CA 92111
(619) 565-2131

American Champion Aircraft
P.O. Box 37
Rochester, WI 53167
(414) 534-6315

American Eurocopter Corp.
2701 Forum Drive
Grand Prairie, TX 75053
(214) 641-0000

Angel Aircraft Corp.
1410 Arizona Place, SW
Orange City, IA 51041
(712) 737-3344

Arctic Aircraft Co.
Box 6-141
Anchorage, AK 99502
(907) 243-2580

Aviat, Inc.
P.O. Box 1149
Afton, WY 83110
(307) 886-3151

Avions Marcel Dassault-Breguet
27, rue du Professeur Pauchet
F-92420
Vaucresson, France
1 474-17921
Dassault-Falcon Jet Corp.
Teterboro Airport, Box 2000
South Hackensack,, NJ 07606
(201) 440-6700

Ayres Corporation
P.O. Box 3090
Albany, GA 31708
(912) 883-1440

Bell Helicopter Textron
P.O. Box 482
Fort Worth, TX 76101
(817) 280-2011

Bellanca, Inc.
P.O. Box 964
Alexandria, MN 56308
(612) 762-1501

Boeing Commercial Airplane Company
P.O. Box 3707
Seattle, WA 98124
(206) 655-2121

Boeing Military Airplane Company
P.O. Box 7730
Wichita, KS 67277-7730
(316) 526-2121

Bombardier, Inc.
800 Rene-Levsque Blvd. West
Montreal, Quebec H3B 1Y8
(514) 861-9481

British Aerospace PLC
11 Strand
London, England WC2N
01 930-1020
13850 McLearen Rd.
Herndon, VA 22071
(703) 435-9100

Canadian Airmotive, Inc.
7400 Wilson Ave.
Delta, BC V4G 1E5
(604) 940-9378

Canadair, Inc.
Box 6067, Station A
Montreal, PQ, Canada H3C 3G9
8 Griffin Rd.
North Windsor, CT 06095
(203) 688-7767

Cessna Aircraft Company
P.O. Box 7704
Wichita, KS 67277
(316) 941-6488

CFE Co.
P.O. Box 5274
Phoenix, AZ 85010
(602) 231-7722

Cirrus Design Corp.
4515 Taylor Cr.
Duluth, MN 55811
(218) 727-2737

Commander Aircraft Co.
7200 NW 63rd, Hangar 8
Bethany, OK 73008
(405) 495-8080

de Havilland of Canada
Garratt Blvd.
Downsview, ONT, Canada M3K 1Y5

Embraer-Empresa Brasileira de Aeronautica S.A.
Box 343, Av. Brig. Faria Lima, 2170 12225
SP, Brazil
123 251378

Extra/Flugzeugbau Gmb
c/o Aero Sport, P.O. Drawer 1989
St. Augustine, FL 32085
(904) 824-6230

EMG Engineering Co.
P.O. Box 1368
Apple Valley, CA 92345
(619) 247-8519

Enstrom Helicopter Corp.
P.O. Box 277
Menominee, MI 49858
(906) 863-9971

Fairchild Dornier
22455 Davis Dr.
Sterling, VA 20164
(703) 444-8330

Fiat Aviazione S.p.A.
via Nizza 312,
I-10127 Turin, Italy
011 69311

Fisher Research Corp.
1000 Rt. 386
Scottsville, NY 144546
(716) 889-8280

Fokker Aircraft B.V.
P.O. Box 12222
NL-1100 AE Amsterdam-Zuidoost, Netherlands
020 5647015

Fuji Heavy Industries, Ltd.
7-2, Nishi-Shinjuki 1-chrome
Shinjuku-ku, Sabaru Bldg.
Tokyo, Japan 160
03 347-2525

Garrett Corporation
P.O. Box 5217
Phoenix, AZ 85010
(602) 231-1000

General Avia S.r.L
via Trieste 24
I-20096 Pioltello (Milan), Italy
2 926-6774

General Dynamics Corporation
Pierre Laclede Center
St. Louis, MO 63105
(314) 889-8200

General Electric Company
GE Aircraft Engines
1 Neumann Way
Evendale, OH 45215-6301
(513) 243-2000

Gulfstream Aerospace
P.O. Box 2206
Savannah, GA 31402
(912) 965-3000

Hawker Siddeley Group PLC
18 St. James's Square
London, England SW 1Y
01 930-6177

Hughes Aircraft Corporation
7200 Hughes Terrace
Los Angeles, CA 90045-0066
(213) 568-7200

IAME S.p.A. KFM Aircraft Motor Div.
Via Lisbona 15
I-24040 ingonia, Italy
035 883022

IHI Co., Ltd.
Shin Ohtemachi Bldg.
2-1 Ohtemachi, 2-chome, Chiyoda-ku
Tokyo, Japan 100
03 244-6496

In-Tech International, Inc.
West 7510 Hall Ave.
Spokane, WA 99204 (509) 455-6116

Israel Aircraft Industries Ltd.
Ben Gurion International Airport
70100 Tel Aviv, Israel
03 971-3111

Javelin Aircraft Co., Inc.
Municipal Airport
Augusta, KS 67101
(316) 733-1011

Kaman Corp.
P.O. Box 1
Bloomfield, CT 06002
(860) 243-7100

Kawasaki Heavy Industries, Ltd.
1-18 Nakamachi-dori
2-chome, Chuo-ku
Kobe, Japan 650-91
078 341-7731

Kestrel Aircraft Co.
P.O. Box 920960
Norman, OK 73069
(405) 573-0090

Lake Aircraft
Laconia, NH 03246
(603) 524-5868

Learjet Corporation*
P.O. Box 7707
Wichita, KS 67277-7707
(316) 946-2200

Light Helicopter Turbine Engine Co.
Paragon Bldg., Suite 400
12400 Olive Blvd.
St. Louis, MO 63141

Lockheed Corporation
4500 Park Granada Blvd.
Calabasas, CA 91399
(818) 712-2000

LTV Corporation
P.O. Box 655003
Dallas, TX 75265-5003
(214) 979-7711

Martin Marietta Corp.
6801 Rockledge Drive
Bethesda, MD 20817
(301) 897-6000

Maule Aircraft
Route 5, Box 319
Moultrie, GA 31768
(912) 985-2045

McDonnell Douglas Corp.
P.O. Box 516
St. Louis, MO 63166 (314) 232-0232

McDonnell Douglas Helicopter Co.
5000 E. McDowell Rd
Mesa, AZ 85205
(602) 891-3000

Melex USA
1221 Front St.
Raleigh, NC 27609
(919) 934-5675

Messerschmitt-Boelkow-Blohm GmbH
Box 801109
D-8000 Munich, W. Germany
089 6000

Mitsubishi Heavy Industries, Ltd.
5-1 Marunouchi 2-chome
Chiyoda-ku
Tokyo, Japan 100
03 212-3111

Mooney Aircraft Corp.
P.O. Box 72
Kerrville, TX 78028
(512) 896-6000

Mundry Aviation Ltd.
Flagler County Airport
Bunnell, FL 32110
(904) 437-9700

Northrop Grumman Corp.
Commercial Aircraft Div.
P.O. Box 655907
Dallas, TX 75255-5907

Partenavia Costruzioni Aeronautiche S.p.A.
Piazzale Techio 51A
I-80100 Naples, Italy
081 759-6311

Rinaldo Piaggio S.p.A.
via Cibario 4
I-16154 Genoa-Sestri, Italy
010 60041
1802 W. 2nd
Wichita, KS 67203
(316) 264-2414

Pilatus Flugzeugwerke AG
CH-6370
Stans, Switzerland
041 636111

Pilatus Aircraft Ltd.
3740 20th St., Suite A
Vero Beach, FL 32960
(303) 465-9099

Pilatus Britten-Norman Ltd.
Bembridge Airport
Bembridge, Isle of Wight, England PO35 5PR
0983 872511
2101 Wilson Blvd.
Arlington, VA 22201
(703) 351-6620

New Piper Aircraft Corporation
P.O. Box 1328
Vero Beach, FL 32961-1328
(305) 567-4361

Pratt & Whitney
400 Main Street East Hartford, CT 06108
(203) 565-4321

Pratt & Whitney Canada
1000 Marie Victora Blvd.
Longueuil, QE J4G 1A1
(514) 677-9411

Raytheon Aircraft Corporation
9709 East Central
Wichita, KS 67201-0085
(316) 676-7603

Robinson Helicopter
24747 Crenshaw Blvd.
Torrance, CA 90505
(213) 539-0508

Rolls-Royce PLC
Civil Engine Group
Box 31
Derby, England DE2 8BJ
0332 242424

Rotorway Aircraft, Inc.
300 S. 25th Ave.
Phoenix, AZ 85009
(602) 278-1199

Saab-Scania AB
S-581 Linkoping, Sweden
4613 180000

Sabreliner Corporation*
18118 Chesterfield Airport Road
Chesterfield, MO 63005-1121
(314) 537-3660

Schweizer Aircraft
P.O. Box 147
Elmira, NY 14902
(607) 739-3821

Sherpa Aircraft Mfg.
13000 SE Mountain Gate Rd.
Portland, OR 97236
(503) 658-7374

Shin Meiwa Industry Co., Ltd.
5-25 Kosone-cho
1-chome, Nishinomiya
Hyogo, Japan 663
0798 47 0331

Short Brothers PLC
Berkley Square House
London, England W1
01 629-9541

SIAI-Marchetti S.p.A.
Via Indipendenza 2
I-21018 Sesto Calende, Italy
0331 924421

Sikorsky Aircraft North Main Street
Stratford, CT 06601
(204-386-4000

Sino Swearingen Aircraft Co.
Sky Place Blvd.
San Antonio, TX 78216
(210) 258-3900

Socata Aircraft
7501 Pembrook Rd.
Pembroke Pines, FL 33029
(800) 999-1110

Soko Aerospace Industry
YU-79000 Mostar Yugoslavia
088 22121

Soloy Conversions, Ltd.
450 Pat Kennedy Way SW
Olympia, WA 98501
(206) 754-7000

Sport Plane Power, Inc.
3659 Arnold Ave.
Naples, FL 33942
(813) 775-2214

Taylorcraft Aircraft Corp.
P.O. Box 480
Lock Haven, PA 17745
(717) 748-8262

TBM North America
2701 Forum Dr.
Grand Prairie, TX 75053

Teledyne Continental Motors
P.O. Box 90
Mobile, AL 36601
(205) 438-3411

Textron Lycoming Recip. Engine Div.
652 Oliver Street
Williamsport, PA 17701
(717) 327-7047

Textron Lycoming Turbine Engine Div.
550 Main St.
Stratford, CT 06497
(203) 385-2754

Turbomeca
Bordes, F-64320
Bizanos, France
59 328437

Utva Aircraft Factory
Utwe Zlatokrite br 9
YU-26000 Pancevo, Yugoslavia
013 44755

United Technologies Corp.
Pratt & Whitney Canada
1000 Marie-Victorin Boulevard
Longueuil, Quebeck J4G 1A1
(514) 677-9411

Westland PLC
Yeovil, Somerset, England BA20 2YB
0935 75222

Williams International
2280 W. Maple Road
Walled Lake, MI 48088
(313) 624-5200

*Civil and military major accessory
and electronics manufacturers*

Aeroquip Corp.
300 S. East Ave.
Jackson, MI 49203-1972
(517) 787-8121

Aire-Sciences, Inc.
216 Passaic Ave.
Fairfield, NJ 70006
(201) 228-1880

Aviation Simulation Technology, Inc.
Hanscom Field-East
Bedford, MA 01730
(617) 274-6600

Bendix/King
400 North Rogers Road
Olathe, KS 66601
(913) 782-0400

Century Flight Systems
P.O. Box 610
Mineral Wells, TX 76067
(817) 325-2517

Champion Spark Plugs
P.O. Box 910
Toledo, OH 43661-0001
(419) 535-2461

Collins Avionics
400 Collins Road NE
Cedar Rapids, IA 52498
(319) 395-1000

EDO Corporation
14-04 111th St.
College Point, NY 11356-1434
(718) 445-6000

Elano Corportion
P.O. Box 217
Alpha, OH 45301-0217
(513) 427-7035

Global Wulfsberg Systems
2144 Michelson Dr.
Irvine, CA 92715
(714) 851-0119

BFGoodrich Corp. Aerospace Div.
250 N. Cleveland-Massillon Rd.
Akron, OH 44313-0501
(216) 374-2000

Goodyear Tire and Rubber Co.
1000 Tradeport Blvd., Suite 1010
Atlanta, GA 30354-2911
(404) 362-3602

Hartzell Propeller Co.
1800 Covington Ave.
Piqua, OH 45356
(512) 778-4200

Honeywell Inc.
P.O. Bopx 29000
Phoenix, AZ 85038
(602) 863-8900

Marathon Power Technologies
P.O. Box 8233
Waco, TX 76714-8233
(817) 776-0650

McCreary Tire & Rubber Co.
1600 Washington St.
Indiana, PA 15701
(412) 349-9010

Parker Hannifin Corporation
18231 Jamboree Blvd.
Irvine, CA 92715
(714) 833-3000

Precision Aerospace Corporation
3220 100th St. SW #E
Everett, WA 98204
(206) 353-8181

Sperry Commercial Flight Systems
P.O. Box 21111
Phoenix, AZ 85038
(602) 863-1030

Unison Industries, Inc.
7575 Baymeadows Way
Jacksonville, FL 32256
(904) 739-4000

Univar Aircraft Corp.
2500 Himalaya Rd.
Aurora, CO 80011
(303) 364-7661

Woodward Governor Co.
P.O. Box 7001
Rockford, IL 61125-7001
(815) 877-7441

GENERAL AVIATION AIRCRAFT
Specifications and Performance

Source: *Business & Commercial Aviation Planning & Purchasing Handbook, 1997;*
Aviation Week & Space Technology Aerospace Source Book, 1997, Manufacturers
Powerplant manufacturers: 250=Allison; AL=Lycoming; ATF=Garrett; BR=BMW/Rolls-Royce; C=General
Electric; Cont= Continental; FJ=Williams; J=General Electric; JT=Pratt & Whitney; LTS & Lyc.=Lycoming;
PT, PW & PWC=Pratt & Whitney; RR=Rolls-Royce; T=GE; TFE=Garrett; TM=Turbomeca; TPE=Garrett
** indicates aircraft not yet flown as of 1/1/97; + indicates aircraft in certification as of 1/1/97*

Model	Price, (x000)	Seats	GW, lb.	Powerplant (hp or lb. thrust)	Cruise, mph (unless noted)
AASI, Long Beach, CA					
Jetcruzer 500+	$1,300	6	5500	PT6-A-42 (1142)	315kt
Jetcruzer 650*	$1,700	13	6500	PT6-A-42 (1142)	261kt
Stratocruzer 1250-ER*	$3,500	13	12500	(2) FJ44-2 (2300#)	445kt
Costruzioni Aeronautiche G. Agusta, Milan, Italy					AGUSTA
Siai Marchetti SF260D		3	2866	Lyc. O-540/IO-540 (260)	187
Siai Marchetti SF260TP		3	2866	250-B17D (350)	216
A109 Power	$3,000	1+7	6000	2 PWC 206C (640)	156 kt
A109K2	$3,400	1+7	5997	2 TM Arriel 1K1 (737)	147 kt
Air Tractor, Olney, TX					AIR TRACTOR
AT-401B	$215.9	1	7860	PW R-1340 (600)	165
AT-402A	$369.5	1	7860	PT6A-11AG (550)	162
AT-402B	$481	1	7860	PT6A-15AG (680)	
AT-502A	$267.5	1	8500	PT6A-45R (customer-furnished)	180
AT-502B	$491	1	8500	PT6A-15AG (680)	180
AT-503A	$553.9	1+1	10,500	PT6A-34AG	225
AT-602	$699.5	1		PT6A-60AG (1050)	
AT-802A	$894	1	16,000	PT6A-67AG (1350)	191
AT-8002AF	$1,015	1	16,000	PT6A-67AG (1350)	
AmericanChampion, Rochester, WI					
Aurora	$57.9	2	1650	Lyc. O-235-K2C (118)	101
Super Decathlon	$85.9	2	1800	Lyc. AEIO-360-H1B (180)	131
Scout	$83.9	2	2150	Lyc. O-360-C2E/T1G (180)	109
Citabria Explorer 7GCBC	$67.9	2	1800	Lyc. O-320-B2B	114
Adventure+	$65.9	2	1650	Lyc. O-320-B2B (160)	147
Angel Aircraft, Orange City, IA					ANGEL
Model 44	$599	8	5800	2 Lyc. IO-540 (300)	175
Astra Jet Corp.					
1125 Astra SP	$8,752	9	16,100	2 TFE 731-3A-200G (3,700#)	M0.85
Astra SPX	$9,995	9	24,650	2 TFE 731-40R (4,700#)	M0.87
Galaxy*		19	33,400	2 PW306A	M0.85
Aviaexport USSR					
Sukhoi Su-26M	$208	1	1841	Vedenyev M-14	280 mph (Vne)
Sukhoi Su-29	$175	2	1800	Lyc. O-360-C1G	140
Aviat, Inc., Afton, WY					
Husky A-1	$92.5	2	1800	Lyc. O-360-C1G (180)	126
Pitts S2B	$137	2	1700	Lyc. AEIO-540-D4A5 (260)	134
Pitts S1T	$99.9	1	1150	Lyc. AEIO-360-A1E (200)	134
Pitts S2S	$115.9	1	1500	Lyc. AEIO-540-D4A5 (260)	134

Model	Price, (x000)	Seats	GW, lb.	Powerplant (hp or lb. thrust)	Cruise, mph (unless noted)
Avions Pierre Robin, Darios, France					ROBIN
ATL		2	1,278	Limbach L2000 (70)	
DR 400/120 Dauphin 2+2		4	1,984	Lyc O-235-L2A	130
DR 400/140B Dauphin		4	2,205	Lyc O-320-D2A	143
DR 400/160 Major		4	2,315	Lyc O-320	176
DR 400/180 Regent		4	2,425	Lyc O-360-A3A	150
DR 400/180R Remorquer		4	2,205	Lyc O-360-A3A	146
DR 400/200R		4	2,425	Lyc IO-360-A186	166
R200		2	1,720	Lyc O-235-L2A	160
R2160		2	1,964	Lyc O-320-D2A	
R3000		4	2,535	Lyc O-360	
Ayres Corporation, Albany, GA					AYRES
S2R-1340/400		1	6900	PW R-1340(600)	136
S2R-R1820/510		1	10,000	Wright R-1820	136
S2R-T11/400		1	8200	PT6A-11AG	136
S2R-T15/400		1	8200	PT6A-15AG	136
S2R-T15/510		1	8200	PT6A-15AG	136
S2R-T34/400		1	8200	PT6A-34AG	136
S2R-T65/400		1+1	10,500	PT6A-54AG	205
S2R-T65/510		1+1	10,500	PT6A-65AG	205
Beech (see Raytheon)					
Bell Helicopter Textron, Fort Worth, TX					BELL
206B-3 Jet Ranger III	$758	1+4	3200	250-C20J TS (420)	118 kt
206 L4 LongRanger IV	$1,135	1+6	4450	250-C30P TS (650)	110 kt
212	$4,743	2+13	11,200	2 PT6T-3B TS (1800)	111 kt
407	$1,349	1+6	5000	250-C47B TS (813)	128 kt
412EP	$5,042	2+13	11,900	2 PT6T-3D TS (1800)	124 kt
430	$4,378	1+9	9000	2 250-C40 TS (808)	140 kt
Bellanca, Inc., Alexandria, MN					BELLANCA
Super Viking	$175.3	4	3325	Cont. IO-520K (300)	178
Boeing Co., Seattle, WA					BOEING
Boeing Business Jet	$66,150	4+50	256,000	2 RR RB211 (43100#)	M0.80
British Aerospace					BAe
Jetstream Super 31	$4,600	19	16,204	2 TPE331-12 (1,020)	264 kt
Jetstream 41	$7,995	16	24,000	2 TPE331-14GTR/HR (1,650)	295 kt
Burkhardt Grob, Palm Beach, FL					GROB
G115C Bavarian	$109.5	2	3260	Lyc. IO-320/AEIO-360 (160)	130
Cadmus Corp. Northfield, IL					PZL
PZL Koliber II 150A	$79.5	4	1874	Lyc. O-320-E2A (150)	106†
PZL Wilga 104-80	$89.5	4	2866	PZL AI-14RA (260)	95
PZL Koliber 150B+		2	1670	Lyc. O-320-E2A (150)	106
PZL Koliber 235+		4	2635	Lyc. O-540 (235)	132
PZL Orlik 130TD+		2	5953	PT6A-25C	270

GA AIRCRAFT SPECIFICATIONS AND PERFORMANCE

Model	Price, (x000)	Seats	GW, lb.	Powerplant (hp or lb. thrust)	Cruise, mph (unless noted)
Canadair div. Bombardier					CHALLENGER
Regional Jet 200		2+50	47,450	2 GE CF34-3B1	
Regional Jet 200ER		2+50	51,000	2 GE CF34-3B1	
Regional Jet 200LR		2+50	53,000	2 GE CF34-3B1	
Canadair SE		19	53,250	2 CF34-3B1	M0.85
Challenger CL-600W		19	41,400	2 ALF502L	M0.85
Challenger CL-601-1A		19	43,250	2 CF34-1A	M0.85
Challenger CL-601-3A		19	43,250	2 CF34-3A	M0.85
Challenger CL-601-3A/ER		2+19	44,750	2 GE CF34-3A	M0.85
Challenger 601A-3R	$15,700	19	42,250	2 CF 34-1A 1(8,729#)	M0.85
Corporate Jetliner	$20,000	30	51,250	2 CF34-3A1	M0.85
Global Express+	$37,500	4+19	93,750	2 BMW/RR BR710	M0.9
Challenger 604	$20,000	2+19	48,300	2 CF-34-3B (8,729#)	468 kt
Cessna Aircraft Co., Wichita, KS					CESSNA
172R Skyhawk	$124.5	4	2457	Lyc. IO-360-L2A (160)	122 kt
182S Skylane	$190.6	4	3110	Lyc. IO-540-AB1A5	142 kt
206 Stationair+		6	3600	Lyc.	
208 Caravan I	$1,076	14	8035	PT6A-114 (600)	184 kt
208B Grand Caravan	$1,159	14	8750	PT6A-114A (675)	184 kt
208 Caravan Super					
Cargomaster	$1,229	2	8750	PT6A-114A (675)	175 kt
208 Caravan II	$2,510	14	9925	2 PT6A-112 (500)	231 kt
CitationJet	$3,150	8	10,400	2 FJ-44 TF (1,900#)	M0.7
Citation Bravo	$4,395	13	14,500	2 PW 530A (2,792#)	M0.7
Citation Ultra	$5,955	11	16,500	2 PWC JT 15D-5D (3045#)	M0.75
Citation Excel+	$6,795	13	18,900	2 PW545A (3,785#)	M0.75
Citation VII	$8,660	15	22,450	2 TFE 731-4R-2S (4,140#)	M0.85
Citation X	$15,595	14	35,700	2 AE30007C (8,600#)	M0.91
Cirrus Design, Duluth, MN					CIRRUS
Cirrus SR20+	$144.5	4	2900	Cont. IO-360-ES (200)	180
Cirrus ST50+	$1,000	5	5000	PT6A-135/7 (500)	
Classic Aircraft, Lansing, MI					WACO
Waco YMS Super	$235	2	2950	Jacobs 755B (275)	105
Commander Aircraft, Bethany, OK					COMMANDER
114B	$322.5	4	3260	Lyc. IO-540-T4B5 (260)	160 kt
114TC	$376.5	4	3305	Lyc. TSIO-540-AG1A (270)	170
Daetwyler					
MD3-160	$120	2	2028	Lyc. O-320-D2A (160)	114
de Havilland					de HAVILLAND
Corporate Dash 8	9,975	2+40	36,500	2 PW120A (2,000)	265 kt
Corporate Dash 8-300	$14,500	2+50	43,000	2 PW123 (2,380)	286 kt
Diamond Aircraft, Englewood, CO					KATANA
DA-20 Katana+	$106.7	2	1610	Rotax 912 (81)	119
EH Industries Ltd., Farnborough, England					EH
EH101	$17,000	2+30	31500	3 GE CT7-6 (1,920)	160 kt

Model	Price, (x000)	Seats	GW, lb.	Powerplant (hp or lb. thrust)	Cruise, mph (unless noted)
Empresa Brasileria de Aeronautica SA, Sao Paulo, Brazil					**EMBRAER**
EMB-720D Minuano		7	3,600	Lyc IO-540-K1GS	175
EMB-810D		7	4,750	2 Cont L/TSIO-360DB	226
Enstrom Corporation, Menominee, MI					**ENSTROM**
F280FX Falcon	$275.4	1+2	2600	Lyc. HIO-360-F1AD (225)	102 kt
480 Turbine	$550	4	2850	250-C20W (420)	106 kt
Eurocopter, Paris, France					
EC120 Colibri+	$815	1+4	3417	TM Arrius 1F (500)	125 kt
AS350 BA Astar	$999	1+5	4,630	TM Arriel 1B (641)	126 kt
AS350 B2 AStar	$1.165	1+5	4960	TM Arriel 1D1 (732)	133 kt
AS35 B3 Astar	$1,300	1+5	4960	TM Arriel 2B (847)	133 kt
BO 105 CBS Super 5	$2,107	1+4	5511	2 250-C20B (420)	131 kt
BO 105 LS	$2,286	1+5	5732	2 250-C2BC (500)	129 kt
AS355N TwinStar	$2,480	1+5	5600	2 TM Arrius 1A (479)	120 kt
AS355 N2 Dauphin 2	$5,300	1+12	9369	2 TM Arriel 1C2 (737)	151 kt
AS 332 L1 Super Puma	$12,900	2+20	18,960	2 TM Makila 1A1 (1,819)	141 kt
AS 332 L2 Super Puma	$13,800	2+20	20,504	2 TM Makila 1A2 (1,841)	150 kt
Eurocopter Deutschland					
EC 135	$2,780	1+6	5800	2 PWC206B (620)	134 kt
BK 117-B2	$4,175	1+7	7385	2 LTS101-750B1 (550)	133 kt
BK117 C1	$4,475	1+7	7385	2 TM Arriel 1E (737)	133 kt
Extra Aircraft, St, Augustine, FL					
Extra 200	$170.0	2	1914	Lyc. AEIO-360-A1E (200)	150
Extra 300	$237.5	2	2095	Lyc. AEIO-540-L1B5 (300)	178
Extra 300S	$235	1	295	Lyc. AIEIO-540-L1B5 (300)	185
Fairchild Aircraft, San Antonio, TX					**FAIRCHILD**
Merlin 23	$3,850	16	16,600	2 TPE 331-11U-611G (1,000)	295 kt
Dornier 328-110	$10,000	35	31,019	2 PW119B (2,180)	335 kt
Falcon Jet Corporation (Dassault)					**FALCON**
Falcon 900S		19	45,500	3 TFE 731-58R (4,750)	M0.87
Falcon 900EX	$27,950	19	48,300	3 TFE731-60 (5,750#)	M0.87
Falcon 50EX	$16,250	11	39,700	3 TFE731-40 (4,700#)	M0.84
Falcon 2000+	$15,595	19	35,800	2 CFE738-1-1B (5,725#)	M0.85
FFA Flugzeugwerke Altenrhein AG					
AS 202 Bravo		2	2381	Lyc. AEIO-360-B1F (180)	114
General Avia					
F22B	$129.9	2	1875	Lyc. O-320 (160)	125
F22C	$169	2	1985	Lyc. O-360 (180	160
Grob Aircraft Div., International Aero Club, Palm Beach, FL					**GROB**
Grob G115C	$119.5	2	2183	Lyc. O-320-AE (160)	125
Grob G115D	$153	2	2028	Luc. IO-360 (180)	130
Gulfstream Aerospace, Savannah, GA					**GULFSTREAM**
Gulfstream IV/SP	$28,000	2+19	74,600	2 RR Tay Mk 611-8 (13,850#)	M0.88
Gulfstream V	$34,500	19	90,500	2 BR710 (14,900#)	M0.90

Model	Price, (x000)	Seats	GW, lb.	Powerplant (hp or lb. thrust)	Cruise, mph (unless noted)
Hiller Aircraft					
UH-12E3	$315	3	3100	Lyc. VO-540 (305)	84 kt
UH-12E3T	$430	3	3100	250-C20B (420)	84 kt
Japan Aerospace					
Tempest 184+	$75	4	2650	Lyc. O-360 (180)	100
Kaman Aerospace, Bloomfield, CT					KAMAN
K-1200 K-Max	$3,500	1	6000	250-C47B (1,500)	128 kt
Kestrel Aircraft Company, Norman, OK					KESTREL
KL-1A+	$89.5	4	2400	Lyc O-320-D26 (160)	127
KL-1B+	$107.7	4	2700	Cont. IO-360-RS (190)	145
KL-1C+	$135	4	2850	Cont. IO-520 (230)	148
KL-1D+	225	6	3600	Cont. TSIO-550-B (325)	157
KL-1R+	125	4	2750	Cont. IO-360-ES (190)	152
Lake Aircraft, Laconia, NH					LAKE
Renegade	$484	4	3140	Lyc. IO-540-C4B5 (270)	132 kt
Turbo Renegade	$528	4	3140	Lyc. TIO-540 AA1AD (250)	155 kt
Seafury	$398	4	3140	Lyc. IO540-AA1AD (250)	132 kt
Turbo Seafury	$435	4	3140	Lyc. TIO-540 (270)	155 kt
Lancair Co., Redmond, OR					LANCAIR
LC-40+		4	3400	Cont. IO-550-N1B (300)	191
Learjet, Inc., Wichita, KS					LEARJET
31A	$5,175	12	15,500	2 TFE 731-2-3B (3,500#)	M0.81
35A	$5,99	10	18,300	2 TFE 731-2-2B (3,500#)	M0.81
60	$9,995	12	22,750	2 PW 305 (4,600#)	M0.81
45+	$5,900	11	20,000	2 TFE731-20 (3,500#)	M0.78
Maule Air, Inc., Moultrie, GA					MAULE
MX-7-160	$99.0	4	2200	Lyc. O-320-B2D (160)	113
MXT-7-160	$108.6	4	2200	Lyc. O-320-B2D (160)	113
MX-7-180C	$121.4	4	2500	Lyc. O-360-C1F (180)	126
MX-7-180A	$104.5	4	2400	Lyc. O-360-C4F (180)	122
MX-7-180B Star Rocket	$116.4	5	2500	Lyc. O-360-C1F (180)	126
MXT-7-180 Star Craft	$126.8	4-5	2500	Lyc. O-360-C1F (180)	122
MXT-7-180A	$114.0	4	2400	Lyc. O-360-C4F (180)	117
MX-7-235	$129.8	4-5	2500	Lyc. O-540-J3A5D (235)	139
MT-7-235 Super Rocket	$149.2	5	2500	Lyc. IO-540-W1A5D (235)	139
McDonnell Douglas Helicopter Co., Mesa, AZ					
MD 500D, E	$900	5	3000	250-C20B TS (375)	135 kt
MD 520N	$1,011	5	3350	250-C30R TS (425)	135 kt
MD 530F	$1.086	5	3100	250-C30 TS (425)	135 kt
MD 600N+	$1,250	8	4100	250-C47M (808)	135 kt
MD 902 Explorer	$4,300	8	6250	2 PWC PW206E (621)	123 kt
Melex USA, Raleigh, NC					
Wilga-80	$85.9	4	2866	Pezetel A114RA (256)	88
M-18 Dromader	$174.9	1	11,700	Pezetel ASZ62-IR-M18 (967)	110
M-26 Air Wolf	$180	2	3086	Lyc. AEIO-540-L1B5 (300)	184
Wilga 80-550+		4	2866	Cont. IO-550 (300)	100

GA AIRCRAFT SPECIFICATIONS AND PERFORMANCE

Model	Price, (x000)	Seats	GW, lb.	Powerplant (hp or lb. thrust)	Cruise, mph (unless noted)
(New) Meyers Aircraft Co., Ft. Pierce, FL					MEYERS
SP20+	$135	2	2350	Lyc. IO-360 (200)	160
M300*	$255	4	3400	Lyc. IO-580 (300)	200
Mooney Aircraft Corp., Kerrville, TX					MOONEY
M20J MSE/ATS	$219.9	4	2740	Lyc. IO-360-A3B6D (200)	168 kt
M20L TLS	$364.9	4	3368	Lyc. TIO-540AF1A (270)	220
M20R Ovation	$299.9	4	3368	Cont. IO-550-G (280)	187 kt
Mundry Aviation, Bunnell, FL					
Cap 10B	$165	2	1830	Lyc. AEIO-360-B2F (180)	135
Cap 232	$250	1	1870	Lyc. AEIO-540L1-B-5D (300)	190
New Piper Aircraft Corp., Vero Beach, FL					PIPER
PA-28-161 Warrior III	$134.9	4	2440	Lyc. O-320-D3G (160)	135
PA-28-181 Archer III	$149.9	4	2,550	Lyc. O-360-A4M (180)	153
PA-28-201 Arrow	$204.7	4	2750	Lyc. IO-360-C1C6 (200)	167
PA-32R-302 Saratoga II	$359.8	4	3167	Lyc. IO-540-K1G5D (300)	165 kt
PA-46-350P Malibu Mirage	$768.7	6	4300	Lyc. TIO-540-AE2A (350)	225 kt
PA-44-1180 Seminole	$319.6	4	3800	Lyc. O-360-A1H6	193
PA-34-220T Seneca V	$472.9	6	4750	2 Cont. TSIO-360-KB(220)	200 kt
Optica					
Scout OA-7	$275	2	2900	Lyc. IO-540 (260)	89
Piaggio Aviation, Inc.					PIAGGIO
Avanti P-180	$4,680	11	10,810	2 PT6A-66 (850)	395
Pilatus Aircraft Ltd., Stans, Switzerland					PILATUS
PC-6		11	6195	PT6A-27 (550)	125
PC-7		2	4189	PT6A-25A (550)	220
PC-XII	$2.190	11	9,039	PT6A-67B (1200)	270
Pilatus-Britten-Norman					
BN2B-20 Islander	$611	10	6600	2 Lyc. IO-540-K1B5 (300)	143
NB2B-26 Islander	$570.5	10	6600	2 Lyc. IO-540-E4C5 (260)	138
BN-2T Turbine Islander	$1,500	10	7000	2 250B-17C (320)	170
Islander 4000	$1,800	10	8800	2 250-B17F (400)	170
Precision Airmotive Corp., Everett, WA					COMMANDER
840 Commander		10	10,375	2 TPE 331-5-254K	291
900 Commander		11	10,775	2 TPE 331-5-254K	290
980 Commander		10	10,375	2 TPE 331-10-501K	309
1000 Commander		11	11,250	2 TPE 331-10-501K	308
Quicksilver Enterprises, Temecula, CA					
GT 500	$29.9	2	1000	Rotax 528 (65)	70
Rans, Inc.					
S-7 Courier	$55	2	1200	Rotax 912 (78)	93
Raytheon Aircraft Corp., Wichita, KS					BEECH
A-36 Bonanza	$425.9	6	3650	Cont. IO-550-B (300)	174 kt
B-36TC Bonanza	$471	6	3850	Cont. TSIO-520-UB (300)	178 kt

Model	Price, (x000)	Seats	GW, lb.	Powerplant (hp or lb. thrust)	Cruise, mph (unless noted)
Raytheon cont.					
58 Baron	$600.8	6	5500	2 Cont. IO-550-C (300)	203 kt
C90B King Air	$2,378	8	10,100	2 PT6A-21 TP (550)	247 kt
C90SE King Air	$1,747	13	10,100	2 PT6A-21 (550)	247 kt
B200 Super King Air	$3,634	10	12,500	2 PT6A-42 TP (850)	289 kt
B200 SE	$2,995	8	12,500	2 PT6A-42 (850)	289 kt
Super King Air 350	$4,509	10	15,000	2 PT6-60A (1,050)	311 kt
Executive 1900D	$4,775	21	16,950	2 PT6A-67D TP (1,279)	278 kt
Beechjet 400A	$5,762	9	16,100	2 JT15D-5 (2,900#)	M0.78
Hawker 800 XP	$10,295	8	28,000	2 TFE 731-5 (4,660#)	M0.8
Hawker 1000	$12,900	9	31,000	2 PW305 (5,225#)	M0.8
Robinson Helicopter, Inc., Torrance, CA					**ROBINSON**
R22 Beta II	$154	2	1370	Lyc. O-320-B2C (131)	96 kt
R44 Astro	$280	4	2400	Lyc. HO-540-F1B (225)	113 kt
Rogerson Hiller Corp., Port Angeles, WA					**HILLER**
UH-12E		3	2800	Lyc. VO-540-C2A	96
RH-1100A		5	2750	250-C20B TS	127
Ruschmeyer Aircraft Production, Galesburg, IL					
R 90-230 RG	$232.5	4	2976	Lyc. IO-540-C4D5 (230)	168
R-90-180 FG		4	2535	Lyc. O-360 (180)	120
R-90-230 Turbo FG+		4	2976	Lyc. IO-540-C4D5 (230)	150
R-90-300 Turbo RG+		4	3414	TBA	220
R-90-420-AT+		4	3194	250-B20 (470)	243
Saab					
2000	$14,500	2+50	50,700	2AZE2100A (4,152)	373 kt
Schweitzer Aircraft, Elmira, NY					**SCHWEITZER**
450B AgCat		1	7020	PW R-985 (450)	100
600B Super B AgCat		1	7020	PW R-1340 (600	100
G164B Super-B Turbine		1	7020	PT6-11/15/34AG (500-750)	113
300CB	$198.5	2	1750	Lyc. HO-360 (180)	85 kt
300C	$248.5	3	2050	Lyc. HIO-360-D1A (190)	88 kt
330	$560	4	2230	250-C20W (235)	100 kt
Sikorsky Aircraft United Technologies Corp., Stratford, CT					**SIKORSKY**
S-76C	$7,000	2+12	11,700	2 TM Arriel 2S1 (860)	155 kt
S-76B	$7,250	2+12	11,700	2 PT6B-36B (1,033)	155 kt
Sherpa Aircraft Mfg. Co., Aloha, OR					**SHERPA**
Sherpa+		5	4750	Lyc. IO-720 (400)	140
Sino-Swearingen					**SWEARINGEN**
SJ30-1+	$3,000	8	10,700	2 FJ44-1A (1,900#)	M0.77
SJ30-2+	$3,500	8	12,300	2 FJ44-2A (2,300)	M0.83
SOCATA					**SOCATA (AEROSPATIALE)**
TB-9 Tampico Club		4	2337	Lyc. O-320-D2A (160)	107
TB-10 Tobago XL		4-5	3086	Lyc. O-360-A1AD (200)	125
TB-200 Tobago XL		4-5	2535	Lyc. IO-360-A1B6 (200)	130
TB-20 Trinidad	$345	4-5	3086	Lyc.IO-540-C4-D5-D(250)	164 kt
TB-21 Trinidad TC	$442	4-5	3086	Lyc. TSIO-540-AB-1-AD(250)	164 kt
TBM 700	$2,610	5-6	6579	PT6A-64 (700)	300 kt
TB360 Tangara	$450	5	3800	(2) Lyc. O-360-A1G6 (180)	168 kt

GA AIRCRAFT SPECIFICATIONS AND PERFORMANCE

Model	Price, (x000)	Seats	GW, lb.	Powerplant (hp or lb. thrust)	Cruise, mph (unless noted)
Swiss Trainer Corp., Huntersville, NC					
MD-160	$120	2	2337	Lyc. O-320-O2A (160)	130
Taylorcraft Aircraft Corp., Morgantown, WV					TAYLORCRAFT
F-22 (A-C)	$62.9	2	1750	Lyc. O-235-L2C (118)	87
	$72.9	2	1750	Lyc. O-360-A4M (180)	104
Visionaire, Chesterfield, MO					
Vantage	$1,700	6	7000	(2) JT15D-5	350
Weatherly Aviation Co., Hollister, CA					WEATHERLY
620		1	5700	PW R-985 (450)	130
620TP		1	6300	PT6A-11AG	155
620A		1	5700	PW R-985 (450)	148
620A-TP		1	6300	PT6A-15AG	165
Yakolev					
Yak-112 Owl	$140	4	2310	Vedenev M-17 (180)	133
Yak-48		12	23,540	2 turbofan	448
Yak-58		6	4576	Vedenev M-14PR (360)	154
Yak-55M	$100	1		Vedenev M-14P (360)	
Yak-54	$160	2		Vedenev M-14X (360)	240
Zenith Aircraft Co., Mexico, MO					
Zenith CH 2000	$69.9	2	1606	Lyc. O-235-N2C (116)	104
Zivko, Guthrie, OK					
Zivko Edge 540+		1		Lyc. IO-540 (327)	
Zlin Aerospace					
Z-242L	$105	2	2400	Lyc. AEIO-360-A1B6 (200)	110
Z-143L		4	2976	Lyc. O-540-J3145 (235)	140

U.S. REGISTERED FIXED-WING AIRCRAFT
BY MAKE AND MODEL

Source: Airpac, Inc., Edmond, OK USA, (800) 654-2006
FAA registrations of 5/97 (some additional antique/warbird listings from *The Vintage Airplane Catalogue*)
(Year type introduced in parenthesis)

AERONCA
Chief KCA (1938)................................9
Chief KM (1938)................................1
Chief 50-C (1938)................................8
Chief 50-F (1938)................................5
Chief 50-L (1938)................................7
Tandem 50-TL (1940)................................1
Chief 65-C (1939)................................89
Chief 65-CA (1938)................................214
Chief 65-LA (1938)................................33
Chief 65-LB (1939)................................45
Grasshopper 65 [L3B](1941)................................1
Grasshopper 65 [L3C](1941)................................3
Grasshopper 65 [O-58A](1941)................................3
Grashopper 65 [O-58B](1942)................................163
KS (1937)................................2
Low Wing LB (1936)................................5
Model K (1937)................................33
Scout CF (1939)................................1
Scout KC (1939)................................5
Tandem TCA (1941)................................1
Tandem 60-TF (1940)................................4
Tandem 65-TAC (1941)................................55
Tandem 65-TAF (1941)................................6
Tandem 65-TAL (1941)................................9
Tandem 65-TC (1939)................................55
Tandem 65-TF (1940)................................13
Tandem 65-TL (1940)................................31
C-2 (1930)................................14
C-3 (1931)................................66
7AC Champ................................2,616
7ACA Champ................................61
7BCM................................261
7CCM Champ................................146
7DC................................187
7EC................................252
7ECA Citabria................................985
7FC................................272
7GC................................60
7GCB................................69
7GCBC................................803
7GCAA................................310
7HC................................21
7JC................................6
7KCAB Citabria................................482
8KCAB Decathlon................................607
8KCAB-180 Decathlon................................5
8GCBC Scout................................256
11AC Chief................................33

11BC Chief................................932
11CC Super Chief................................100
15AC Sedan................................225
ALSO SEE BELLANCA

AEROSPATIALE
TB-10 Tobago (1986)................................63
TB-20 Trinidad (1984)................................174
TB-21 TC Trinidad (1986)................................26
TBM 700................................54
MS760 Paris Jet................................17
601 SN Corvette................................3

AIRCOUPE
415-C (1941)................................31
415-D (1945)................................2

ALEXANDER
Eaglerock A-1 (1928)................................7
Eaglerock A-2 (1928)................................2
Eaglerock A-3 (1928)................................1
Eaglerock A-4 (1928)................................1
Eaglerock A-12................................1

ALLIANCE
Hess Argo (1929)................................1

AMERICAN AIR
Pilgrim 100B (1931)................................1

AMERICAN EAGLE
Eaglet (1926)................................21
Eaglet B-31 (1931)................................1

ARROW
Sport (1929)................................10

BARNARD
Standard (1929)................................1
Standard D-25 (1929)................................1
Standard D-29A (1929)................................1
Standard D-31 (1941)................................1

BEAGLE
B.121 Series 1 Pup................................1
B.121 Series 2 Pup................................1
B.206 Series 1................................11
B-206 Series 2 (1966)................................18

BRITTEN

BN-2 Islander	4
BN-2A Islander	14
BN-2A-03 Islander	13
BN-2A-06 Islander	2
BN-2A-08 Islander	20
BN-2A-09 Islander	8
BN-2A-20 Islander	2
BN-2A-21 Islander	11
BN-2A-26 Islander	26
BN-2A-27 Islander	6
BN-2AMK3 Tri Islander	9
BN-2B-20 Islander	2
BN-2B-21	3
BN-2B-26 Islander	5
BN-2B-27 Islander	1
BN-2T Islander	1

BUCKER

Bu-131 Jungmann (1938)	2
Bu-133 Jungmeister (1938)	3

CALL-AIR

Model A (1943)	4
S-1A (1941)	105
L-6 (1943)	1
S1B1 (1942)	27

CANADAIR

CL-600 Challenger (1981)	332
CL-601 Challenger (1983)	10

CESSNA

AW (1928)	6
DC-6A (1929)	1
C-34 Airmaster (1935)	9
C-37 Airmaster (1937)	12
A37	2
C-38 Airmaster (1937)	5
C-145 Airmaster (1938)	7
C-165 Airmaster (1938)	25
T-50 Bobcat (1940)	107
T-50 UC78 Bobcat (1943)	5
120 (1946)	941
140 (1946)	2,273
140A (1949)	283
150 (1959)	610
150A (1961)	187
150B (1962)	184
150C (1963)	255
150D (1964)	410
150E (1965)	495
150F (1966)	1,813
150G (1967)	1,670
150H (1968)	1,330
150J (1969)	1,155
150K (1970)	573
150K Reims	4
A150K Aerobat	109

150L (1971)	2,523
150L Reims	3
A150L Aerobat (1973)	126
150M (1975)	2,533
A150M Aerobat	113
152 (1978)	4,466
152 Reims	2
A152 Aerobat	106
170 (1948)	384
170A (1949)	672
170B (1952)	1,611
172 Skyhawk (1956)	2,537
172 (T-41B)	8
172A Skyhawk (1960)	596
172B Skyhawk (1961)	534
172C Skyhawk (1962)	486
172D Skyhawk (1963)	605
P172D Skyhawk	37
172E Skyhawk (1964)	786
R172E Skyhawk	143
R172E Reims Skyhawk	3
172F Skyhawk (1965)	1,029
172G Skyhawk (1966)	960
R172G Skyhawk	6
172H Skyhawk (1967)	1,053
172H Reims Skyhawk	4
172I Skyhawk (1968)	453
172K Skyhawk (1969)	1,434
172K Reims Skyhawk	3
R172K Hawk XP (1977)	864
172L Skyhawk (1971)	1,091
172M Skyhawk (1973)	4,764
172N Skyhawk (1977)	4,448
172N Reims Skyhawk	4
172P Skyhawk (1981)	1,668
172Q Skyhawk (1983)	27
172R Skyhawk (1996)	18
172RG Cutlass (1980)	790
175 Skylark (1958)	810
175A Skylark (1960)	371
175B Skylark (1961)	147
175C Skylark (1962)	64
177 Cardinal (1968)	677
177A Cardinal (1969)	111
177B Cardinal (1970)	1,004
177RG Cardinal (1977)	996
177RG Reims Cardinal	2
180 Skywagon (1953)	1,364
180A Skywagon (1957)	241
180B Skywagon (1959)	108
180C Skywagon (1960)	64
180D Skywagon (1961)	59
180E Skywagon (1962)	43
180F Skywagon (1963)	51
180G Skywagon (1964)	49
180H Skywagon (1965)	403
180J Skywagon (1973)	293
180K Skywagon (1977)	313
182 Skylane (1956)	614

Cessna cont.

182A Skylane (1957)	1,078
182B Skylane (1959)	537
182C Skylane (1960)	393
182D Skylane (1961)	347
182E Skylane (1962)	492
182F Skylane (1963)	383
182G Skylane (1964)	490
182G 460 Wren	5
182H Skylane (1965)	522
182H 460 Wren	3
182J Skylane (1966)	586
182K Skylane (1967)	514
182K 460 Wren	3
182L Skylane (1968)	522
182M Skylane (1969)	515
182N Skylane (1970)	412
182P Skylane (1972)	2,565
182Q Skylane (1977)	1,840
182R Skylane (1981)	582
182R Reims	2
182S (1997)	9
T182R Turbo Skylane (1981)	58
182RG Skylane (1976)	907
T182RG Turbo Skylane (1979)	579
185 (1961)	107
185A (1962)	45
185B (1963)	20
185C (1964)	22
185D (1965)	34
185E (1966)	15
A185E (1972)	295
185F (1973)	2
A185F (1973)	1,206
190 (1947)	93
195 (1947)	276
195 LC126C	3
195A (1947)	139
195B (1952)	133
205 (1963)	199
205A (1964)	49
206 Super Skywagon (1964)	108
P206 (1965)	101
U206 (1965)	85
P206A (1966)	58
PT206A (1968)	18
U206A (1966)	49
TU206A (1966)	40
P206B (1967)	46
206B (P)	14
U206B (1967)	74
TU206B (1967)	29
P206C (1968)	40
206C (P) (1968)	15
U206C (1968)	78
206C (U) (1968)	39
P206D (1969)	39
206D (P) (1969)	8
U206D (1969)	39
TU206D (1969)	17
P206E (1970)	11
TP206E (1970)	5
U206E (1970)	65
TU206E (1970)	17
U206F Stationair (1972)	430
TU206F Turbo Stationair (1972)	196
U206G Stationair (1977)	591
206G (U) Turbo Stationair (1977)	489
207 Skywagon (1969)	90
T207 Turbo Skywagon (1.969)	35
207A Skywagon (1977)	152
T207A Turbo Skywagon (1977)	71
208 Caravan I (1985)	86
208A Caravan	9
208B Super Cargomaster (1987)	404
210 Centurion (1960)	414
210-5 Centurion	70
210-5A Centurion	17
210A Centurion (1961)	172
210B Centurion (1962)	157
210C Centurion (1963)	96
210D Centurion (1964)	193
210E Centurion (1965)	132
210F Centurion (1966)	61
T210F Turbo Centurion (1966)	131
210G Centurion (1967)	69
T210G Turbo Centurion (1967)	75
210H Centurion (1968)	67
T210H Turbo Centurion (1968)	56
210J Centurion (1969)	81
T210J Turbo Centurion (1969)	37
210K Centurion (1970)	91
T210K Turbo Centurion (1970)	65
210L Centurion (1972)	536
T210L Turbo Centurion (1972)	631
210M Centurion (1977)	244
T210M Centurion (1977)	700
210N Centurion (1979)	205
P210N Press. Centurion (1978)	580
T210N Turbo Centurion (1979)	861
210R Centurion (1985)	5
P210R Press. Centurion (1985)	30
T210R Turbo Centurion (1985)	48
303 Crusader (1982)	3
T303 Crusader (1982)	108
305 Bird Dog	1
305 L-19 Bird Dog	6
305 L-19A Bird Dog	7
305 L-19E Bird Dog	32
305 O-1A Bird Dog	10
305 O-1DT Bird Dog	1
305 O-1E Bird Dog	1
305A Bird Dog	245
305B Bird Dog	4
305C Bird Dog	15
305E Bird Dog	3

PQ-14A (1944)..............................3
PQ-14B..4
PQ-14TD2C1..................................1

CUNNINGHAM HALL
PT-6F (1938).................................1

CURTISS AEROPLANE
Fledgling J-1 (1929).......................5

CURTISS-ROBERTSON
Robin C-1 (1928)..........................38
Robin J-1 (1929)............................3
Robin C-2 (1929)...........................2
Robin 4C-1A (1930)........................1

CURTISS-WRIGHT
Jenny JN4D (1918)..........................4
Junior CW1 (1931).........................26
Sedan 15-D (1931)..........................3
Travel Air 4 (1928)..........................9
Travel Air 6 (1928)..........................3
Travel Air 10 (1930)........................1
Travel Air 12 (1931).......................10
Travel Air 14 (1931)........................4
Travel Air 16 (1931)........................4
Condor 22A (1938)..........................2
Condor T32C..................................1
Falcon SNC-1.................................1
Hawk P1..1
Hawk F6C-4...................................1
Helldiver SB2C5 (1944).....................1
Commando C-46 (1940).....................9
Commando C-46A...........................11
Commando C-46C.............................1
Commando C-46D.............................4
Commando C-46F............................20
Commando C-46R.............................4
Warhawk P-40 (1939).......................2
Warhawk P-40E (1941)......................8
Warhawk P-40K................................1
Warhawk P-40N (1942).....................11
O-52 (1941)...................................1

DASSAULT
Falcon 10 (1973).........................149
Falcon 20 (1966).........................205
Falcon 50 (1980).........................156
Mystere Falcon 200 (1983)...............21
Mystere Falcon 900 (1986)...............81
Mystere Falcon 900 EX....................12
Falcon 2000...................................32

DAVIS
D-1 (1929).....................................1
D-1 K (1930)..................................2
D-1 W (1929).................................4
D-1 66 (1929).................................2
V-3 Special (1929)...........................5

DE HAVILLAND
Gypsy Moth DH-60 (1929).................6
Moth 60-GM (1929)..........................6
Tiger Moth DH-82 (1939)..................98
Puss Moth DH-80..............................1
Dragon DH-84 (1934)........................1
Horne Moth DH-87 (1935)..................3
Dragon Rapide DH-89 (1941)..............4
Moth Minor DH-94 (1937)..................2
Mosquito CH-98MK35 (1946).............13
Vampire DH-100 (1947)....................10
Sea Venom DH-112 (1956)..................7
DHC-3 Otter..................................54
DHC-6 Twin Otter...........................64
DHC-6 100 Twin Otter.......................3
DHC-6 200 Twin Otter (1968).............15
DHC-6 300 Twin Otter (1970).............61

DEWOITNE
D26 (1931)....................................1

DOUGLAS
Dragon B-23 (1940)..........................6
Dauntless SBD-4 (1941)......................1
Dauntrless SBD-5..............................2
Dauntless A-24B (1943)......................1
Havoc A-20B...................................1
Havoc A-20G (1941)..........................2
Havoc A-20J (1943)...........................1
Invader A-26 (1941)...........................5
Invader A-26B (1941)........................25
Invader A-26C.................................13
Invader B-26...................................6
Invader B-26B (1941)........................15
Invader B-26C (1941).........................9
Invader RB-26C (1944)........................3
Invader TB-26B (1945)........................2
Invader TB-26C.................................1
DC-2 (1935)....................................2
DC-3 (1936).................................241
DC-3 C-R.......................................1
DC-3 S...3
DC-3 SCG (1937)............................75
DC-3 SC3G.....................................2
DC-3 S1C3G...................................68
DC-3 S4C4G...................................16
DC-3 R4D-6....................................1
DC-3 C-47.....................................57
DC-3 C-47A...................................14
DC-3 C-47B.....................................5
DC-3 C-47D.....................................4
DC-3 C-47D V..................................5
DC-3 C-47H.....................................1
DC-3 C-47J.....................................6
DC-3 C-47 R4D-8..............................5
DC-3 C-47 G102...............................1
DC-3 C-47 G102A.............................5
DC-3 C-117B....................................1
DC-3 C-117D...................................10

JAMIESON
J-1 (1943)..3

JOHNSON
Rocket 125 (1942)...............................1

JUNKERS
JU-52 (1934).......................................2

KARI-KLEEN
Sioux J-4 Coupe (1929)......................2

KINNER
Playboy (1935)....................................1
Sportster (1933)...................................7
Sportster Model B1 (1934).................1

LAIRD
LC-B (1926)..2
Special...1
LC-B-200 (1928)................................1
LC-DW500..1
LC-1B-300 (1930)..............................1

LAKE
C-1 Skimmer (1957).........................12
C-2 Skimmer (1958).........................10
LA-4 Buccaneer (1960)...................593

LEARFAN
2100...3

LEAR JET
23 (1965)...45
24 (1966)...56
24A...10
24B (1969)...31
24B-A...2
24D (1970)...55
24E (1976)...12
24F (1976)..8
25 (1968)...54
25B (1970)...75
25C (1970)..8
25D Century III (1976)...................114
28 Longhorn (1979)............................3
31...22
31A...79
35 (1974)...54
35A Century III (1976)...................404
36 (1974)...14
36A Century III (1976)......................23
55 (1981)...82
55B (1987)..6
55C..8
60 (1994)...87

LOCKHEED
Vega 2D (1929)....................................1
Vega 5C (1929)....................................2
Vega 1 DL-1B (1933)...........................1
Electra 10A (1935)...............................5
Electra 10E...2
Electra JR 12 (1936)...........................23
Super Electra 14-H2.............................1
Ventura PV-1 (1943)...........................27
Ventura B-34 (1942).............................2
Harpoon PV-2 (1943)..........................66
Lightning F-5 (1944)............................2
Lightning P-38 (1944).........................13
Constellation 049 (1945)......................2
Constellation 49 (1946)........................1
T-33 Shooting Star (1952)..................96
Starfighter F-104..................................7
1329 JetStar (1961)............................46
1329-23D JetStar..................................2
1329-23E JetStar.................................18
1329-25 JetStar...................................27
1329-731 JetStar...................................2

LUSCOMBE
Silvaire 8 (1938)..................................8
Silvaire 8A (1938)............................231
Silvaire 8B (1939)..............................23
Silvaire 8C (1939)..............................62
Silvaire 8D (1939)..............................22
Silvaire 8E (1939)............................457

MAHONEY-RYAN
B-1 (1928)..3
B-5 (1929)..1
B-7...1

MARTIN
AM-1 Mauler..1
B-26CT Marauder (1940).....................1

MAULE
M-4 Jetasen..73
M-4--180 Astro Rocket (1970)............3
M-4-210 (Rocket 1973)......................84
M-4-220 Strata Rocket (1971)..........130
M-5-180 Lunar Rocket........................71
M-5-200...2
M-5-210 Lunar Rocket(1974)............130
M-5-220 Strata Rocket (1974)............42
M-5-235 Lunar Rocket (1977)..........258
M-6-180 (1985).....................................3
M-6-235 (1981)...................................88
M-7-180 (1985).................................116
M-7-235 Super Rocket (1984)..........175
M-8-235...4

MBB
HFB 320 Hansa Jet.............................23

MERCURY
S-1 (1932)..1
T-2 Chic (1929)..1

MESSERSCHMITT
Me108 Taifun (1942)...................................4
Me109 (1935)..15

MEYERS (ALSO SEE ROCKWELL)
OTW (1935)..22
OTW-145 (1940)..13
OTW-160 (1940)..20

MILES
MIIA (1937)..1

MILLER P D
Y-1S (1937)..2

MITSUBISHI
A6M Zero...1
A6M2-21 Zero..1
MU-2B (1967)..26
MU-2F (1969)..46
MU-2K (1972)..35
MU-2M...17
MU-2P..29
MU-2G (1970)..15
MU-2J (1972)..53
MU-2L (1975)..27
MU-2N (1977)..18
MU-2B-40 Solitaire (1979)..........................43
MU-2B-60 Marquise (1979)........................94
MU-300 Diamond (1982).............................79
MU-300-10 Diamond....................................3

MONO AIRCRAFT
Monocoupe 70 (1928)..................................6
Monocoupe 90 (1930)................................78
Monocoupe 110 (1930)...............................18
Monocoupe 113 (1928)................................5
Monocoupe 125 (1932)................................1
Monocoupe D-145 (1934).............................5
Monoprep (1929)...1

MOONEY
A-2A Cadet !968)...6
M10 Cadet..52
M18 Mite (1949)..1
M18C Mite..75
M18C 55 Mite...16
M18L Mite (1948).......................................44
M18LA Mite (1953).....................................21
M20 Mk 20 (1955).....................................100
M20A Mk 20A (1958).................................226
M20B Mk 21 (1961) 139

MOSWEY
III (1943)...1

NAVAL AIRCRAFT FACTORY
N3N-3 (1939)...167

NAVION
Navion (1946)...980
L17A Navion (1948).....................................5
L17B Navion (1951).....................................6
B Navion (1951).......................................167
D Navion (1958)..11
E Navion (1959)..2
F Navion (1960)..7
G Range Master (1961)...............................82
H Range Master (1967)...............................42
Twin Riley...2

NEW STANDARD
D-25 (1930)...3

NIEUPORT
28C-1 (1917)...4

NORTH AMERICAN
BT-14 (1941)...1
Mitchell B-25 (1941)...................................67
Mustang A36A (1942)..................................2
Mustang P-51 (1941).................................106
Mustang P-51F...66
Sabre F-86 (1948).......................................45
Super Sabre F-100 (1954).............................8
Texan AT-6 (1939).....................................681
Texan P-64 (1940)..1
Trojan T-28 (1950)....................................230
Twin Mustang F-82 (1944).............................1

NORTHROP
Black Widow P-61 (1942)..............................1
Scorpion F-89 (1953)....................................1
Talon T-38 (1961).......................................29

M20C Ranger (1963)...............................1,658
M20D Master (1963)..................................125
M20E Chaparral (1965)............................1,113
M20F Executive 21 (1967)..........................913
M20G Statesman (1968).............................155
M20J 201 (1977)....................................1,585
M20K 231 (1979)......................................880
M20L (1988)...32
M20M...171
M20R..83
M22 Mustang (1967)...................................20
M30...1

PARKS
P-1-T (1929)..1
P2A (1929)...2
P-2-A..1

PASPED
W1 Skylark (1935)..............................1

PHEASANT
H-10 (1929)...1

PIAGGIO
P-136-L1...10
P.136-L2..4
P.149D...5
P.166...3

PILATUS
PC-12...44

PIPER
J-2 Cub (1938)...................................118
J-3 Cub (1938)................................4,592
J-4 Coupe (1939).............................302
J-5 Cruiser (1940)............................417
J-5 PT1 Cub...1
J-5 L-14...1
AE-1...3
PA-11 Cub Cruiser (1947).................494
PA-12 Cub Super Cruiser (1946)....1,527
PA-14 Cub Family Cruiser (1948)....118
PA-15 Cub Vagabond (1948)............195
PA-16 Cub Clipper (1950)................398
PA-17 Cub (1949)..............................117
PA-18 Super Cub (1950)................4,036
PA-19 Super Cub (1949)....................27
PA-20 Pacer (1950)...........................526
PA-22 Colt (1951)...........................1,037
PA-22 Tri Pacer (1951)..................4,242
PA-23-150 Apache (1954).................610
PA-23-160 Apache (1958).................453
PA-23-180 Apache................................2
PA-23-235 Apache..............................54
PA-23-235 UO Apache..........................2
PA-23-250 Aztec (1960)..................2,477
PA-24-180 Comanche (1958).............760
PA-24-250 Comanche (1958)..........1,652
PA-24-260 Comanche (1965).............710
PA-24-400 Comanche (1964).............104
PA-25-235 Pawnee.............................895
PA-28-140 Cherokee (1964)...........6,359
PA-28-150 Cherokee..........................186
PA-28-151 Warrior (1974)..............1,247
PA-28-160 Cherokee (1962).............373
PA-28-161 Warrior (1977)..............2,024
PA-28-180 Cherokee (1963)...........4,300
PA-28-181 Archer (1976)................2,257
PA-28R-180 Arrow (1967).................733

PA-28R-200 Arrow (1969)..............1,604
PA-28R-201 Arrow (1977)..................361
PA-28R-201T Turbo Arrow (1977)....524
PA-28RT-201 Arrow...........................240
PA-28RT-201T Arrow.........................334
PA-28-201T Dakota (1979)..................79
PA-28-235 Pathfinder (1964)..........1,061
PA-28-236 Dakota (1979)..................511
PA-28R-300 (1982)................................1
PA-29 (1962)..1
PA-30 Twin Comanche (1963).......1,216
PA-30B-160 Turbo Comanche.............2
PA-31-300 Navajo (1968).....................8
PA-31-310 Navajo (1967)..................524
PA-31-325 Navajo CR (1975)............260
PA-31-350 Chieftan (1973)...............884
PA-31-P350 Mojave..............................37
PA-31-P425 Navajo............................132
PA-31-500 T Cheyenne I...................158
PA-31-620 T Cheyenne II..................337
PA-32-260 Cherokee 6.......................931
PA-32-300 Cherokee 6....................1,333
PA-32R-300 Lance..............................679
PA-32-300RT Lance............................271
PA-32R-300T Lance............................296
PA-32-301 Saratoga...........................202
PA-32R-301 Saratoga.........................380
PA-32RT-301 Saratoga.......................266
PA-32-301T Saratoga...........................86
PA-34-200 Seneca...............................457
PA-34-200T Turbo Seneca...............1,105
PA-34-220T Turbo Seneca.................285
PA-38-112 Tomahawk......................1,130
PA-39 Twin Comanche CR...................83
PA-40 Flag Ship....................................1
PA-41P...1
PA-42-720 R Cheyenne.........................7
PA-42-720 Cheyenne III......................82
PA-42-1000LS Cheyenne IV...............21
PA-44-180 Seminole...........................269
PA-44-180T Turbo Seminole...............49
PA-46-310P Malibu.............................284
PA-46-350P..214
PA-48..2
600 Aerostar.......................................167
600A Aerostar (1967)............................2
601 Aerostar (1968)...........................106
601B Aerostar..5
601P Aerostar.....................................339
602P Aerostar (1967)...........................83
700P Aerostar..20

PITCARIN
PA 4 Fleetwing......................................1
PA 5 (1927)..4
PA 6 (1929)..2
PA 7..3
PA 7S (1930)..1
PA 8 (1931)..2

PORTERFIELD

35 (1940)	4
35W	1
40 [CP] (1938)	2
50 [CP] (1939)	9
55 [CP]	1
65 [CP] (1940)	30
65 [FP] (1941)	9
65 [LP] (1940)	32
75C (1941)	2
35-70 (1935)	16

REARWIN

Skyranger 175 (1940)	15
Skyranger 180 (1941)	8
Skyranger 185 (1945)	5
Sportster 6000 (1936)	3
Sportster 7000 (1935)	10
Sportster 8000 (1939)	1
Sportster 8125 (1939)	9
Sportster 8135 (1939)	17
Sportster 8135T (1941)	6
Sportster 8500 (1936)	7
Sportster 9000 (1935)	16

REPUBLIC

AT12 (1940)	1
P-47 Thunderbolt (1942)	3
P-47D Thunderbolt (1944)	5
F-84 Thunderjet (1952)	2
F-84F Thunderstreak (1951)	6

ROCKWELL

100 Darter (1965)	184
100 Lark (1969)	137
112 (1972)	153
112A (1974)	125
112B (1977)	22
112TC (1976)	78
112TCA (1977)	57
114 (1976)	216
114A (1979)	22
114B (1992)	62
114TC	12
200	2
200A (1959)	8
200B (1961)	13
200C (1963)	7
200D (1965)	72
500 Shrike (1958)	53
500A Shrike (1960)	42
500B Shrike (1960)	118
500S Shrike (1968)	109
500U Shrike (1966)	9
520 Commander (1952)	73
560 Commander (1954)	35
560A Commander (1955)	49
560E Commander (1957)	40
560F Commander (1961)	35
680 Super (1955)	105
680E (1958)	57
680F Grand Commander (1960)	48
680FL Courser Commander (1963)	66
680FLP Courser Liner (1964)	18
680FP Grand Commander (1962)	3
680T Turbo Commander (1966)	28
680V Turbo II (1967)	23
680W Turbo II (1968)	27
681 Hawk Commander (1970)	19
681B Hawk Commander (1971)	3
685 Commander (1972)	37
685A Commander	1
690 Commander (1972)	30
690A Commander (1973)	111
690B Commander (1977)	115
690C Jet Prop 840 (1980)	61
690D Commander 900 (1982)	16
695 Jet Prop 980 (1980)	37
695A Commander 1000 (1982)	46
695B Commander 1000	11
700 (1972)	20
720 AltiCruiser (1958)	8
Sabreliner 40 (1964)	75
Sabreliner 60 (1967)	75
Sabreliner 65 (1980)	70
Sabreliner 75 (1972)	4
Sabreliner 75A (1974)	46

ALSO SEE GRUMMAN, MEYERS

ROSE

Parakeet (1936)	9

RYAN

M-2	1
SCW-145 (1938)	4

RYAN AERONAUTICAL

ST-A Special (1935)	35
STM (1938)	3
PT-22 (1941)	181
SCW-145 (1935)	5

SAAB

Lansen 32 [A] (1956)	2
Lansen 32A [A]	1
350 Draken	2
91D Safir (1961)	1

SAVOIA-MARCHETTI

S-56-B (1930)	1

SHORTS

SC7 Series 3 Skyvan	18

SIAI-MARCHETTI

S.220 Vela	1

SPARTAN
Executive 7W (1937)............21
Executive 12 (1945)............1
C2 (1931)............2
C3 (1929)............5

ST. LOUIS
Cardinal C2 (1929)............1
Cardinal C2-110 (1930)............2

STAR
Cavalier (1936)............1
Cavalier 1
Cavalier E (1930)............2

STATE
Arrow F (1936)............7

STEARMAN
C2A (1927)............1
C3B (1927)............10
C3R (1929)............8
4C (1929)............2
4E (1930)............6
6L (1930)............2
4CM-1 (1931)............1
70............1
Kaydet PT-13 (1936)............5
Kaydet PT-13B............4
Kaydet PT-13D............18
Kaydet PT-17 (1941)............50
Kaydet PT-17A............1
Kaydet N2S-1 (1938)............1
Kaydet N2S-2............2
Kaydet N2S-3............16
Kaydet N2S-4............4
Kaydet N2S-5............4
Kaydet 75 (1941)............40
Kaydet A75............131
Kaydet B75............17
Kaydet E75............338
Kaydet IB75A............33
Kaydet A75J1............1
Kaydet A75N1............1,033
Kaydet B75N1............271
Kaydet D75N1............41
Kaydet E75N1............95
Kaydet A75L3............62
Kaydet A75L300............36

STINSON
SM-1B Detroiter............1
SM-1F Detroiter (1929)............1
Junior R............2
Junior S (1931)............12
Junior W (1931)............32
Junior SM-2 (1928)............1

Junior SM-2AA (1929)............2
Junior SM-7A (1930)............1
Junior SM-7B............2
Junior SM8A............19
SM6000 Airliner (1931)............3
Reliant SR (1933)............10
Reliant SR-5............2
Reliant SR-5A............5
Reliant SR-5B............1
Reliant SR-5C............2
Reliant SR-5E............3
Reliant SR-6............4
Reliant SR-6A............1
Reliant SR-7B............4
Reliant SR-7C............3
Reliant SR-8B............3
Reliant SR-8C............6
Reliant SR-8D............1
Reliant SR-8E............4
Reliant SR-9............2
Reliant SR-9B............2
Reliant SR-9C............15
Reliant SR-9E............7
Reliant SR-9E M............1
Reliant SR-9F............4
Reliant SR-10B............1
Reliant SR-10C............2
Reliant SR-10E............2
Reliant SR-10G............2
Reliant SR-10J............6
Reliant V77 (1942)............123
Reliant SR-10J3............3
Sentinel L-1 (1941)............1
Sentinel L-1F............2
Sentinel OY-2............1
Sentinel L-5 (1942)............80
Sentinel L-5B............8
Sentinel L-5C............1
Sentinel L-5E............24
Sentinel L-5G............28
Sentinel L-5E-1............5
Voyager 10 (1940)............41
Voyager 10A............119
Voyager 10L9B............2
108............421
108-1 (1946)............611
108-2 (1947)............529
108-3 (1947)............775

SUPERMARINE (See Vickers)

SWALLOW
Swallow (1928)............3
TP (1917)............5

SWEARINGEN (FAIRCHILD)
SA-26T Merlin IIA (1967)..............................17
SA-226AT Merlin IIB (1968)........................59
SA-226T Merlin III (1970)........................104
SA-226AT Merlin IV (1971)........................31
SA-226TC Metro (1970).............................82
SA-227AC Metro (1982)...........................150
SA-227AT Merlin IVA (1982)....................15
SA-227TT Merlin III (1982)......................19
SA-227DC..39

TAYLORCRAFT
Model A (1937)......................................27
Model B (1938)....................................755
Model D (1941)....................................338
L-2M...1
TG-6 (1936)..2
F-19 (1974)..148
F-21 (1980)..20
F-21A (1983)..5
F-21B (1986)..12
F-22A...9
F-22C...2

TIMM
Collegiate (1928)....................................2
N2T-1 (1943)..8

TRAVEL AIR
Model 2000 (1927)................................37
Model 3000 (1928)..................................3
Model 4000 (1927)..............................133
Model 6000 (1929)..................................7
R Mystery Ship (1929)............................1

VICKERS
Spitfire (1936)...8
Seafire..1

VIKING
Kitty Hawk B-4 (1929)............................2
Kitty Hawk B-8 (1931)............................2

WACO
A Model series (1932).............................5
C Model series (1931)...........................68
D S3HD (1934)......................................1
E Model series (1927) 24
F Model series (1930)..........................271
G Model series (1930).............................2
M Model series (1929)............................3
O Model series (1928)..........................78
S Model series (1934)...........................87
Model W...1
N AVN-8 (1938)....................................5
Model 9 (1927)......................................4
Model 10 (1928)...................................13
Model 12 (1928)....................................1

WALLACE
B-330...1

WILEY POST AIRCRAFT
Model A (1935).......................................2

Aircraft Sales Activity (new and used)

based on FAA new owner registration Source: Airpac, Inc., Edmond, OK USA, (800) 654-2006

Month	LSE	HPSE	Twin	TP	Jet	Total
1993						
July	803	585	331	118	104	1941
August	1429	1014	534	119	121	3217
September	1307	900	462	166	158	2993
October	1130	727	368	127	118	2470
November	1087	719	375	118	136	2435
December	1039	743	371	148	172	2473
1994						
January	1201	817	414	147	168	2747
February	1304	886	377	139	168	2974
March	1035	738	419	139	140	2471
April	688	493	279	105	110	1675
June	1201	779	400	150	118	2648
July	1368	907	467	161	127	3030
September	851	571	288	111	131	1952
October	1067	637	331	111	143	2289
November	1113	724	366	151	152	2506
December	787	572	323	157	145	1982
1995						
January	928	624	314	147	156	2169
February	697	540	287	138	137	1799
March	994	615	397	129	160	2295
April	976	635	387	115	129	2242
May	1156	863	407	169	138	2733
June	1147	865	372	139	170	2693
July	835	562	316	107	110	1930
August	1482	1007	445	154	159	3247
September	930	630	334	148	137	2179
October	1036	734	372	128	170	2440
November	986	654	368	126	145	2279
December	703	485	249	111	155	1703
1996						
January	1021	706	339	120	188	2374
February	1066	792	403	164	188	2613
March	1079	834	391	142	166	2612
April	912	707	358	128	166	2271
May	1315	940	407	147	166	2975
June	1136	757	412	134	167	2606
July	1166	821	436	372	220	3015
August	1083	727	379	178	205	2572
September	1119	733	389	149	161	2551
October	1322	861	440	171	180	2974
November	843	620	367	196	178	2204
December	745	560	270	234	164	1973
1997						
January	1136	768	398	168	231	2701
February	801	596	326	123	202	2048
March	870	637	316	143	167	2133
April	942	649	364	291	160	2406
May	849	681	331	129	163	2153
June	746	502	269	103	144	1764
July	740	549	277	144	173	1883

All the Pipers and Cessnas
Production aircraft and prototypes

Cessna Aircraft Co. production
(through 1988)

Year	Model	No.
1911	First Cessna	1
1912	Silver Wings	1
1913-17	1-palce monoplane	4
1926	4-place monoplane	1
1927	4-place monoplane	1
1928	AA	14
	AS	3
1928-30	AW	48
1928-29	BW	13
1929	AC	1
	CW-6	1
	CPW-6 (Goebel Spl.)	1
	CM-1	1
	DC-6	5
1929-34	DC-6A	22
	DC-6B	22
1930	GC-1 racer	1
	GC-2 racer	1
	CG-1 glider	1
	CG-2 glider	84
	CS-1 sailplane	1
	CPG-1 power glider	1
	Baby Cessna	1
	EC-1	2
	FC-1	1
	EC-2	1
1935-36	C-34 Airmaster	42
1937	C-37 Airmaster	46
1938	C-38 Airmaster	16
1938-41	C-145 Airmaster	42
	C-165 Airmaster	38
1940-41	T-50	40
1941	P-7	1
	P-10	1
	AT-8	33
1941-42	Crane I	640
1941-43	AT-17	1140
1942	Crane Ia	190
	CG-4A Waco glider	750
1942-44	UC-78	3356
1943	C-106/106A	2
1944	P780	1
1945-1986		
1946-49	120	2171
1946-50	140/140A	5430
1947-53	190	230
1947-54	195	866
1948-56	170	5173
1949	305 (L-19)	1

Year	Model	No.
1950-52	LC-126	83
1950-54	L-19A	2360
1951	X210	1
	308	1
1951-53	OE-1	62
1952	XL-19B	1
1953	L-19A-1T	68
	319	1
1953-55	305A	4
1953-67	310	5
1953-81	180 Skywagon	6193
1954	321	1
1954-62	CH-1Skyhook	33
1954-77	T-37	1272
1955	XL-19C	2
	OE-2	27
1955-70	180	4
1955-81	310	5242
1956	620	1
	325	2
1956-57	TL-19D	310
1956-63	L-19E	591
1956-86	172 Skyhawk	33629
	182 Skylane	19643
1957-58	L-27	160
	YH-41	10
1958-62	175 Skylark	2119
1957-68	210	4
1959	175	1
1959-77	150	22082
1960	336	2
1960-65	185	3
	320	2
1960-86	210 Centurion	8453
1961	U-3B	36
1961-64	205	2
1961-85	185 Skywagon	3859
1962	160	1
	411	1
1962-68	320 Skyknight	575
1962-70	U-17C	17
1963-64	205	574
1963-67	U-17A	265
1964-65	336 Skymaster	195
	337	2
1964-72	188	4
1964-86	206	7652
1965	P411	1
	360	1

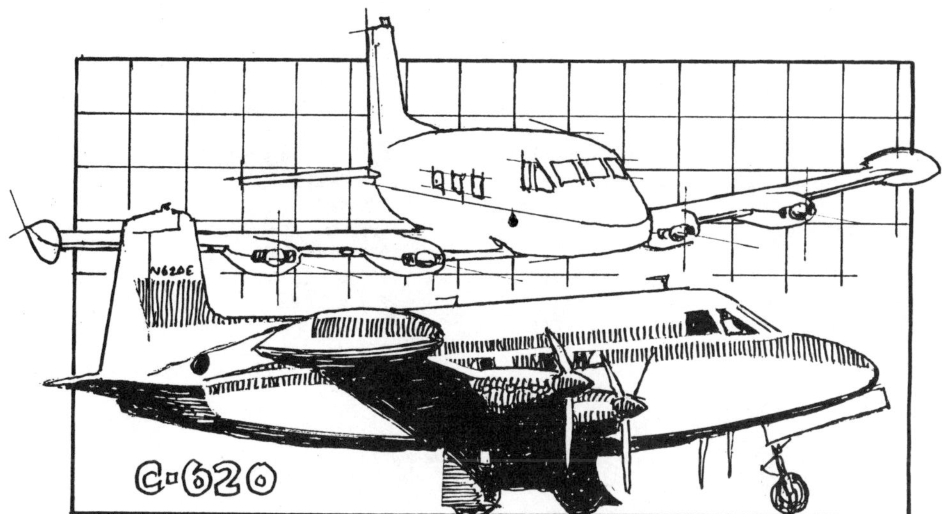

The four-engined, pressurized Model 620 was intended for executive transport, but Cessna discovered that potential buyers could get surplus military or airline aircraft for far less money.

Cessna cont

1965-68	411	301
1966	330	1
	172J	2
1966-67	T-41A	237
1966-77	F150	1758
1966-80	337 Skymaster	1857
1966-86	F172	2144
1967	T-41B	255
	210L	2
	182M	1
	207	1
	327	1
	187	1
1967-72	401	404
1967-73	U-17B	215
1967-70	O-2A	479
1967-85	402	1540
1968-83	188 Ag Wagon	3977
1968-69	T-41C	52
1968-77	414	2
1968-78	A-37B	577
	Reims Rocket	591
	177 Cardinal	2751
1968-83	T-41D	320
1968-84	421 Golden Eagle	1909
1969	1008	1
1969-70	500	2
1970	340	2
	1014	1
1970-85	414/Chancellor	1067
1971	TU206	1
	T337G	1
1971-78	Cardinal RG	1314
1972-76	Citation 500	349
1972-77	Reims Skymaster	67
1972-82	340	1298
1973-80	441	2
	Pressurized Skymaster	332
1973-82	207	790

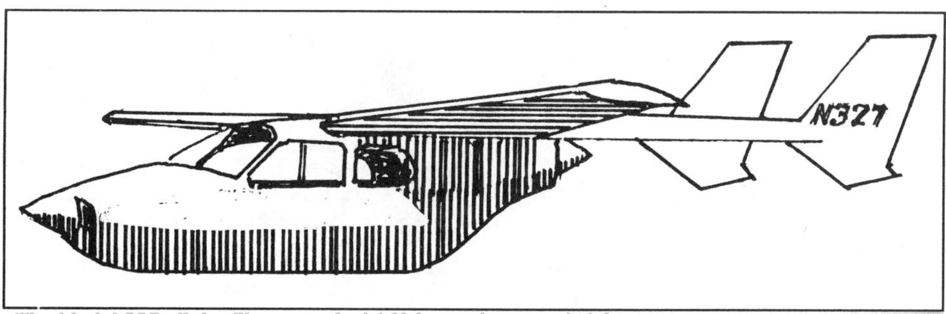

The Model 327   Baby Skymaster had 160 hp engines, carried four.

Cessna cont.

Year	Model	Qty
1974	404	1
1974-78	Reims P-Skymaster	27
1976	P210	1
	550	1
	182RG	1
1977	402C	1
	303	1
1977-81	R172 Hawk XP	1452
	Reims Hawk XP	85
	Reims 182	169
	404 II/Titan	395
1977-85	Citation I 501	342
1978	172RG	2
	425	1
1978-85	152	6860
	F152	640
	441 Conquest II	360
1978-86	182RG Skylane	2029
	Reims 182 RG	73
	P210	851
1979	T303	2
1979-80	650	2
1980	335	65
1980-84	172RG Cutlass	1159
1981-86	425 Conquest I	232
1982	208	1
1982-84	T303 Crusader	297
1983	406	1
1983-84	172Q Cutlass	36

Piper Aircraft Co. production
(through 1985)

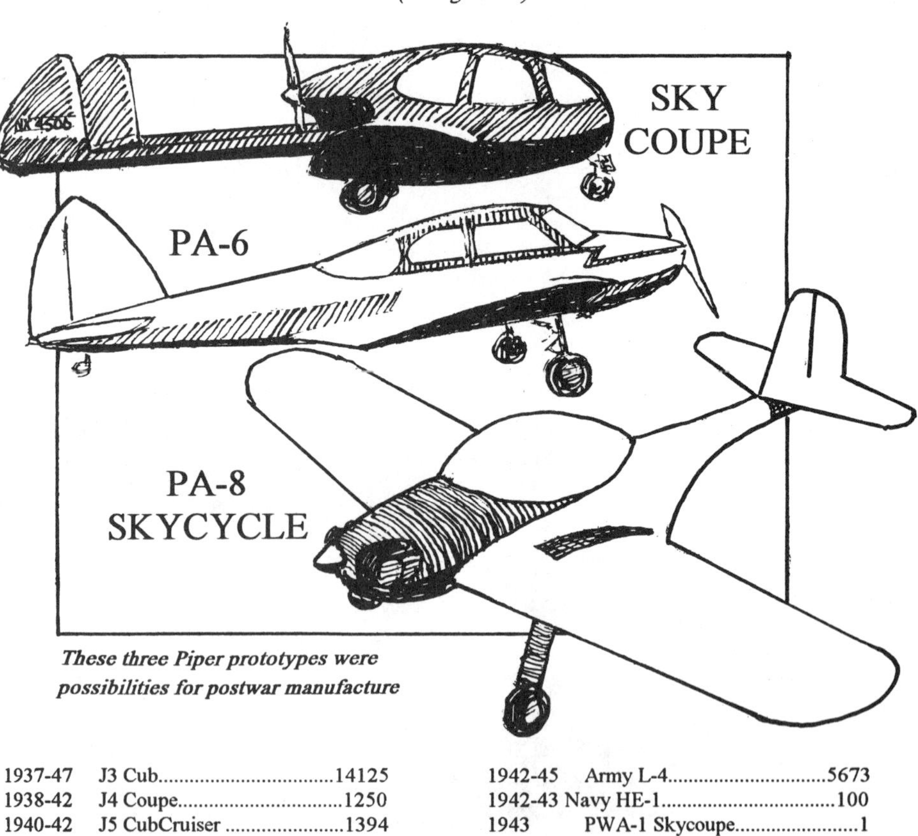

*These three Piper prototypes were
possibilities for postwar manufacture*

Year	Model	Qty
1937-47	J3 Cub	14125
1938-42	J4 Coupe	1250
1940-42	J5 CubCruiser	1394
1941	P-2 Cub	1
	P-4 Cub	1
1941-50	Commercial J5C	10
1942	PT-1 Trainer	1
1942-45	Army L-4	5673
1942-43	Navy HE-1	100
1943	PWA-1 Skycoupe	1
1943-44	TG-8 Glider	253
1944	PWA-8 (PA-8)	1
1945	Army L-14	14
	PWA-6 (PA-6)	1

Piper cont.

Year	Model	Qty
1946-49	PA-12 Super Cruiser	3758
1947	PA-11 Cub Special	1428
1948-52	PA-14 Family Cruiser	232
1948	PA-15/17 Vagabond	585
1948-50	Stinson 108-3	325
1949-54	L-18	972
1949-52	PA-16 Clipper	726
1949-85	PA-18 Super Cub	5792
1950-55	PA-20 Pacer	1120
1951	Twin Stinson	1
1951-60	PA-18A Super Cub	2650
1951-63	PA-22 Tripacer	7668
1951-58	Army L-21	716
1952-53	Air Force 18T	243
1954-57	PA-23 Apache 150	1231
1957-61	PA-23 Apache 160	816
1958-64	Comanche 180	1143
	Comanche 250	2537
1959-62	PA-25 Pawnee 150	731
1960	Navy UO-1	20
1961-67	Cherokee 150	300
	PA-28 Cherokee 160	810
1962-65	PA-23 Apache 235	119
1962-79	PA-25 Pawnee 235	3773
1962-75	PA-28 Cherokee 180	6338
1963-69	Twin Comanche 160	1998
1963-77	PA-28 235	2094
1964-70	PA-23 Aztec	4771
1964-78	PA-28 Cherokee 140	10086
1964-65	PA-24 Comanche 400	148
1965-72	PA-24 Comanche 260	1028
1965-78	PA-32 Cherokee Six 260	1493
1966-79	PA-32 Cheokee Six 300	2379
1967-83	PA-31 Navajo 300/310	1322
1967-77	PA-25 Pawnee 260	646
1967	PA-33 Pressurized Comanche	1
1967-71	PA-28R	1161
1967-	PA-28R-200	2851
1967-80	PA-25 Pawnee 260	646
1968	PA-35 Pocono	1
1968-81	Aerostar 600	265
	Aerostar 601B	157
1970-77	PA-31P Navajo 425	248
1970-72	PA-39 Twin Comanche C/R	145
1972-74	PA34-200 Seneca	918
1973	PA-40 Arapaho	3
1973-84	PA-31-350 Chieftan	1825
1973-77	PA-36 Pawnee Brave 285	372
1973-77	PA-28-151 Warrior	577
1974	PA-41P Pressurized Aztec	1
1974-83	PA-31T-620 Cheyenne II	524
1974-81	Aerostar 601P	454
1975-	PA-28-181 Archer	3376
1975-81	PA-34-200T Seneca	2602
1976-	PA-32R-300 Lance	1138
1978-	PA-32RT-300 Lance	388
1978-84	P-31T-500 Cheyenne I	211
1977-81	PA-36 Pawnee Brave 300	312
1977-78	PA-28R-201T Turbo Arrow	798
1978-81	PA-36 Pawnee Brave 375	239
1978-82	PA-38 Tomahawk	2512
1979-	PA-28-RT-201 Arrow	474
	PA-28RT-201T T. Arrow	870
	PA-28-236 Dakota	92
1979-82	PA-44 Seminole	467
1980-	PA-32-301 Saratoga	382
	PA-32R-301T Saratoga	378
1980-87	PA-42-720 Cheyenne III	
1981-84	PA-31T-620XL	76
1981-	PA-34-220T Seneca	739
1981-84	Aerostar 602P	110
1982-	PA-36 Pawnee Brave 400	25
1983-84	Aerostar PA-60-700P	25
1984-87	PA-42000 Cheyenne 400LS	29
1984-88	PA-46-310P Malibu	212
1984	PA-31P-350 Mojave	50

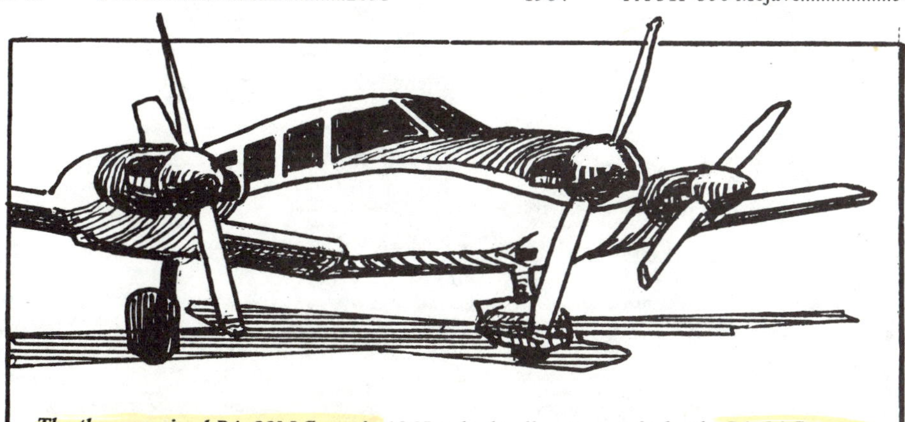

The three-engined PA-32M flown in 1965, udoubtedly a test mule for the PA-34 Seneca

new fiberglas **TAYLORCRAFT**

featuring **DESIGNED SAFETY**

WITH SIX SAFETY FIRSTS

1  **DIE-MOLDED FIBERGLAS FUSELAGE**

An entirely new concept in aircraft design. Tough in substance, yet lightweight and non-conductive. Resin impregnated fiberglas is formed to make it stronger than most metals. Will not dent, puncture, or corrode.

2  **STRONG, WELDED STEEL FRAMEWORK**

Passengers and pilot are surrounded by a protective, reinforced frame of steel tubing. All control cables are between frame and fiberglas fuselage giving an unobstructed interior.

3  **SLOTTED FIBERGLAS WINGS**

Leading edge of wing is slotted to give greater flight stability. This feature assures ease in handling controls in slow flight under heavy load conditions. Permits slower, safer landing speeds.

4 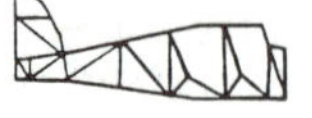 **SLOTTED FLAPS AND AILERONS**

Full control at and near stalling speeds. Greater performance features of shorter takeoff in smaller areas, quicker climb and smoother landings.

5  **HIGHER "FLIGHT STRENGTH"**

A rugged design to permit safer flying even in "rugged weather." Additional built-in strength for maximum passenger safety. TAYLORCRAFT has been tested under highest operational category by CAA.

6  **THREE DOORS**

No more sliding seats, cramped exit and entrance. Ease and comfort for pilot and passengers through three-door accessibility. Roomy interior even with full baggage load.

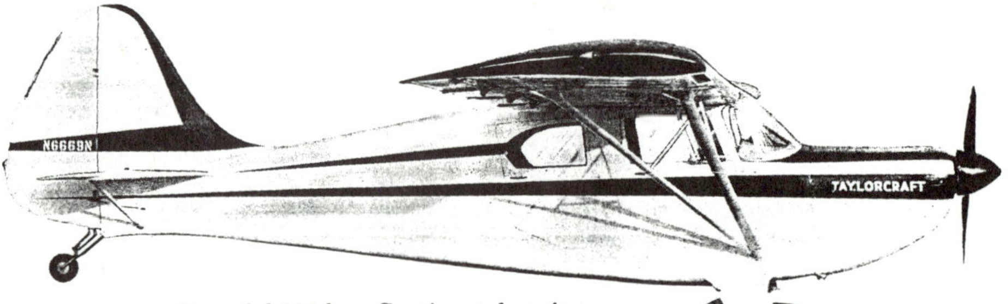

- Powerful 225 h.p. Continental engine
- 150 m.p.h. cruising speed

You fly the finest when you fly the new FOUR-PLACE fiberglas TAYLORCRAFT. Named by many as the "strongest, safest" plane in use today. Years ahead in safety and performance. Write today for full information. Price: $10,865*

Sales Outlets still available for interested airport operators

*f.a.f. Conway, Pa.

FLYING IS SAFER — SAFEST IN A TAYLORCRAFT

II. GENERAL AVIATION PILOTS

General Aviation pilots will introduce more than a million young people to the flight experience by 2003, the 100th anniversary of the first flight, in the EAA's "Young Eagles" program (Photo by Phillips 66)

ACTIVE U.S. PILOT CERTIFICATES HELD
(Dec. 31 of each year)

Source: FAA

Year	Stud't	Pvt. (only)	Comml.	ATP	Heli. (only)	Sail (only)	LTA*	Rec.	Total
1975	176,978	305,863	189,342	45,592	4,932	5,348	3,132		728,187
1976	188,801	309,005	187,801	45,072	4,804	5,789	2,974		744,246
1977	203,510	327,424	188,763	50,149	4,819	6,208	3,059		783,932
1978	204,874	337,644	185,833	55,881	4,874	6,541	3,186		789,833
1979	210,180	343,276	182,097	63,652	5,218	6,796	3,448		814,667
1980	199,833	357,479	183,442	69,569	6,030	7,039	3,679		827,071
1981	170,912	328,562	168,580	70,311	6,453	7,388	2,976		764,182
1982	156,361	322,094	165,093	73,471	7,034	7,842	1,360		733,255
1983	147,197	318,643	159,495	75,938	7,237	8,157	1,337		718,004
1984	150,081	320,086	155,929	79,192	7,532	8,390	1,166		722,376
1985	146,652	311,086	151,632	82,740	8,123	8,168	1,139		709,540
1986	150,273	305,736	147,798	87,186	8,122	8,411	1,133		709,118
1987	146,016	300,949	143,645	91,287	8,702	7,901	1,153		699,653
1988	136,913	299,786	143,030	96,968	8,608	7,600	1,111		694,016
1989	142,544	293,179	144,540	102,087	8,863	7,708	1,089		700,010
1990	128,663	299,111	149,666	107,732	9,567	7,833	*	87	702,659
1991	120,303	293,306	144,365	112,167	9,860	8,033	*	161	692,095
1992	114,597	288,078	146,385	115,855	9,652	8,205	*	187	682,959
1993	103,583	283,700	143,014	117,070	9,166	8,326	*	206	665,069
1994	96,254	284,236	138,728	117,434	8,719	8,476	*	241	654,088
1995	101,279	261,399	133,980	123,877	7,183	11,234	*	232	639,184
1996	94,947	254,002	129,187	127,484	6,961	9,413	*	265	622,261

*Lighter than air--not issued after 1989

AGE DISTRUBITION OF ACTIVE AIRMEN BY CLASS AND SEX
(Dec. 31, 1989)

Source: FAA

Age	First Class male/female	Second Class male/female	Third Class male/female	Total male/female
to 15	3/1	7/2	69/14	79/17
15-19	3,544/403	5,978/619	16,168/2,361	25,690/3,383
20-24	11,315/919	15,305/1,075	25,456/2,690	52,076/4,684
25-29	18,166/1,189	19,977/1,191	35,036/3,752	73,179/6,132
30-34	23,102/1,000	20,701/1,029	42,902/4,125	86,705/6,154
35-39	18,075/528	23,260/1,004	44,084/3,913	85,419/5,445
40-44	20,050/271	26,595/816	40,874/3,555	87,519/4,641
45-49	16,288/104	19,753/571	33,262/2,445	69,303/3,120
50-54	12,213/49	14,872/342	26,132/1,642	53,217/2,033
55-59	7,306/22	11,866/291	23,575/1,130	42,747/1,443
60-64	1,426/12	8,314/229	18,583/688	28,323/929
65-69	680/5	5,841/128	10,531/303	17,052/436
70-74	143/1	1,835/43	3,902/89	5,880/133
75-79	24/1	394/13	1,024/25	1,442/39
80-84	3/0	72/1	200/3	275/4
85+	0/0	10/0	22/0	32/0

ESTIMATED ACTIVE PILOTS AND FLIGHT INSTRUCTORS
BY FAA REGION AND STATE
(Dec. 31, 1993)

Source: FAA

	Stud't	Pvt.	Comml.	ATP	Misc.	Inst.	Total
Alaska							
Alaska	1124	4293	2451	1476	129	984	7472
Central							
Iowa	1024	3665	1330	549	65	726	6633
Kansas	1224	4546	1950	1027	120	988	8867
Missouri	1899	5698	2443	1882	224	1415	12146
Nebraska	715	2449	1118	436	31	444	4749
	4962	16358	6841	3894	440	3573	32395
Eastern							
Delaware	259	657	336	315	45	248	1612
District of Columbia	105	232	123	51	33	51	544
Maryland	1517	4282	1817	1450	315	1150	9435
New Jersey	2317	5697	2427	2201	463	1597	13105
New York	4144	9495	4192	2423	990	2552	21244
Pennsylvania	3461	9102	3898	3454	546	2643	20461
Virginia	2301	5521	3616	3244	438	1835	15120
West Virginia	415	1196	459	228	59	241	2357
Armed Forces Europe	214	268	260	99	58	71	899
	14733	36460	17182	13465	2947	10388	84777
Great Lakes							
Illinois	3934	10582	4828	4227	448	3067	24019
Indiana	2048	6054	2380	1414	155	1372	12051
Michigan	3212	9464	3525	2125	405	2150	18731
Minnesota	2459	7622	3414	2644	203	1856	16342
North Dakota	125	1481	998	171	73	324	3098
Ohio	3319	9908	4013	2643	405	2382	20288
South Dakota	389	1148	576	265	19	233	11761
Wisconsin	1950	6177	1093	1420	121	1230	11761
	`7736	52436	21827	14909	1779	12614	108687
New England							
Connecticut	1147	3217	1361	1692	255	848	7072
Maine	660	1761	757	401	47	342	3626
Massachusetts	1919	5354	2018	1295	442	1250	11028
New Hampshire	597	1776	830	1166	103	558	4472
Rhode Island	241	697	267	170	35	150	1410
Vermont	296	790	326	234	68	187	1714
	4860	13696	5559	4958	950	3336	29922
NW Mountain							
Colorado	2407	6387	3636	4123	496	2360	17049
Idaho	692	2286	976	527	94	421	4575
Montana	557	1921	932	372	49	397	3831
Oregon	1602	5309	2106	973	267	1080	10257
Utah	1086	2605	1104	1248	122	606	6165
Washington	3356	8964	4475	4770	530	2390	22115
Wyoming	250	991	382	237	33	196	1893
	9960	28493	13611	12250	1591	7460	65885

Pilots by region and state cont.

Southern							
Alabama	1484	3451	1978	902	390	1024	8206
Florida	6675	17800	11278	10786	770	5714	47309
Georgia	2588	6207	3464	5247	326	1731	17812
Kentucky	1111	2542	1106	1363	152	657	6274
Mississippi	798	1732	1288	463	113	448	4394
North Carolina	2295	6044	2957	2925	378	1485	14599
South Carolina	1109	2830	1497	913	135	733	6484
Tennessee	1998	4998	2373	2980	307	1517	12656
Armed Forces							
Europe	27	31	35	17	17	14	127
	18065	**45848**	**25976**	**25586**	**2588**	**13223**	**117860**
Southwest							
Arkansas	932	2502	1449	612	78	562	5573
Louisiana	1143	2723	2027	975	268	860	7136
New Mexico	676	2055	11228	679	138	565	4676
Oklahoma	1803	4652	2248	1128	125	1210	9956
Texas	6757	18835	10563	12059	1072	5886	49286
	11311	**30767**	**17415**	**15463**	**1681**	**9083**	**75627**
Western-Pacific							
Arizona	2196	6124	3350	3129	523	2172	15324
California	12800	29708	17265	12936	3263	9025	96972
Hawaii	446	840	689	863	250	360	3088
Nevada	685	2192	1088	1576	162	707	5703
Armed Forces							
Pacific	265	213	193	65	31	48	767
	16394	**49077**	**22585**	**18569**	**4228**	**12312**	**110854**
Other U.S. territories	4548	6806	9567	8601	1368	2059	28590
Total	**99035**	**277094**	**133447**	**110569**	**16334**	**72962**	**636479**

ACTIVE PILOTS BY TYPE OF CERTIFICATE
(Thousands)

Source: FAA

Year	Student	Pvt.	Comm.	ATP	Heli	Glider	Total
1990	139.6	300.7	145.0	101.5	8.7	7.7	704.3
1991	142.0	302.8	146.5	104.7	9.0	8.3	714.5
1992	144.1	303.7	148.0	108.6	9.1	8.4	723.1
1993	145.9	304.3	149.4	112.6	9.2	8.5	731.1
1994	147.3	304.9	150.9	116.7	9.4	8.6	739.1
1995	148.4	305.8	152.4	121.0	9.5	8.7	747.2
1996	149.1	306.2	154.0	124.4	9.6	8.8	754.2
1997	149.7	307.7	155.5	127.8	9.7	8.9	760.9
1998	150.2	308.6	157.1	131.3	9.8	9.0	767.7
1999	150.7	309.5	158.6	133.9	9.9	9.1	773.5
2000	151.2	310.5	160.2	136.6	10.0	9.2	779.6
2001	151.7	311.4	161.8	139.3	10.1	9.3	785.6

Corporate Pilot Salary Survey
Business & Commercial Aviation survey , April 1997

SALARY BY JOB TITLE

Job title	Salary range	Average Salary	Median Salary
Director of Operations	$20,000-126,000	$62,033	$48,950
Flight Dept. Manager	$12,000-225,000	$83,500	$85,756
Chief Pilot	$15,000-128,000	$60,927	$57,250
Captain	$18,000-155,000	$59,765	$58,000
Pilot	$16,500-78,400	$45,349	$47,500
First Officer	$15,000-68,000	$35,040	$31,022

SALARY BY EQUIPMENT TYPE

Job title	Salary range	Average Salary	Median Salary
HEAVY JETS (30,000+ lb. MTOW)			
Director of Operations	$66,000-225,000	$111,231	$105,000
Flight Dept. Manager	$100,700-126,000	$117,233	$125,000
Chief Pilot	$42,000-109,000	$78,625	$82,000
Captain	$26,000-225,000	$73,291	$73,500
Pilot	$55,000-67,700	$59,892	$57,000
First Officer	$26,000-68,000	$43,524	$46,000
MEDIUM JETS (20,000-29,999 lb. MTOW)			
Director of Operations	$38,000-68,000	$48,000	$47,000
Flight Dept. Manager	$43,000-120,000	$94,717	$100,000
Chief Pilot	$42,000-128,000	$80,926	$77,000
Captain	$24,000-101,000	$62,606	$63,000
Pilot	$28,000-67,675	$44,154	$42,950
First Officer	$26,000-60,000	$36,206	$29,023
LIGHT JETS (to 19,999 lb. MTOW)			
Director of Operations	$46,124-107,000	$76,148	$77,900
Flight Dept. manager	$38,000-72,000	$46,400	$40,000
Chief Pilot	$34,000-128,000	$64,747	$60,000
Captain	$24,000-101,000	$53,429	$52,000
Pilot	$16,500-65,000	$45,363	$49,500
First Officer	$15,000-68,000	$35,040	$31,022
TURBOPROPS			
Director of Operations	$37,124-100,000	$61,836	$65,998
Flight Dept. Manager	$20,000-45,000	$37,400	$40,000
Chief Pilot	$24,000-85,000	$52,255	$51,000
Captain	$18,000-155,000	$47,987	$43,661
Pilot	$16,500-66,500	$42,092	$45,000
First Officer	$15,000-34,000	$27,761	$31,023
PISTON TWINS			
Director of Operations	$12,000-105,000	$62,501	$62,950
Flight Dept. Manager	$20,000-50,400	$35,200	$35,200
Chief Pilot	$15,000-75,000	$38,815	$36,000
Captain	$18,000-57,200	$32,658	$31,900
Pilot'	$20,000-66,500	$36,400	$31,500

III. GENERAL AVIATION AIRPORTS

U.S. CIVIL AND JOINT-USE AIRPORTS, HELIPORTS AND SEAPLANE BASES ON RECORD BY TYPE OF OWNERSHIP

Dec. 31, 1996

FAA Region and state

	Total		Airports open to public				
	Public	Pvt.	Paved		Unpaved		Total
			Lighted	Not	Lighted	Not	Airports
Alaska							
Alaska	383	163	45	5	99	156	546
Central							
Iowa	136	169	96	1	13	14	305
Kansas	132	252	101	9	17	21	384
Missouri	137	370	113	5	6	10	507
Nebraska	94	203	76	2	5	9	297
	474	932	382	15	63	72	1,493
Eastern							
Delaware	4	31	5	0	3	1	35
DC	8	9	2	0	0	0	19
Maryland	21	179	28	1	3	3	200
New Jersey	47	301	36	2	4	7	348
New York	91	444	83	13	22	31	535
Pennsylvania	74	681	89	5	14	28	755
Virginia	73	292	63	1	1	3	365
West Virginia	31	73	26	4	1	9	104
	349	2,010	332	26	48	82	2,359
Great Lakes							
Illinois	122	774	92	0	19	5	896
Indiana	87	516	78	4	5	21	603
Michigan	134	337	124	7	43	56	471
Minnesota	150	325	107	0	23	16	475
North Dakota	96	338	66	5	12	11	434
Ohio	136	603	124	2	17	24	739
South Dakota	77	79	54	1	14	5	156
Wisconsin	103	382	99	2	12	18	485
	905	3,354	744	21	145	156	4,259
New England							
Connecticut	16	120	17	2	0	4	136
Maine	46	104	30	7	2	7	150
Massachusetts	35	191	36	4	1	5	226
New Hampshire	16	78	16	3	1	6	94
Rhode Island	9	17	7	0	0	0	26
Vermont	20	53	10	1	0	5	73
	142	563	116	17	4	27	705
NW Mountain							
Colorado	92	299	64	4	3	8	391
Idaho	131	96	44	9	1	64	227
Montana	123	120	73	7	8	31	243
Oregon	103	308	58	17	4	18	411
Utah	58	67	41	5	0	1	125
Washington	127	310	93	8	4	17	437
Wyoming	49	55	34	1	1	5	104
	683	1,255	407	51	21	144	1,938

Airports by type of ownership cont.

	Total		Airports open to public				
	Public	Pvt.	Paved		Unpaved		Total
			Lighted	Not	Lighted	Not	Airports
Southern							
Alabama	103	137	86	5	4	6	240
Florida	163	606	100	5	8	15	769
Georgia	136	265	97	9	2	2	401
Kentucky	73	106	53	9	0	3	179
Mississippi	90	131	70	9	2	2	221
North Carolina	95	263	87	4	5	16	358
Puerto Rico	17	15	10	1	0	0	32
South Carolina	69	98	57	1	4	6	167
Tennessee	88	170	71	6	0	2	258
Virgin Islands	6	3	2	0	0	0	9
	840	1,794	633	49	25	52	2,634
Southwest							
Arkansas	115	150	90	5	0	5	265
Louisiana	110	316	71	2	1	7	426
New Mexico	73	91	43	9	0	10	164
Oklahoma	160	250	112	13	6	17	410
Texas	358	1,326	321	18	5	35	1,684
	816	2,133	637	47	12	74	2,949
Western-Pacific							
Arizona	88	188	54	8	0	14	276
California	319	614	207	42	1	11	933
Hawaii	18	28	11	2	0	0	46
Nevada	60	59	26	7	3	19	119
S. Pacific	27	8	11	2	0	13	35
	512	897	309	61	4	57	1,409
U.S. Total	5,129	13,163	3,609	294	399	802	18,292

NOISE LEVELS FOR BUSINESS JETS

Multiple dBA levels indicate quietest and noisest readings for aircraft; weights are maximum gross takeoff; stage levels from FAA Advisory Circular (AC) 36-F1, Appendix 1; dBA ratings are from AC 36-3F, Appendix 2 and 3. *EPNdB=Effective Perceived Noise Decibel system.

Aircraft	Takeoff dBA	Approach dBA	Flaps (°)	Takeoff EPNdB*	Approach EPNdB*	Stage
Beechjet						
400	71.8	83.0		88.6	91.4	3
400A				89.0	91.7	3
Canadair						
Challenger 600	66.9	81.7		81.6	91.2	3
Challenger 601	66.4	80.4		79.4	89.4	3
Challenger IAII			45	87.4	91.6	3
Challenger 2A12			45	79.9	89.8	3
Challenger 2B16			45	79.9	89.4	3
Challenger RJ 600-2B19 (@51,000#)	62.7	81.4		65.6	81.4	3
Challenger RJ600-2B19 (@53,000#)	67.2	81.4		78.6	92.1	3
Cessna						
500	67.0	77.7	40	76.4	87.7	3
501	67.3	77.7	40	78.0	87.9	3
550	67.4	79.8	40	80.1	90.5	3
650	69.3	84.8	37	80.1	93.8	3
560	69.4	80.5	35	84.6	88.9	3
660	65.4	81.6	40	77.1	90.8	3
525	60.3	81.7	40	73.4	92.1	3
Dassault Falcon						
10	69.4	81.8	30	81.6	95.4	3
20 (CF700)	77.0	90.1	25	90.0	103.0	2
20 (CF700-20-2)	70.6	79.4	25	80.3	90.7	3
200	71.7	84.1	20	83.9	93.9	3
50	70.9	82.0	20	84.8	97.1	3
900	69.2	81.0	40	81.9	91.7	3
900B	69.9	82.5	40	79.8	91.7	3
2000	64.0	83.7	40	79.4	93.1	3
Gulfstream						
G-II (@62,000#)				90.0	98.4	2
(@65,500#)	84.2	83.9	20	92.5	98.4	2
G-IIB/G-III	82.8	82.5	20	91.1	97.3	2
G-IV (@71,700#)	66.9	80.6	39	76.8	91.0	3
(@73,200#)	64.2	80.7	39	79.0	91.0	3
G-IV SP	64.9	81.3	39	77.5	92.0	3
Hawker						
HS-125-1A/3A	70.4-72.4	83.3-85.5	25	84.2	96.0	3
125-400	72.4	83.0	25	85.5	95.7	3
HS-125-600	81.9	96.0	45	92.3	102.9	2
HS-125-700	75.8	83.6	25	88.0	96.3	3

Business jet noise levels, cont.

Aircraft	Takeoff dBA	Approach dBA	Flaps (°)	Takeoff EPNdB*	Approach EPNdB*	Stage
Hawker cont.						
800	69.7	85.0		80.9	96.5	3
1000	71.8	82.2	25	81.8	91.6	3
IAI						
1121 Commodore	89.7	100.0		100.9	107.0	1
1123 Westwind	89.7	99.0		100.9	106.0	1
1124/1124A	70.3	84.2	40	85.4	92.8	3
1125 Astra						
(@23,500#)	70.3	80.4		82.3	89.8	3
(@24,700#)	72.1	80.4	40	84.1	89.8	3
Learjet						
23 (CJ610)	84.7	89.7		88.0	98.0	2
24 (CJ610,						
@13,500/11,900#)	73.1-80.6	88.3-94.7		83.6-91.9	95.3-101.7	2
25 (CJ610,						
@16,300/a5,000#)	79.7-82.8	88.2-93.8		90.9-94,0	95.2-102.7	2
28/29	87.0	101.7		87.0	99.7	2
31	68.9	82.9		81.0	87.0	3
35/36	65.6-71.5	81.6-81.9		84.0	92.2	3
55	67.0-68.4	81.5-81.9		84.2=86.7	90.6-92.4	3
60	60.9	77.4		70.8	87.7	3
Lockheed Jetstar						
1329-23 (JT12A)	88.7	101.0		106.6	107.5	1
1329-23 (TFE731)	82.3	88.3	50	92.7	96.0	2
1329-23 (@44,250#)	74.7	88.3	20	85.2	96.9	3
1329-25 (@43,800#)	82.3	88.3		93.1	96.9	2
1329-25 (@44,500#)	75.0	88.3	20	85.4	96.6	3
Sabreliner						
40	83.4	92.0		89.7-94.5	97.5-98.4	2
60	83-8-84.7	92.0-95.4		94.4-95.0	98.5-102.2	2
65	70.8	81.7		82.3-84.0	90.6	3
75A	77.7	90.3	25	90.7	100.2	2
80	79.6-80.5	90.3-91.0		90.7-91.2	100.2-101.1	2

FAA AIR ROUTE FACILITIES 1972-1996 (Dec. 31)

	VOR/ VORTAC	NDB	ARTC center	ATC tower	FSS	ILS	ASR
1972	991	706	27	355	324	403	125
1973	995	739	27	403	315	467	142
1974	1000	793	27	417	320	490	156
1975	1011	848	26	487	321	580	177
1976	1020	920	25	488	321	640	175
1977	1021	959	25	495	319	678	182
1978	1020	988	25	494	319	698	185
1979	1028	1015	25	499	318	753	192
1980	1037	1055	25	502	317	796	192
1981	1033	1123	25	501	316	840	199
1982	1029	1143	25	492	316	884	197
1983	1032	1183	25	494	316	934	197
1984	1035	1211	25	497	310	955	197
1985	1039	1222	25	500	302	968	198
1986	1043	1239	25	686	293	977	312
1987	1045	1262	25	686	253	1111	312
1988	1043	1287	24	692	220	1130	311
1989	1044	1263	24	693	199	1147	312
1990	1044	1309	24	691	180	1114	311
1991	1045	1325	24	691	175	1188	318
1992	1044	1314	24	691	179	1177	312
1993	1041	1344	24	684	165	1231	310
1994	1041	1342	24	685	156	1245	303
1995	1039	1333	24	683	91	1280	298
1996	1027	1165	21	664	94	1197	228

TOP 20 U.S. AIRPORTS

Rank by total traffic, FY 1995; Source: FAA

	Total operations	General Aviation as % of total
1. Chicago O'Hare, Illinois	892,330	4.1
2. Dallas/Fort Worth, Texas	873,510	1.8
3. Atlanta Hartsfield, Georgia	747,136	3.1
4. Los Angeles Int'l, California	716,293	3.8
5. Miami Int'l, Florida	576,609	12.6
6. Phoenix Sky Harbor, Arizona	522,634	17.6
7. St. Louis Intl, Missouri	516,021	7.5
8. Las Vegas McCarran, Nevada	508,077	20.9
9. Detroit Metro Wayne, Mich.	501,402	14.3
10. Denver Stapleton, Colorado	487,225	6.1
11. Boston Logan, Massachusetts	478,253	8.2
12. Charlotte Douglas, N. C.	474,338	12.9
13. Minneapolis/St. Paul, Minn.	466,916	12.9
14. Newark, New Jersey	428,703	4.7
15. Pittsburgh Greater Intl, Penn.	452,900	5.1
16. San Francisco Int'l, California	436,907	5.9
17. Seattle-Tacoma Int'l, Wash.	382,100	2.5
18. Houston Intercontinental, Tex.	375,246	6.5
19. JFK, New York	352,494	3.9
20. LaGuardia, New York	346,869	5.4

TOP 20 U.S. GENERAL AVIATION AIRPORTS

FY 1995, rank by total traffic; Source: FAA

Airport	Operations	Air carrier as % of total
1. Van Nuys, California	500,959	0.001
2. Oakland Int'l, California	502,952	32.7
3. Long Beach, California	493,391	1.4
4. Santa Ana/Orange Co., California	491,464	15.5
5. Denver Centennial, Colorado	400,103	--
6. Seattle Boeing, Washington	357,266	2.3
7. Tulsa Riverside, Oklahoma	282,527	--
8. Morristown, New Jersey	277,066	0.0007
9. Daytona Beach, Florida	254,442	3.5
10. Meacham Ft. Worth, Texas	248,137	0.06
11. Ft. Lauderdale Exec., Florida	234,675	--
12. San Diego Montgomery, California	233,298	--
13. Caldwell, New Jersey	225,611	--
14. Atlanta DeKalb-Peachtree, Georgia	224,156	--
15. Carlsbad Palomar, California	215,112	--
16. Teterboro, New Jersey	195,853	0.06
17. Chicago Palwaukee, Illinois	194,798	--
18. Santa Monica, California	191.918	--
19 Dallas Addison, Texas	158,166	--
20. Pontiac, Michigan	148,882	0.11

CIVIL AVIATION ACCIDENT DATA--1984-1996

Accident rates per 100,000 flight hours--fixed wing
Source: NTSB

Year	Air Taxi total/fatal	Commuter total/fatal	Airlines total/fatal	Corporate total/fatal	G.A.. total/fatal
1984	5.14/0.81	1.26/0.41	0.208/0.012	0.52/0.08	10.36/1.87
1985	5.99/1/36	1.209/0.403	0.253/0.08	0.88/0.31	9.66/1.75
1986	4.35/1.15	0.87/0.116	0.231/0.02	0.53/0.08	9.54/1.75
1987	3.65/1.13	1.644/0.514	0.329/0.038	0.56/0.12	9.25/1.65
1988	3.84/1.06	0.908/0.096	0.251/0.018	0.27/0.05	8.69/1.68
1989	3.68/0.83	0.803/0.223	0.248/0.098	0.29/0.11	7.96/1.53
1990	4.71/1.24	0.642/0.128	0.198/0.049	0.21/0.09	7.78/1.55
1991	3.88/1.2	1.013/0.368	0.218/0.034	0.23/0.08	7.98/1.58
1992	3.78/1.19	1.008/0.321	0.144/0.032	0.21/0.08	8.71/1.87
1993	3.81/1.06	0.66/0.165	0.178/0.008	0.23/0.07	9.05/1.77
1994	4.26/1.3	0.429/0.129	0.159/0.030	0.18/0.07	9.09/1.83
1995	3.75/1.2	0.465/0.078	0.266/0.022	0.25/0.11	10.31/2.05
1996*	4.63/1.42	0.40/0.04	0.30/0.039	0.14/0.06	8.06/1.51

*preliminary

Accident rates per 100,000 flight hours--rotorcraft
Source: NTSB

	Piston engine	Turbine engine	All rotorcraft
1984	3.5	10.4	14.9
1985	3.0	9.0	12.0
1986	2.8	9.0	11.8
1987	3.4	10.4	13.8
1988	5.4	4.6	10.0
1989	4.2	6.7	10.9
1990	4.0	8.2	12.2
1991	5.8	2.7	8.5
1992	6.8	4.2	11.0
1993	5.3	5.2	10.5
1994	4.4	5.2	9.6

Phases of flight in which accidents occurred--rotorcraft

	Single-engine	Multi-engine
Cruise	26.8%	29.3%
Descent	2.6%	1.5%
Approach	5.6%	6.6%
Hover	8.1%	--
Maneuver	16.7%	--
Hover/maneuver	--	32.3%
Landing	13.1%	14.1%
Parked/taxi	4.0%	6.6%
Takeoff	12.2%	13.1%
Climb	3.4%	4.0%

NATIONAL AVERAGE FUEL COSTS
July 1996-May 1997
Compiled by Fillup Flyer Fuel Finder (800) 333-7900, www.fillupflyer.com/ff

	1996						1997				
	July	Aug.	Sep.	Oct.	Nov.	Dec.	Jan.	Feb.	Mar.	April	May
Jet-A	1.91	1.91	1.96	2.00	2.01	2.02	2.03	2.00	1.99	2.03	2.02
100LL	2.06	2.07	2.07	2.07	2.07	2.08	2.11	2.10	2.10	2.12	2.09
80 Oct.	2.03	2.03	2.04	2.05	2.05	2.05	1.98	1.98	1.98	1.55	2.05
Augogas	1.56	1.56	1.56	1.56	1.56	1.56	1.55	1.55	1.56	1.55	1.57

IV. CIVIL & MILITARY AIRCRAFT

CIVIL & MILITARY AIRCRAFT
Specifications and Performance

Source: *Aviation Week & SpaceTechnology Aerospace Source Book,* 1997

Powerplant key: TF=Turbofan; TJ=Turbojet; TP=Turboprop; TS=Turbohaft
Powerplant manufacturers: 250=Allison; AL=Lycoming; ATF=Garrett; C=General Electric;
Cont.=Teledyne Continental piston; FJ=Williams; J=General Electric; JT=Pratt & Whitney; LTS=Lycoming;
Lyc.=Lycoming piston; PT & PW=Pratt & Whitney; P&W Pratt & Whitney
piston; RR=Rolls-Royce; T=GE; TFE=Garrett; TM=Turbomeca; TPE=Garrett
* indicates production terminated. Other models may be listed under General Aviation Aircraft

Model	Crew +pax	GW, lb.	Powerplant (hp or lb. thrust)	Cruise, mph (unless noted)
Aero Boreo, Cordoba, Argentina				AERO BOREO
115BS	1+2	1,543	Lyc. 235-C2 A	137
150-RV	1+2	1,861	Lyc. O-320 A/E	152
150-AG	1+2	2,207	Lyc. O-320 A/E	142
180-RV	1+2	1,861	Lyc. O-360 A1A	153
180-RVR	1+2	1,861	Lyc. O-360 A1A	153
180-AG	1+2	2,207	Lyc. O-360 A1A	143
260-AG	1+1	2,977	Lyc. O-540 H2B5D	156
Aero Industry Development Center, Taipei, Republic of China				
Transport & Utility				
XC-2	3+30-38	27,500	2 T53-L-701A TP	244
Fighters				
IDF Ching-Kuo	1	20,000	2 ITEC F1225 TF	M1.7
Trainers				
TCH-1 Chung-hsing	2	9,200	T53-L-701 TP	368
PL-1B Chieh-shou	2	1,440	Lyc. O-320-E2A	205
AT-TC-3	2	16,467	2 TFE 731-2-2L TF	M0.85
Aero Vodochody, Prague, Cxech Republic				
L-39ZA Albatros	2	12,346	Progress AI-25TL TF	469
Aeromacchi, Varese, Italy				MACCHI
Trainers				
AMX-T	2	28,660	RR Spey Mk 807	500
MB.339B	2	14,000	RR Viper 680-43 TJ	496
MB.339D/FD	2	14,000	RR Viper 632-43 TJ	496
M-290TP Rediogo	2	4,190	250-B17F TP	190
Aerospace Technologies of Australia Pty. Ltd., Port Melbourne				
Utility				
N22B Nomad 22C	2+12	8,950	2 250-B17C TP	200
Turboprop Commuters				
N24A Nomad 24A	2+16	9,400	2 250-B17C TP	200
Aerospatiale, Paris, France				AEROSPATIALE
N 262 C Fregate	2+29	23,920	2 TM Bastan 7	260
N 262 Mohawk 298	2+29	23,480	2 PT6A-45 TP	250
Helicopters				
SA 315B Lama	5	4,207	TM Artouste 3B	120
AS 332B1 Super Puma	23	19,840	2 TM Makila 1A1	173
AS 332L1 Super Puma	26	18,960	2 TM Makila 1A1	173
AS 332M1 Super Puma	27	19,840	2 TM Makila 1A1	173
AS 332F1 Super Puma	23	19,840	2 TM Makila 1A1	173

Model	Crew +pax	GW, lb.	Powerplant (hp or lb. thrust)	Cruise, mph (unless noted)
Aerospatiale cont.				
SA 342L1 Gazelle	5	4,409	TM Astazou 14M	174
SA 365N2 Dauphin 2	14	9,369	2 TM Arriel 1C1	178
SA 365F Dauphin 2	14	9,039	2 TM Arriel 1M	184
SA 365K Panther	12	9,369	2 TM Arriel 1M1	178
AS 550U2	6	4,960	2 TM Arriel 101	178
AS 550 UR	6	5,600	2 250-C20R	178
AS 565UA Panther	12	9,369	2 TM Arriel 1M1	178
AS 565MA Dauphin 2	12	9,369	2 TM Arriel 1M1	178
AS 532UC	23	19,384	2 TM Makila 1A1	173
SA 366G1 Dauphin 2	8	8,950	2 LTS101-750B2	190
AS 350B Ecureuil AStar	6	4,207	TM Arriel 1B	169
AS 350D AStar	6	4,299	LTS 191-600A	169
AS 350L1 Ecureuil	6	4,850	TM Arriel 1D	173
AS 355F2R Ecureuil 2/ TwinStar	6	5,600	2 250-C20R	173
AS 350B2 Ecureuil AStar	6	4,960	TM Arriel 1D1	178
AS 355M2 Ecureuil 2	6	5,600	2 250-C20F	173
Socata (fixed wing)				
TB30 Epsilon	1+1	2,755	Lyc. AE1540-L1B5D	
TB31 Omega	2	3,638	Turbomeca TP-319	287

Aerospatiale/Daimler-Benz

Model	Crew +pax	GW, lb.	Powerplant (hp or lb. thrust)	Cruise, mph (unless noted)
Transport & Utility				
Transall NG	4	108,356	2 RR Tyne Mk 22 TP	319

Aero Vodochody, Prague, Szech Republic

Model	Crew +pax	GW, lb.	Powerplant (hp or lb. thrust)	Cruise, mph (unless noted)
Trainers				
L-39ZA Albatros	1-2	12,346	Progress AI-25L TF	469

Agusta/SIAI-Marchetti, Sesto Calende, Italy _____ SIAI-MARCHETTI

Model	Crew +pax	GW, lb.	Powerplant (hp or lb. thrust)	Cruise, mph (unless noted)
Trainers				
S.211	2	7,165	JT15D-4C TF	414
S.211A	1-2	8,820	JT15D-5C TF	414
S.F.260E	1-2	2,866	Lyc AIO 540-D4AS	187
S.F.260F	2	2,866	Lyc AIO 540-D4AS	187
S.F.260TP	2	2,866	250-B17 TP	230
Turboprops				
SF.600A Canguro	2+9	7,974	2 250-B17F TP	194

Airbus Industrie GIE, Blagnac, France _____ AIRBUS

Model	Crew +pax	GW, lb.	Powerplant (hp or lb. thrust)	Cruise, mph (unless noted)
Narrow Body Turbofans				
A319	2+124	141,000	2 XCFM56-5 or IAE V2500	M0.82
A320-100	2+150	149,000	2 CFM56-5 or IAE V2500	M0.82
A320-200	2+179	162,000	2 CFM56-5 or IAE V2500	M0.82
A321-100	2+220	181,200	2 CFM56-5 or IAE V2500	M0.82
A321-200	2+185	187,400	2 CFM56-5 or IAE V2500	M0.82
Wide Body Turbofans				
A300 B2-100	3+251	302,000	2GE CF6 or PW JT9D	M0.86
A300 B2-200	3+251	313,000	2GE CF6 or PW JT9D	M0.86
A300 B4-100	3+251	347,200	2GE CF6 or PW JT9D	M0.82
A300 B4-200	3+251	363,800	2GE CF6 or PW JT9D	M0.82
A300 C4-200	3+251	363,800	2GE CF6 or PW JT9D	M0.82
A300 F4-200	3+	363,800	2GE CF6 or PW JT9D	M0.82
A300-600C	2+	375,900	2 GE CF6-80C2 or PW 4000	M0.82

Model	Crew +pax	GW, lb.	Powerplant (hp or lb. thrust)	Cruise, mph (unless noted)
Airbus cont.				
A300-600	2+375	363,760	2 GE CF6-80C2 or PW 4000	M0.82
A300-600F	2+	375,890	2 GE CF6-80C2 or PW 4000	M0.82
A300-600R	2+375	375,890	2 GE CF6-80C2 or PW 4000	M0.82
A310-200	2+280	313,100	2 GE CF6-80C2 or PW 4000	M0/84
A310-300	2+280	330,700	2 GE CF6-80C2 or PW 4000	M0.84
A310-200C	2+280	313,100	2 GE CF6-80C2 or PW 4000	M0.82
A330-300	2+335	467,400	2 GE CF6-80 or PW4000	M0.86
A330-300	2+440	478,400	2 GE CF6-80 or PW4000	M0.86
A340-200	2+263	556,600	4 CFM56-5C	M0.86
A340-300	2+295	597,500	4 CFM 56-5C	M0.86
A340-800	2+232	285,100	4 CFM 56-5C	M0.86
Alenia, Naples, Italy				
Transport & Utility				
C-27A Spartan	3	56,878	2 GE T64/P4D	250
G.222	3	61,700	2 T64-P4D TP	300
Trainers				
G.91T/3	2	11,995	RR Orpheus 803 D-1`1	M1.0
Alenia/Aermacchi/Embraer				
Attack				
AMX	-	28,660	RR Spey Mk 807	560
Atlas Aircraft Corp., Ltd., Kempton Park, South Africa				ATLAS
Trainers				
MB 326M Impala	2	9,500	RR Viper 11 Mk 22A TJ	512
Avions Marcel Dassault-Breguet Aviation, Paris, France				DASSAULT
Attack				
Mirage 5	1	30,200	Snecma Atar 9C TJ	M2+
Alpha Jet NGEA	2	17,600	2 Snecma-TM Larzac 04-C20	M0.86
Bombers				
Mirage 2000N	2	36,375	Snecma Atar M53 TF	M2.2+
Mirage 4	2	70,000	2 Snecma Atar 9K TJ	M2+
Fighters				
Super Etendard	1	26,000	Snecma Atar 8K50 TJ	M1+
Mirage 3E	1	29,800	Snecma Atar 9C TJ	M2+
Mirage 3NG	1	32,400	Snecma Atar 9K50 TJ	M2+
Mirage F1	1	33,500	Snecma Atar 9K50 TJ	M2.2
Mirage 50	1	30,200	Snecma Atar 9K50 TJ	M2.2
Mirage 2000	1	36,375	Snecma Atar M53 TF	M2.2+
Mirage 2000B	2	36,375	Snecma Atar M53 TF	M2.2+
Mirage 2000D	1	32,6400	2 Snecma Atar M53 TF	M2.2
Mirage 2000-5	1	36,375	Snecma Atar M53P2 TF	
Mirage 4000	1	32,400	2 Snecma Atar M53 TF	
Rafale C/B	2	44,550	2 Snecma M88-2	
Rafale M	1	44,550	2 Snecma M88-2	
Patrol ASW				
1150 Atlantic/ATL2	12	100,000	2 RR Tyne R. Ty. 20 TP	380
Falcon Guardian	5	32,000	2 ASE ATF3 TF	M0.86
Reconnaissance				
Mirage 3R	1	29,000	Snecma Atar 9C TJ	M2+
Mirage F1CR	1	33,500	Snecma Atar 9K50 TJ	M2+

Model	Crew +pax	GW, lb.	Powerplant (hp or lb. thrust)	Cruise, mph (unless noted)
Dassault cont.				
Trainers				
Mirage 3B	2	29,800	Snecma Atar 9C TJ	M2+
Avions de Transport Regional (Aero Inaternational), Toulouse, France				**ATR**
Turboprop Commuters				
ATR-42-400	2+50	39,462	PW121A (2,150)	266
ATR-42-500	2+50	41,005	PW127E (2,750)	304
ATR 72-200	2+74	47,400	PW 124 (2,160)	284
ATR 72-210	2+74	47,400	PW 127 (2,480)	284
Jetstream 32	2+19	16,204	2ASE TPE331-12 UAR	287
Jetstream 41	2+29	24,000	2 ASE TPE331-14 GR/HR	240
Narrow Body Turbofans				
Avro RJ70	2+82	95,000	4 ASE LF507	M0.70
Avro RJ85	2+100	97,000	4 ASE LF507	M0.70
Avro RJ100	2+100	101,500	4 ASE LF507	M0.70
Avro RJ115	2+116	101,500	4 ASE LF507	M0.70
Beech Aircraft Corp. (see Raytheon)				
Bell Helicopter Textron, Fort Worth, TX				**BELL**
206 (OH-58A Kiowa)	2+3	3,000	250-C18 TS	130
206 (OH-58C Kiowa)	2+3	3,200	250-C20B TS	135
206B-JR3 (TH-57B/C)				
Jet Ranger	1+4	3,200	250-C20J	140
212	15	11,200	2PT6T-3B TS	115
230	10	8,400	2 250-C30G/2	172
407	7	5,500	2 250-C47 TS	161
412 EP	15	11,900	2 PT6T-3D TS	160
430	10	9,300	2 250-C40 TS	172
UH-1H Iroquois	2+14	9,500	T53-L-13 TS	127
UH-1N Twin Huey	2+13	10,000	2 T400-CO-400 TS	116
209 (AH-1G Cobra)	2	9,500	T53-L-13 TS	219
209 (AH-1S E Cobra)	2	10,000	T53-L-702 TS	195
209 (AH-1S F Cobra)	2	10,000	T53-L-703 TS	195
209 (AH-1S P Cobra)	2	10,000	T53-L-703 TS	195
209 (AH-1J Sea Cobra)	2	10,000	2 T400-CP-400 TS	210
209 (AH-1T Sea Cobra)	2	14,000	2 T400-CP-402 TS	218
209 (AH-1W Super Cobra)	2	14,750	2 T700-GE-401 TS	218
406 (OH-58D Kiowa Scout)	2	5,000	250-C30R	137
406CS CombatScout	1+4	4,500	250-C30V	136
901 V-22 Osprey	3+24	32,821	2 T406-AD-400	250
Boeing Aircraft Co., Seattle, WA				**BOEING**
Bombers				
B-52H Stratofortress	6	488,000	8 TF33-P-3 TF	650
Transport & Utility				
C-22B	4	161,000	3 P&W JT80-9A	470
KC-135 A/E	4	297,000	4 J57-P-59W TJ	530
KC-135R	4	322,500	4 F108-CF-100 TF	
VC-137C	4	328,000	4 JT3D-3B TF	M 0.84
VC-25A	23	836,000	4 CF6-80C2B1	484
Early Warning				
E-3A/B/C AWACS	23	325,000	4 TF33-P-100 TF	

Model	Crew +pax	GW, lb.	Powerplant (hp or lb. thrust)	Cruise, mph (unless noted)
Boeing Early Warning cont.				
E-4A/B NEACP	31	800,000	4 CF6-50E TF	
E-6A TACAMO	18	342,000	4 CFM 56 TF	
Narrow Body Turbofans				
707-320B/C		3+165	336,000 4 JT3D-7	M0.80
727-100	3+94	170,000	3 JT8D-1, 7 or 9	M0.80
727-200	3+145	191,500	3 JT8D-15A	M0.80
727-200F	3+	204,000	3 JT8D-15A	M0.80
737-100	2+103	111,000	2 JT8D-7 or 9	M0.78
737-200	2+120	116,000	2 JT8D-15A	M0.73
737-200	2+120	128,600	2 JT8D-17A	M0.73
737-200C	2+120	116,000	2 JT8D-15A	M0.73
737-300	2+141	124,500	2 CFM 56-3-B1	M0.73
737-400	2+146	150,000	2 CFM 56-3C	M0.74
737-400HGW	2+146	150,000	2 CFM 56-3C	M0.74
737-500	2+108	115,500	2 CFM 56-3-B1/3C-1	M0.74
737-600	2+108	143,500	2 CFM56-7B	M0.79
737-700	2+128	153,000	2 CFM56-7B	M0.79
737-800	2+162	172,500	2 CFM56-7B	M0.79
757-300	2+240	255,000	2 RR RB.211-535E4 or PW2037	M0.80
Wide Body Turbofans				
747-100B	3+452	750,000	4 GE CF6-45A2	M0.84
747-200B	3+452	833,000	4 PW JT19D-7R4G24	M0.84
747-200B Combi	3+452	833,000	4 PW JT19D-7R4G2	M0.84
747-200C	3+452	833,000	4 PW JT9D	M0.84
747-200F	3	833,000	4 PW JT19D-7R4G2	M0.84
747-300B	3+490	833,000	4 PW JT19D-7R4G2	M0.84
747-300 Combi	3+496	833,000	4 PW JT19D-7R4G2	M0.84
747-400	2+420	875,000	4 PW4056	M0.84
747-400 Combi	2+266	875,000	4 PW4056	M0.84
747-400D	3+568	600,000	4 GE CF6-80C2B1F	M0.84
747-400F	2	811,000	4 PW4056	M0.84
767-200	2+290	315,000	2 JT9D-7R4 or CF6-80A	M0.80
767-200ER	2+290	351,000	2 JT9D-7R4 or CF6-80A	M0.80
767-300	2+290	351,000	2 JT9D-7R4 or CF6-80A	M0.80
767-300ER	2+290	412,000	2 PW 4000 or CF6-80C2	M0.80
767-300F	2	412,000	2 PW 4056	M0.80
777-200	2+375	535,000	2 PW4000 or GE90 or RR Trent	M0.84
777-200 IGW	2+305	632,500	2 PW4000 or GE90 or RR Trent	M0.84
777-300	2+368	660,000	2 PW4000 or GE90 or RR Trent	M0.84

Model	Crew +pax	GW, lb.	Powerplant (hp or lb. thrust)	Cruise, mph (unless noted)
Boeing Helicopter, Philadelphia, PA				BOEING HELICOPTER
CH-46E Sea Knight	3+25	24,300	2 T58-16 TS	166
CH-46D Sea Knight	3+25	23,000	2 T58-10 TS	166
CH-47D Chinook	3+33	50,000	2 T55-L-712 TS	183
MH-47E Chinook	4+44	54,000	2 T55-L-714 TS	190
414-100 Chinook	3+55	54,000	2 T55-L-712S/SB TS	187
234-LR Boeing 234	3+44	48,500	2 AL5512 TS	172
234-UT Boeing 234	2	51,000	2 AL5512 TS	166
234-ER Boeing 234	2+18	48,500	2 AL5512 TS	172
234 MLR Boeing 234	2+3	48,500	2 AL5512 TS	172
Transport & Utility				
Dash 8 Series 200	2	36,300	2 PW123D	

Model	Crew +pax	GW, lb.	Powerplant (hp or lb. thrust)	Cruise, mph (unless noted)
Bombardier, Inc. Canadair Division, St. Laurent, Quebec				CANADAIR
Electronic Warfare				
E-9A Dash 8 Series 200	2+2	36,300	2 PW123D	
Patrol ASW				
Dash 8 Series 200	2+2-4	36,300	2 PW123D	
Trainers				
CT-142 Dash 8	2-6	36,300	1 PW123D	
Narrow Body Turbofans				
Regional Jet 200	2+50	47,450	2 CF34-3B1	M0.81
Regional Jet 200ER	2+50	51,000	2 CF34-3B1	M0.81
Regional Jet 100LR	2+50	53,000	2 CF34-3B1	M0.81
Turboprop Commuters				
DHC-7-100	2+54	44,000	4 PT6A-50 TP	266
DHC-8-100	2+40	34,500	2 PW120A	312
DHC-8-200	2+39	36,300	2 PW123C/D	340
DHC-8-300	2+56	43,000	4 PW123B/E	330
DHC-8-400	2+78	62,500	3 PW150	403
British Aerospace, London, England				BAe
Attack				
GR7 Harrier	1	32,000	RR Pegasus Mk 105/F402-RR406A	M 0.91
Hawk 200	1	29,061	RR Adour Mk 871 TF	M 0.82
Transport & Utility				
146-100*	2+94	84,000	4 Lyc. ALF502R-5 TF	490
C-29A (Bae 125-800)	5	27,520	2 TFE731-5R-1H	397
HS.780 Andover C	2+58	51,000	2 RR Dart 12 TP	270
Fighters				
F/A2 Sea Harrier	1	26,200	RR Pegasus Mk 104/106 TF	635 kt
Patrol ASW				
HS.801 Nimrod MR Mk 1/2	12	177,500	4 RR Spey Mk 250 TF	575
Nimrod 2000	10		4 RR 710	
Trainers				
Series 120 Bulldog	2	2,350	Lyc. AEIO360-A1B6	150
Hawk 60	2	20,061	RR Adour 861 TF	M0.82
Hawk 100 Trainer	2	20,061	RR Adour Mk 871 TF	M0.82
T. Mk 3 Harrier	2	25,300	RR 402 103 TF	
T. Mk 4 Harrier	2	25,300	RR Pegasus Mk 103 TF	M0.85
T. Mk 10 Harrier 2	2	25,300	RR Pegasus Mk 105 TF	M0.85
T. Mk 60 Harrier	2	25,300	RR Pegasus Mk 152-32 TF	M0.85
T. Mk 4/Mk 54 Harrier	2	26,200	RR Pegasus Mk 103 TF	M0.85
Narrow Body Turbofans				
Concorde	3+100	408,000	4 RR Olympus 593 Mk 610	M2.0
BAe 146-200	2+112	97,500	4 ASE ALF502R-5	500
Bae 146-300	2+128	97,500	4 ASE ALF502R-5	500
Turboprop Commuters				
HS 748 Series 2A	2+58	46,500	2 RR Dart 7 Mk 535 TP	278
HS 748 Series 2B	2+58	46,500	2 RR Dart Mk 536-2 TP	280
BAe ATP	2+72	50,550	2 PW124/126 TP	315
Caproni Vizzola Costruzion Aeronautiche SpA, Milan, Italy				CAPRONI
C22J	2	2,760	2 MT TRS 18-1 TJ	330

Model	Crew +pax	GW, lb.	Powerplant (hp or lb. thrust)	Cruise, mph (unless noted)
Cessna Aircraft Co., Wichita, KS				CESSNA
Attack				
A-37B Dragonfly	1	14,000	2 J85-GE-17A TJ	M 0.82
Trainers				
T-37B	2	6,618	2 J69-T-25	425
Construcciones Aeronauticas, S.A., Madrid, Spain				CASA
Transport & Utility				
C-212M Aviocar	2+1	17,857	2 TOE-331-10R-531C	
Trainers				
C-101 Aviojet	2	13,889	TFE 731-5 TF	517
Turboprop Commuters				
C.212-300 Aviocar	3+26	16,975	TPE 331-10R-513C	230
Helicopters				
A129 Mangusta	2		2 LHTEC T800 TS	150
Dornier GmbH, Friedrichshafen, Germany				DORNIER
Turboprop Commuters				
DO 228-212	2+20	14,109	2 TPE 331-5A TP	
DO 328-100	2+33	27,558	2 PW119	397
EH Industries, Ltd., Farnborough, England				
EH101	2+30	32,188	3 RTM 322	192
Embraer-Empresa Brasileria de Aeronautica S.A., Sao Paulo, Brazil				EMBRAER
Attach				
EMB-312H Super Tucano	2	7,618	PWC P6A-68 TP	320
Patrol ASW				
EMB-111 Bandeirante	6	15,432	2 PT6A-34 TP	286
Utility				
EMB-312 Tucano	2	7,000	PT6A-25C TP	278
EMB-110K1 Bandeirante	2	12,500	2 PT6A-34 TP	286
Trainers				
EMB -312A Tucano	2	7,000	PT6A-25C	512
Narrow Body Turbofans				
EMB-145 Amazon	2+50	43,328	2 AE3007A	M0.78
Turboprop Commuters				
EMB-110K Banderante*	2+19	12,500	2 PT6A-34	225
EMB-120 Brasilia	2+30	26,433	2 PW118A TP	313
Eurocopter (also see Aerospatiale)				
HAP/UHU/HAC Tiger	2	13,062	2 MTR 390	200
Eurocopter Deutschland				
BO 105CBS-5	1+5	13,063	2 250-C20B	150
Eurofighter, Munich, Germany				
EF 2000	1	46,200	2 Eurojet 200 TF	M2.0
Fairchild Dornier, San Antonio, TX				FAIRCHILD
Transport & Utility				
C-26A	2	16,000	2 TPE-331	275
SA-227-AT Expediter	2	16,000	2 TPE 331-11-612G TP	324
Turboprop Commuters				
SA-227AC Metro 3	2+19	14,000	2 TPE 331-11U-612G TP	315
SA-227-AT Metro 3	2+19	15,500	2 TPE 331-11-612G TP	324

CIVIL & MILITARY AIRCRAFT

Model	Crew +pax	GW, lb.	Powerplant (hp or lb. thrust)	Cruise, mph (unless noted)
Fairchild Republic				
Attack				
A-10A Thunderbolt II	1	48,560	2 GE FT34-GE-100 TF	400
T-46A	2	7,295	2 F109-GA-100 TF	425
FAA Flugzeugwerke Altenrhein AB, Altenrhein, Switzerland				FFA
Trainers				
AS 202/18A Bravo	2	2,315	Lyc. AEIO-360	150 kt
AS 202/32P Bravo	2	2,380	250-B17C	166 kt
Fokker, Amsterdam, Netherlands				FOKKER
Narrow Body Turbofans				
Fokker 70	2+79	81,900	2 RR Tay 620	M 0.77
Fokker 100	2+100	98,000	2 RR Tay 650 TF	M 0.77
Turboprop Commuters				
Fokker 50	2+58	45,900	2 PW125B	325
Fokker 50 High Performance	2+58	45,900	2 PW127B	327
Fokker 60 Utility	2+60	50,600	2 PW127B	322
Fuji Heavy Industries, Tokyo, Japan				FUJI
Trainers				
TL-1	4	3,350	Lyc. IGSO480-A1A6	210
T-5	4	3,493	250-B17D	236
T-3	2	3,322	Lyc. IGSO480-A1F6	211
Helicopters				
UH-1H (Bell)	13	9,500	T53-K-13B	127
FB-204B2 (Bell)	10	8,500	T53-K-13B	127
AH-1S (Bell)	2	10,000	T53-K-703 TS	195
Gulfstream Aerospace, Savannah, GA				GULFSTREAM
Transport & Utility				
C-20A/B/D/E (G-III)	2	69,700	2 RR Spey Mk 511-6 TF	M0.77
C-20H (G-IV)	5	73,5600	2 RR Tay Mk 611-8 TF	M0.8
Hindustan Aeronautics, Ltd., Bangalore and Kanpur, India				
Fighters				
Jaguar	1	33,075	2 Adour Mk 811	841
Trainers				
HJT-16 Kiran Mk 2	2	10,890	RR Orpheus 701 TJ	432
Ajeet Trainer	2	10,000	RR Orpheus 701 TJ	679
HPT-32	4	3,490	Lyc. AEIO-540	175
HTT-34	4		Allison 250	205
Turboprop Commuters				
H.S. 748-2	3+20	44,402	2 RR Dart 7 Mk 531 TP	312
Israel Aircraft Industries Ltd, Ben-Gurion Intl. Airport				IAI
Fighters				
Kfir C2	1	32,340	J79 J1E TF	M 2.2
Kfir C7	1	36,300	J79 J1E TF	M 2.2
Kfir 2000	1	35,700	J79E TJ	M2.2
Patrol ASW				
1125 Astra Sea Scan	5	24,650	2 TFE 731-3-A-2006 TF	353

Model	Crew +pax	GW, lb.	Powerplant (hp or lb. thrust)	Cruise, mph (unless noted)
IAI cont.				
Research				
Lavi	1	22,000	PW 1120 TF	M 1.8
Trainers				
Kfir TC2	2		J79 J1E TF	M 2.2
Kfir TC7	2		J79 J1E TF	M 2.2
Utility				
IAI 202 Arava	2+30	17,000	2 PT6A-34 TP	185
Turboprop Commuters				
IAI 101B Arava	2+19	15,140	2 PT6A-36 TP	191
IAI 201-202 Arava	2+24	15,000	2 PT6A-34 TP	191
IAI 202 Arave	2+30	17,000	2 PT6A-36	185

Jetstream Aircraft Ltd., Prestiwck Airport, Scotland____JETSTREAM

Model	Crew +pax	GW, lb.	Powerplant	Cruise
Turboprop Commuters				
Jetstream 32 (Super 31)	2+19	16,204	2 TPE331-UAR	285
Jetstream 41	2+30	24,000	2 TPE331-GR/HR	228

Kaman Aerospace Kaman Corp, Bloomfield, CT____KAMAN

Model	Crew +pax	GW, lb.	Powerplant	Cruise
K-888 Seasprite	1+3	13,500	2 T58-GE-8F TS	160
K-894 Super Seasprite 6		13,500	2 T700-GE-401 TS	173

Kawasaki Heavy Industries, Kobe, Japan____KAWASAKI

Model	Crew +pax	GW, lb.	Powerplant	Cruise
Transport & Utility				
C-1	3+60	85,320	2 JT8D-M-9 TF	490
Trainers				
T-4	2	12,125	2 F-3 TF	M 0.9
KV-107-2A-5	7	21,400	2 CT58-IHI-140-1	137
HK-500 (369HS)	6	2,550	C250-M-5A	152
OH-6D (500D)	6	3,000	C250-C20B	175
BK-117	11	7,055	2 LTS101-750B1	173
CH-47J	47	54,000	2 T55-K-712 TS	183
P-3C ASW	10	135,000	4 T56-IHI-14 TP	473

Learjet, Inc., Wichita, KS____LEARJET

Model	Crew +pax	GW, lb.	Powerplant	Cruise
Transport & Utility				
C-21A	2	18,398	2 TFE 731-2-2B	360 kt

Lockheed Martin Aeronautical Systems Co., Marietta, GA_________________________________ LOCKHEED

Model	Crew +pax	GW, lb.	Powerplant	Cruise
Bombers				
FB-111A	2	114,000	2 TF30-P-7 TF	M2.0+
Transport & Utility				
AC-130A Spectre	14	124,200	4 T56-A-9 TP	288
AC-130H Spectre	14	155,000	4 T56-A-15 TP	374
C-5B Galaxy	6	837,000	4 TF39-1C TF	563
C-130H Hercules	4	155,000	4 T56-A-15 TP	374
C-130H-30 Super Hercules	4	155,000	4 T56-A-15 TP	363
C-130J Hercules	3	155,000	4 AE2100D3 TP	340
C-130J-30 Hercules	3	155,000	4 AE2100D3 TP	340
C-130T Hercules	6	155,000	4 T56-A-16 TP	320kt
C-140A Jetstar	5	21,455	4 J60-P-5 TJ	M 0.8
C-141B Starlifter	4	343,000	4 TF33 P-7 TF	563
KC-130-F/R/T Hercules	6	155,000	4 T56-16	305kt
MC-130E Combat Talon	9	155,000	4 T56-A-15 TP	325

Model	Crew +pax	GW, lb.	Powerplant (hp or lb. thrust)	Cruise, mph (unless noted)
Lockheed cont.				
MC-130H Combat Talon II	7	155,000	4 T56-A-15 TP	325
Fighters				
F-22	2		2 PW F119 TF	
F-117A	1	52,500	2 F404 TF	
F-16A/B Fighting Falcon	1	37,500	F100-PW-200 TF	M2.0+
F-16C/D Fighting Falcon	1	18,238	F100-PW-200 TF or F110-GE-100 TF	M2.0+
F-16N Fighting Falcon	1	18,400	F110-GE-100 TF	M2.0+
F-111A/D/E/F	2	100,000	2 TF3--P-100 TF	M2.5
Patrol ASW				
P-3C Orion	12	139,760	4 T56-A-14 TP	
S-3A/B Viking	4	52,539	2 TF34-GE-400A/B TF	370
Reconniassance				
SR-71 Blackbird	2		2 PW J58 TJ	M 3.0+
U-2	2	17,000	PW J75 TJ	
U-2S	1	19,000	GE F118-GE-101 TF	
Wide Body Turbofans				
L-1011 TriStar*	3+400	358,000	3 RR RB.211-22B	M0.83
L-1011-100*	3+400	368,000	3 RR RB.211-22B	M0.83
L-1011-200*	3+400	368,000	3 RR RB.211-524B4	M0.83
L-1011-250	3+400	368,000	3 RR RB.211-524B4	620
L-1011-500	3+330	368,000	3 RR RB.211-524B4	620
Turboprop Utility				
L-100-30 Super Hercules	3	155,000	4 501-D22A	387
<u>Lockheed Aircraft Argentina</u>				
Attack				
IA-58 Pucara	2	15,011	2 TM Astazou	255
Trainers				
IA-63 Pampa	2	11,023	TFE731-2 TF	440

Lovaux, Ltd., Brooklands Aerospace, Bournemouth Intl. Airport, UK

Model	Crew +pax	GW, lb.	Powerplant (hp or lb. thrust)	Cruise, mph (unless noted)
OA7 300 Optica	2	2,899	Lyc. IO540	161

McDonnell Douglas, St. Louis, MO   McDONNELL DOUGLAS

Model	Crew +pax	GW, lb.	Powerplant (hp or lb. thrust)	Cruise, mph (unless noted)
Attack				
A-4F Skyhawk	1	24,500	J52-P-8A TJ	667
A-4M Skyhawk	1	24,500	J52-P-408 TJ	690
A-4N Skyhawk	1	24,000	J52-P-408 TJ	M0.9
AV-8B Harrier 2	1	31,000	RR F402/408 TF	M0.9
F-15C/D Eagle	2	68,000	2 F100-PW-100/200	M2.5
F-15E/F Eagle	2	81,000	2 F100-PW-220	M2.5
Transport & Utility				
KC-10A Extender	4	590,000	3 CF6-50C2 TF	M0.82
C-9A Nightingale	7	108,000	2 JT8D-9 TF	570
C-9B Skytrain II	7	110,000	2 JT8D-9 TF	570
VC-9C	12	110,000	2 JT8D-9 TF	570
C-17		570,000	4 PW2037 TF	403
Fighters				
F-4E Phantom 2	2	58,000	2 J79-GE-17 TJ	M2.0+
F-4G Wild Weasel	2	58,000	2 J79-GE-17 TJ	M2.0+
F-4S/N Phantom 2	2	56,000	2 J79-GE-10 TJ	M2.0
F-15A/B Eagle	1-2	54,000	2 PW F100-PW-100	M2.5
F-15C/D Eagle	1-2	68,000	PW F100-PW-100/220	M2.5
F-15E Eagle	2	81,000	2 PW F100-PW-220/229	M2.5
F/A-18A/B Hornet	1-2	51,900	2 GE F404-GE-402 TF	
F/A-18C/D Hornet	1/2	51,900	2 GE F404-GE-402 TF	M1.8

Model	Crew +pax	GW, lb.	Powerplant (hp or lb. thrust)	Cruise, mph (unless noted)
McDonnell Douglas cont.				
F/A-18E/F Super Hornet	1/2	66,000	2 GE F414 TF	
Trainers				
TAV-88 Harrier 2	2	31,000	RR F402/406A TF	
T-45A Goshawk	2	12,758	RR F405/RR 401 TF	M1.04
Narrow Body Turbofans				
DC-8 Series 30	3+176	315,000	4 JT4A-9, 11	600
DC-8 Series 40	3+117	315,000	4 RR Co. 12	600
DC-8 Series 50	3+189	325,000	4 JR3D-3B	600
DC-8 Super 61	3+259	328,000	4 JT3D-3B	600
DC-8 Super 62	3+189	350,000	4 JT3D-7	600
DC-8 Super 63	3+259	355,000	4 JT3D-7	600
DC-9 Series 10	2+85	90,700	2 JT8D-1 or 7	576
DC-9 Series 20	2+85	98,000	2 JT8D-9 or 11	576
DC-9 Series 30	2+110	108,000	2 JT8D-9	586
DC-9 Series 40	2+120	114,000	2 JT8D-11 or 15	586
DC-9 Series 50	2+135	121,000	2 JT8D-15 or 17	586
MD-81	2+155	140,000	2 JT8D-209 or 217A	576
MD-82	2+155	149,500	2 JT8D-217C	576
MD-83	2+155	160,000	2 JT8D-219	576
MD-87	2+130	140,000	2 JT8D-217C	M0.80
MD-88	2+155	149,500	2 JT8D-217C	M0.80
MD-90-10	2+139	139,000	2 IAE V2500-D5	M0.80
MD-90-30	2+172	156,000	2 IAE V2500-D5	M0.80
MD-90-40	2+208	163,500	2 IAE V2500-D5	M0.80
MD-95-30	2+129	114,000	2 BMW/RR BR715	
Wide Body Turbofans				
DC-10 Series 10	3+380	440,000	3 CF6-6D	M0.82
DC-10 Series 15	3+380	455,000	3 CF6-50C2F	M0.82
DC-10 Series 30	3+380	572,000	3 CF6-50C2	M0.82
DC-10 Series 40	3+380	572,000	3 JT9D-59A	M0.82
MD-11	2+410	618,000	3 PW4360 or CF6-80C2-DF1 or RR Trent-685	M0.82

Model	Crew +pax	GW, lb.	Powerplant (hp or lb. thrust)	Cruise, mph (unless noted)
McDonnell Douglas Helicopter Co., Mesa, AZ				**MDD**
OH-6A	2	2,700	T63-A-5A TS	150
500MD Defender	3	3,000	250-C20B TS	175
500MG Defender	2	3,000	250-C20B TS	150
530MG Defender	2+2	3,100	250-C30 TS	150
600MG Defender	2+2	4,100	250-47 TS	155
77 (AH-64A Apache)	2	17,650	2 T700-GE-701 TS	226

Model	Crew +pax	GW, lb.	Powerplant (hp or lb. thrust)	Cruise, mph (unless noted)
Messerschmitt-Boelkow-Blohm, Ottobrunn-bei-Munchen, Germany				**MBB**
(also see Eurocopter Deutschland)				
BO 105LS	6	5,732	2 250-C28C	149
BO 105CB	5	5,511	2 250-C20B	150
BK 117 B-1	11	7,055	2 LTS101-750B-1	173

Model	Crew +pax	GW, lb.	Powerplant (hp or lb. thrust)	Cruise, mph (unless noted)
Mitsubishi Heavy Industries, Tokyo, Japan				**MITSUBISHI**
Fighters				
F-1	1	30,200	2 RR TF40-IHI-801A TF	M1.6
F-2	1	26,400	GE F110-GE-129 TF	M2.0
F-4EJ Phantom II	2	46,640	2 J79-IHI-17 TJ	M 2.2
F-15J Eagle	1	41,500	2 F100-IHI-100 TF	M 2.5

Model	Crew +pax	GW, lb.	Powerplant (hp or lb. thrust)	Cruise, mph (unless noted)
Mitsuhishi cont.				
Patrol ASW				
MU-2S	4	10,030	2 TPE 331-25A/B TP	363
Reconnaissance				
LR-1 (MU-2K)	7	8,840	2 TPE 331-25A TP	363
Trainers				
T-2	2	25,270	2 RR TF40-IHI-801A TF	M1.6
Helicopters				
SH-60J	3	21,039	2 CT1700-IHI-401C TS	171
Northrop Grumman, Los Angeles, CA				NORTHROP
Attack				
A-10A Thunderbolt 2	1	48,560	2 TF34-GE-100 TF	400
A-7D Corsair II	1	42,000	TF41--A-1 Spey TF	607
A-7K Corsair II	2	42,000	TF41-A-1 TF Spey TF	607
A-6E Intruder	2	60,400	2 P&W J52-P BB TJ	M0.8
Bombers				
B-2A Stealth Bomber	2	336,500	4 F118-GE-100 TF	
Transport & Utility				
C-2A Greyhound	5	54,354	2 All T56-A-425 TP	260 kt
Early Warning				
E-2C Hawkeye	5	51,933	2 T56-A-425A TP	
Electronic Warfare				
EA-6A Intruder	2	54,504	2 J52-P-8 TJ	648
EA-6B Prowler	4	60,000	2 J52-P-408 TJ	374
EF-111A Raven	2	88,950	2 PWTF30-P-9 TF	M2.4
Fighters				
F-5A/B Freedom Fighter	2	20,294	2 J85-GE-13 or -15 TJ	M1.4
F-5E Tiger II	1	24,722	2 J85-GE-21 TJ	M1.6
F-5F Tiger II	2	25,152	2 J85-GE-21 TJ	M1.5
F-14A Tomcat	2	71,000	2 TF30-P-414 TF	M 2.3
F-14B Super Tomcat	2	74,349	2 F110-GE-400 TJ	M 2.3
F-14D Super Tomcat	2	74,349	2 F110-GE-400 TJ	M 2.3
Reconnaissance				
E-8C Joint Stars	18	336,000	4 PW JT3D-3B TJ	M0.84
RF-5E Tigereye	1	25,204	2 J85-GE-21 TJ	M1.5
Trainers				
T-38A Talon	2	12,500	2 J85-GE-5 TJ	M1.2
Omnipol Foreign Trade Corp., Prague, Czechoslovakia				
Turboprop				
L-410UVP Turbolet	2+15	12,786	2 Walter M-601D TP	227
L-410UVPE Turbolet	2+19	14,110	2 Walter M-601E TP	227
Pacific Aerospace Corp., Hamilton Airport, New Zealand				
Fu24-954 Fletcher	4	5,430	Lyc. NG IO 720-A1B	167
600 Cresco	6	7,000	LTP101 700-1A	204
CT4B Airtrainer	3	2,400	Cont. IO-360-HB9	238
CT4CR Airtrainer	2	2,500	250-B170	297
Panavia Aircraft, Munich				
Attack				
Tornado IDS	2	62,000	2 Turbo-Union RB.199-34R Mk 103	M2.0
Electronic Warfare				
Tornado ECR	2	61,600	2 Turbo-Union RB.199-34R Mk 105	M2.0

CIVIL & MILITARY AIRCRAFT

Model	Crew +pax	GW, lb.	Powerplant (hp or lb. thrust)	Cruise, mph (unless noted)
Pamavia cont.				
Fighters				
Tormado ADV	2	61,700	2 Turbo-Union RB.100-34R Mk 104	M2.0
Panstwowe Zaklady Lotnicze, Swidnik, Poland				PZL HELICOPTER
SM-1 W	5	5,636	OKL Lit-4	109
SM-2	5	5,644	OKL Lit-4	118
Mi-2	8	7,826	2 GTD-350	130
Mi-2 Kunia	8	7,826	2 C250-C20B	130
M-2M	11		2 GTD-350	
Partenavia, Naples, Italy				PARTENAVIA
Patrol ASW				
P.68C	6	4,387	2 Lyc. IO360-A1B6	200
Peoples Republic of China				
Attack				
Qiang Ji-5M	1	54,500	2 Wopen-6 TJ	M1.2
Bombers				
Hongsha (B)-7	2	30,800	2 RR Spey Mk 202 TF	M1.7
Fighters				
Jian Ji 7 (MiG-21)	1	19,580	RD-11 TJ	M2.05
Jian Ji 8-2	1	31,526	2 WP13A TJ	M2+
Super 7	1			M2.0
Patrol ASW				
PS-5	8	121,250	4 WJ6 TP	344
Trainers				
Chu Jiao-6	2	3,080	Hou Sai-6	178
Jianjiao-7	2	18,920	WP-7BM TJ	M2.0
K-8 Karakorum 8	2	7,716	TFE 731-2A TF	495
Narrow Body Turbofans				
Yunshu-10	3+124	224,400	4 JT3D-7 TF	569
Turboprop				
Yunshu-7 (An-24)	3+52	46,200	2 WJ5-1TP	297
Yunshu-8 (An-12)	3+96	134,480	4 WJ6 TP	411
Yunshu-12	2+17	12,100	2 PT6A-11 TP	
Pilatus Aircraft Ltd., Stans, Sitzerland				PILATUS
Early Warning				
BN2TR AEW Defender	4	8,500	2 250-B17F TP	96
Trainers				
PC-7 Turbo Trainer	2	4,189	PT6A-25A TP	310
PC-9	2	4,960	PT6A-62 TP	368
Pilatus Britten-Norman Ltd., Isle of Wight, UK				
Early Warning				
BN2TR AEW Defender	4	8,500	2 250-B17F	196
Patrol ASW				
BN2T Maritime Defender	4	7,000	2 250-B17C	197
Reconnaissance				
BN2T-4R MSSA	1-4	8,500	2 250-B17F	
Turboprop Utility				
BN2T Turbine Islander	1+9	7,000	2 250-B17C	197

Model	Crew +pax	GW, lb.	Powerplant (hp or lb. thrust)	Cruise, mph (unless noted)
Polskie Zaklady Lotnicze, Warsaw and Mielec, Poland				PZL
Trainers				
PZL-I-22 Iryda	2	16,519	2 PZL SO-3W22	M0.85
PZL-M26 Model 00 Iskierka	2	2,645	PZL F-6A35OCA	165
PZL-M26 Model 01 Iskierka	2	3,086	Lyc. AEIO-540 L1B5D	211
Raytheon Aircraft, Wichita, KS				BEECH
Transport & Utility				
C-12A/B/C/D/F/R Huron	2	12,500	2 PT6A-38 TP	299
C-12J	2	16,600	2 PT6A-65B TP	312
Reconnaissance				
RC-12D/G/H/K	2	16,000	2 PT6-/67 TP	340
RU-21A-E/G/H	2	10,200	2 T74-CP-700 TP	239
Trainers				
T-1A Jayhawk	3	16,300	2 JT15D-5 TF	
T-34C Turbo Mentor	2	4,300	PT6A-25 TP	243
T-42A Cochise	1	5,100	2 Cont. IO-470L	236
T-44A	2	10,100	2 PT6A-34B TP	278
T-400	3	16,300	2 PWC JT15D-5 TF	538
Utility				
RU-21J	2	15,000	2 PT6A-41 TP	294 kt
U-8D/G Seminole	2	7,300	2 Lyc. GSO-480-A1A6	208 kt
U-8F Seminole	2	7,700	2 Lyc. GSO-480-3	205 kt
U-21A/G Ute	2	9,650	2 T74-CP-700 TP	208 kt
U-21F Ute	2	11,500	2 PT6A-28 TP	248 kt
UC-12F/M	2	12,500	2 PT6A-41 TP	259 kt
Turboprop Commuter				
1900C Airliner	2+19	16,600	2 PT6A-65B TP	308
1900D Airliner	2+19	16,950	2 PT6A-67D TP	331
Rhein-Flugseugbau GmbH, Moenchengladbach, Germany				
Trainers				
FT600 Fantrainer	2	5,070	250-C30 TS	320
Rinaldo Piaggio, Genoa, Italy				PIAGGIO
Patrol ASW				
P.166DL3-SEM	4	9,480	2 LTP101-100A-1A	230
Rockwell International (Boeing North American), Seal Beach, CA				
Attack				
AC-130U Gunship	13	155,000	4 All T56-A15 TP	
Bombers				
B-1B Lancer	4	477,000	4 GE F101-GE-102 TF	
Observation				
OV-10A Bronco	2	12,432	2 ASE T76-G-418/419 TP	305 kt
OV-10D NOS	2	15,000	2 ASE T76-G-420/421 TP	305 kt
Trainers				
T-2C Buckeye	2	13,191	2 GE J85-GE-4/4A TJ	
T-28D Trojan	2	12,104	Wr R-1820-86A	300
Saab-Scania AB, Linkoping, Sweden				SAAB
Attack				
35 XD Draken	1	25,000	Flygmotor RM6C TJ	M2.0

Model	Crew +pax	GW, lb.	Powerplant (hp or lb. thrust)	Cruise, mph (unless noted)
Saab cont.				
1050E	2	14,300	2 J85-17B TJ	585
AJ37 Viggen	1	35,000	Flygmotor RM8A TF	M2.0
Sk60B, C	1-2	10,400	2 TN Aubisque TF	478
Fighters				
J35F, Q Draken	1	23,000	Flygmotor RM6C TJ	M2.0
JA37 Viggen	1	37,500	Flygmotor RM8A TF	M2.0
JAS 39 Gripen	1	17,635	Flygmotor/GE RM12	M2.0
Reconnaissance				
SF37 Viggen	1	35,500	Flygmotor RM8A TF	M2.0
SH37 Viggen	1	35,000	Flygmotor RM8A TF	M2.0
Trainers				
Sk35C Draken	2	20,000	Flygmotor RM6B TJ	M1.8
Sk35X Draken	2	25,000	Flygmotor RM6C TJ	M2.0
Sk37 Viggen	2	35,000	Flygmotor RM8A TF	M2.0
Sk60W	2+3	8,800	2 TM Aubisque TF	478
Safari	2+1	2,590	Lyc. IO360	148
Supporter	2+1	2,590	Lyc. IO360	148
Turboprop Commuter				
340	2+36	29,000	2 CT-7E	322
2000	2+58	50,265	2 All AE2100A	430
Sabreliner Corp., St. Louis, MO				SABRELINER
Attack				
A/T-33 Shooting Star	2	17,250	J-33	M0.80
Trainers				
T-39A/B Sabreliner	2	19,305	2 J60-P3	M0.75
Short Brothers PLC, Belfast, N. Ireland				SHORTS
Transport & Utility				
C-23A Sherpa	3+30	22,900	2 PT6A-45R TP	231
C-23B Sherpa	3	25,600	2 PT6A-65AR TP	233
Trainers				
Shorts Tucano	2	5,842	TPE 331-12B TP	322
Turboprop Commuter				
SD330 Sherpa	3+30	22,900	2 PT6A-45R TP	228
SD360-300	3+39	27,100	2 PT6A-67R TP	249
Sikorsky Aircraft United Technologies Corp., Stratford, CT				SIKORSKY
S-65 (CH-53A Sea Stallion)	3+27	42,000	2 T64-GE-6B TS	195
S-65 (HH-53C)	3	42,000	2 T64-GE-7A TS	186
S-65 (CH-53D)	3+37	42,000	2 T64-413 TS	195
S-65 (RH-53D)	6+37	42,000	2 T64-415 TS	195
S-65 (CH-53E) Super Stallion	3+55	69750	2 T64-416 TS	196
S-65 (MH-53E Sea Dragon	8	69,750	2 T64-416 TS	195
S-70 (MH-60K)	4+12	24,500	2 T700-GE-701C TS	130
S-70C	2	22,000	2 CT7-2C TS	196
HH-60H	4+12	20,138	2 T700-GE-401C TS	135
HH-60J Jayhawk	4+6	21,246	2 T700-GE-401C TS	135
S-70 (UH-60A Black Hawk)	3+11	16,994	2 T700-GE-700/701 TS	184
S-70 (VH-60N)	3+11	15,546	2 T700 TS	150
S-76 Mk II	2+12	10,300	2 250-C30 TS	178
H-76	2+12	10,300	2 250-C34-C30 TS	178
SH-3H Sea King	4	21,000	2 T58-GE-10 TS	120
SH-60B Seahawk	3	21,481	2 GE T700-GE0401 TS	
SH-60F CV-Helo	4	21,800	2 T700-GE-401 TS	177

Model	Crew +pax	GW, lb.	Powerplant (hp or lb. thrust)	Cruise, mph (unless noted)
Sepecat				
Jaguar	1	34,600	2 RR Turbomdca Adour	M 1.5
Westland Helicopters, Ltd., Yeovil, Somerset, UK				WESTLAND
Super Lynx	2+11	11,750	2 RR Gem 42-1	173
Commando	30	21,000	2 RR Gnome H 1400-1	149
Sea King	25	21,000	2 RR Gnome H 1400-1	149
Adv. Sea King	28	21,500	2 RR Gnome H 1400-1	150
AH Mk 1 Lynx	11	9,600	2 RR GEM 2	180
Lynx II(Army)	11	10,000	2 RR GEM 60-3/2	179
Lynx III(Army)	11	12,000	2 RR GEM 60-3/2	173
HAS Mk 2 Lynx	11	9,750	2 RR GEM 2	180
HAS Mk 3/4 Lynx	11	10,500	2 RR GEM 41-1	180
Lynx II (Navy)	11	10,750	2 RR GEM 41-2/43-1	171
Lynx III (Navy)	11	12,000	2 RR GEM 60-3/2	171
W30-140 Westland 3017		12,350	2 RR GEM 41-1 Mk 510	138
W30-160 Westland 3021		12,800	2 RR GEM 60-3	144
W30-200 Westland 3021		12,800	2 CT7-2B	144
W30-300 Westland 3019		15,500	2 CT702B or T700-401	155
EH 101	4	31,500	3 T700-GE-401	

Russian Aircraft

Source: *Aviation Week & Space Technology* Aerospace Source Book, Aeroflot

Powerplant key: TF=Turbofan; TJ=Turbojet; TP=Turboprop; TS=Turboshaft
Powerplant manufacturers: AI=Ivchenko; AM=Mikulin; ASh=Shvetsov; NK=Kuznetsov; Sol=Soloviev;
Tum=Tumansky; Ved=Vedeneev; VK=Klimov; Zap=Zaporozhye

Model	Crew +pax	GW, lb.	Powerplant (hp or lb. thrust)	Cruise, mph (unless noted)
Antanov Scientific Production Complex,Kiev, Ukraine				**An**
Transport & Utility				
An-12 Cub	5	121,500	4 AI-20M TP (4,250)	450
An-14 Clod	1+7	7,607	2 AI-14R TP	143
An-22 Cock	5	500,000	4 NK-12MV TP 470	
An-24V Coke	3+50	46,305	2 AI-24T TP (2,250)	
			+ RU19A-300 TJ (1,760#)	335
An-26 Curl	3	52,920	2 AI-24UT TP (2,820)	
			+ RU19A-300 TJ(1,760#)	270
An-28 Cash	3+15	12,350	2 Isotov TVD-850	205
An-30 Clank	7	51,040	2 AI-24VT TP (2,820)	
			+ RU19A-300 TJ (1,760#)	360
An-32 Cline	3	59,525	2 AI-20M TP	379
An-70T	5	271,170	4 Progress D-27 TP	379
An-72/74 Coaler/B/C	2	66,000	2 Lotarev D-36 378	
An-124 Condor	6+350	800,000	4 Lotarev D-18P 350	
An-225 Cossack	6	1,320,000	6 Lutarev D-18T	525
Reconnaissance				
A-50 Mainstay	5+40	418,870	4 Avia D-30KP TF	
An-71	3+3		1 RD-38A, 2 D-436K	M0.65
Beriev				**Be**
Transport & Utility				
Be-30 Cuff	2+15	12,919	2 TVD-10 TP	290
Ilushyn Design Bureau, Moscow				**Il**
Transport & Utility				
Il-12 Coach	5	38,000	2 ASh-82FNV	252
Il-18V Coot	5+110	134,640	4 AI-20K TP	425
Il-62 Classic	6+186	347,125	4 NK-8-4 TF	560
Il-62M Classic	5+186	363,000	4 Sol. D-30KU TF (24,200#)	560
Il-76M Candid	7	374,000	4 Sil. D-30KP TF (26,400#)	480
IL-76 MF	5	462,960	4 PS-90A	M0.80
Il-86 Camber	4+350	413,600	4 NK-86 TF (28,600#)	560
Early Warning				
A-50 Mainstay	5	418,870	4D-30 KP	
Reconnaissance				
Il-38 May	12		4 AI-20 TH	400
Il-78 Midas			4 Sol. D-30KP TF	480
Kamov				**Ka**
Helicopters				
Ka-26 Hoodlum	2+7	6,600	2 Ved M14V-26 TS	106
Ka-32 Helix	2	24,250	2 TVZ-117 (2,225)	150
Ka-62	2+16	13,230	2 RD-600 (1300)	186
Ka-115	1+5	4,080	PW-Klimov PW206K/2	155
Ka-126	1+7	7,165	TVO-100 (720)	111
Ka-226	1+6	6,835	2 250-C20B	127

Model	Crew +pax	GW, lb.	Powerplant (hp or lb. thrust)	Cruise, mph (unless noted)
Mikkoyan-Gurevich (VPK Mapo)				**MiG**
Trainers				
MiG-AT	2	12,037	2 Snecma Larzac 04-R20 TF	530
Fighters				
MiG-21 Fishbed C	1	16,700	Tum R-11 TJ	M2.0
MiG-21 Fishbed D, F	1	21,000	Tum R-11 TJ	M2.0
MiG-21 Fishbed H/J/K/N	1	21,000	Tum R-13-300 TJ	M2.2
MiG-23 Flogger B	1	35,000	TJ	M2.3
MiG-23 Flogger C/E/F/H/G/K	2	35,000	TJ	M2.3
MiG-25 Foxbat E	1	64,200	2 Tum R-266 TJ	M2.8
MiG-25 Foxbat B	1	64,200	2 TJ	M2.3
MiG-27 Flogger D/J	1	37,000	TJ	M1.7
MiG-29 Fulcrum A/B/C	1	60,000	2 TJ	M2.3
MiG-31 Foxhound A	2	64,200	2 R-15-300 TJ	M2.4
Mil				
Helicopters				
Mi-2	2+8			102
Mi-4 Hound	2	17,199	ASh-82V	130
Mi-6A Hook	5+90	93,712	2 Sol. D-25V TS	186
Mi-8 Hip E	2+27	26,400	2 TV2-117A TS (1,700#)	135
Mi-10 Harke	3	95,807	2 Sol. D-25V TS	214
Mi-12 Homer	6	231,000	4 Sol. D 25V TS	163
Mi-17 Hip	2 +24	28,660	2 TV3-117MT TS	155
Mi-24 Hind A/B/D/E	3	24,200	2 TV2-117A TS	190
Mi-26 Halo	5	123,200	2 D-136 TS	183
Mi-28 Havoc	5		2 TS	190
Mi-38	2+30	34,400	2 TVA-3000	175
Myasischev Experimental Machine Building Factory, Zhukovsky				**Mya**
Transport & Utility				
Mya-4 Bison	8	250,000	2 AM-209 TJ	610
Reconnaissance				
M-17	1	40,565		M0.75
M-55	1	52,910		M0.75
Sukhoi Design Bureau, Moscow				**Su**
Trainers				
G2-A Goleb	2	9,480	RR Viper Mk 22-6	
TJ-1 Jastreb Trainer	2	11,250	RR Viper Mk 531	287
UTVA-75A-21	2	2,116	Lyc IO-360-B1F	116
LASTA Swallow	1-2	3,600	Lyc AEIO-540-L1B5D	210
Su-7 Fitter A	1	29,000	Lyuka AL7F-2 TJ	M2.0
Su-9 Fishpot	1		TJ	M2.0
Su-11 Fishpot C	1		TJ	M1.8
Su-15 Flagon A/B/E/F 1		35,000	2 TJ	M2.5
Su-20 Fitter C/D	1	39,000	TJ	M2.1
Fighters				
Su-24 Fencer A/B/C	2	65,000	2 TJ	M2.0
Su-25 Frogfoot A/B	2	36,000	2 R-13-300 TJ	M0.9
Su-27 Flanker B/C	1-2	63,500	2 TF	M2.3

Model	Crew +pax	GW, lb.	Powerplant (hp or lb. thrust)	Cruise, mph (unless noted)
Sukhoi cont.				
Attack				
Su-25 Frog Foot	1	36,380	2 R195 TJ	M0.95
Su-30	2	72,752	2 AL-35 TF	M2.0
Su-30K	2	67,100	2 Ak-31F TF	
Su-30MK	2	75,555	2 AL-37 w/thrust vectoring	M2.0
Su-33 (Su-27K) Flanker	1	70,550	2 AL-35F TF	M2.0
SU-35	1	74,955	2 AL-31FM TF	M2.3
Su-37	1	56,590	2 AL-37 FU	
Bombers				
Su-24 Fencer	2	87,520	2 AL-21F3 TJ	M1.35
Su-30MK		52,910		M2.0
Su-34	2	99,200	2 AL-31F	M1.8
Tupolev Design Bureau, Moscow				**Tu**
Transport & Utility				
Tu-114 Cleat	5+172	361,620	2 NK 12MV TP	540
Tu-134 Crusty	3+72	98,000	2 Sol. D-30 TF (14,960#)	540
Tu-144 Charger	3+140	396,000	4 NK-144 TF	M2.35
Tu-154 Careless	3+164	198,450	3 NK-8-2U TF (23,100#)	594
Bombers				
Tu-22M Backfire B/C	4	273,270	2 NK 144 TF	M2.0
Tu-95 Bear B/G/H	5	407,800	4 NK-12 TP	550
Tu-160 Blackjack	4	606,260	4 NK-32 TF	M2.3
Fighters				
Tu-28P Fiddler	2	82,000	2 TJ	M1.7
Yakolev Design Bureau, Moscow				**Yak**
Trainers				
Yak-130	2	19,840	2 Klimov RD-35	540
Transport & Utility				
Yak-40 Codling	3+32	35,275	3 AI-25 TF (3,300#)	375
Yak-40M Codling	2+40	39,200	3 AI-25 TF	435
Yak-42 Clobber	2+120	110,000	3 Zap D-36 (14,300#)	496
Bombers				
Yak-28 Brewer	2	30,000	2 Tum R-11 TJ	1,200
Fighters				
Yak-38 Forger A/B	1	26,000	2 TF + 2 TJ	M0.9
Yak-141 Freestyle	1	42,900	R-79 and 2 RD	M1.8

Gas Turbine Engines

Source: *Aviation Week & Space Technology Aerospace Source Book*, 1997
shp=shaft horsepower; #=lbs thrust

Model	Type*	Stages Comp.	turbine	Power @S.L.	Dry wt.,lb.	Notes, use
AlliedSignal Garrett Engine Div., Phoneix, AZ						**GARRETT**
ATF 3-3	ACFF	1,5,1	1,3,2	5440#	1118	HU-25
ATF 3-6A	ACFF	1,5,1	1,3,2	5440#	1125	Falcon 200
F109-GA-100	ACFF	1,2	2,2	1330#	439	
F124-GA-100	ACFF	3,5	1,1	6300#	1100	
T76-G-10, 12, 400 series	CFP	2	3	715/1040 shp	320/360	OV-10
TFE 1042-70	ACFF	3,5	1,1	9250#	1360	
TFE731-2	ACFF	1,4,1	1,3	3500#	734	Falcon100
TFE731-3	ACFF	1,4,1	1,3	3700#	742	Bae 125-700
TFE731-3A	ACFF	1,4,1	1,3	3700#	766	Lear 55
TFE731-3B	ACFF	1,4,1	1,3	3650#	750	Citation 3,6
TFE731-4	ACFF	1,4,1	1,3	4080#	822	Citation 7
TFE731-5	ACFF	1,4,1	1,3	4300#	852	Bae 125-800
TFE731-5AR	ACFF	1,4,1	1,3	4500#	852	Falcon 900
TFE731-5BR	ACFF	1,4,1	1,3	4750#	867	
TFE-731-20	ACFF	1,4,1	1,3	3650#	885	Lear 45
TFE-731-40	ACFF	1,4,1	1,3	4250#	885	
TFE-731-60	ACFF	1,4,1	1,3	5000#	988	
T7G10, 12, 420, 421	CFP	2	3	710/1040shp	20/360	
TPE331-1	CFP	2	3	665shp	335	MU-2, Merlin
TPE331-2	CFP	2	3	715shp	335	Shorts
TPE331-3	CFP	2	3	840shp	355	Merlin 3,4
TPE331-4/5	CFP	2	3	715shp	355	
TPE-331-6	CFP	2	3	840shp	355	Beech B100
TPE331-8/9	CFP	2	3	715/865shp	370	Conquest
TPE331-10	CFP	2	3	900/1000shp	385	Jetstream 31
TPE331-11	CFP	2	3	1000/1100shp	400	Metro 3
TPE331-12B	CFP	2	3	1100shp	420	Tucano
TPE331-12UAR	CFP	2	3	1000/1100shp	390	Cheyenne 400
TPE331-14A/B/F/UA/UB	CFP	2	3	1250/1645shp	585	
TPE331-14 GR/HR	CFP	2	3	1759shp	633	Jetstream 41
TPE331-15AW	CFP	2	3	1645shp	622	Grumman S-2
TPF51-25,-43,-47,55,-62	CFP	2	2,3	575shp	335	MU-2, Cmdr.
Stratford Operations, Stratford, CT						
T53-L-701A	ACFP	5,1	2,2	1400shp	693	
T53-L-13B	ACFS	5,1	2,2	1400shp	544	UH-1H
T53-L-703	ACFS	7,1	2,2	18005shp	545	AH-1
T5313B	ACFS	5,1	2,2	1400shp	544	Bell 205
T55-L-712	ACFS	7,1	2,2	3750shp	750	CH-47D
T55-L-714	ACFS	7,1	2,2	4868shp	832	MH-47E
AL5512	ACFS	7,1	2,2	4075shp	770	Boeing 234
ALF502L/L2	ACFF	3,7,1	2,2	7500#	1311	CL-600
ALF502L-2A/-2C	ACFF	3,7,1	2,2	7800/7500#	1311	
ALF502L3	ACFF	3,7,1	2,2	7500/7800#	1321	
ALF502R-3	ACFF	2,7,1	2,2	6700#	1336	Bae 146
ALF502R-5	ACFF	2,7,1	2,2	6970#		
ALF502R-6	ACFF	3,7,1	2,2	7500#	1375	

*first one or two letters designate compressor: A=Axial, C=Centrifugal; next letter for flow: final letters define output: F=Fan, J=Jet, LJ=Lift Jet, P=Propeller, S=Shaft

Model	Type*	Stages Comp. turbine		Power @S.L.	Dry wt.,lb.	Notes, use
AlliedSignal cont.						
LF507-1F/H	ACFF	3,7,1	2,2	7000#	1385	Avro RJ
LTC4B-8D	ACFS	7,1	2,1	2250shp	605	
LTP101-600A/A-1A	ACFP	1,1	1,1	620shp	335	Air Tractor
LTP101700-A-1A	ACFP	1,1	1,1	700eshp	335	
LTS101-600-A3	ACFS	1,1	1,1	615shp	253	AS 350D
LTS101-650B-1	ACFS	1,1	1,1	650shp	266	BK-117A
LTS101-650C-3/3A	ACFS	1,1	1,1	675shp	241	Bell 222
LTS101-750A-1	ACFS	1,1	1,1	742shp	268	AS 366
LTS101-750B-1	ACFS	1,1	1,1	735shp	271	BK-117-1
LTS101-750B-2	ACFS	1,1	1,1	742shp	268	HH-65A
Allison Engine Co., Indianapolis, IN						**ALLISON**
250-B17C	ACFP	6,1	4	420shp	198	
250-B17D, E, F	ACFP	6,1	4	420shp	202	
250-B17F,F1,F2	ACFP	4,1	1	450shp	212	
250-C20B, F, J, W	ACFS	6,1	4	420shp	161	Bell 206B
250-C20R, R1, R2, R4	ACFS	4,1	1	450shp	173	A109C
250-C20S	ACFS	6,1	4	420shp	162	
250-C28B,C	CFS	1	4	550shp	238	BO-105
250-C30G, G2	CFS	1	4	650shp	258	Bell 222
250-C30L, P, R, S, U	CFS	1	4	650shp	250	MD 530,S-67
250-C30M	CFS	1	4	650shp	250	AS 350G
AE 2100	CFS	2	4	1000shp		C-130J
AE 3007	AFP	1,14	2,3	7150#	1581	Citation X
T406-AD-400	AFS	14	4	6150shp	971	V-22
T56-A-14	AFP	14	4	4591shp	1890	P-3C
T56-A-15	AFP	14	4	4591shp	1848	C-130H
T56-A-425	AFP	14	4	4591shp	1899	E-2C
T56-A-427	AFP	14	4	5250shp	1940	
T63-A-720	ACFS	6,1	4	420shp	158	OH-58C
T703-A-700	CFS	1	4	650shp	252	OH-58D
TF41-A-400, A-2B	AFF	3,2,11	2,2	15,000#	3204	
BMW Rolls-Royce, Oberusel, Germany						
BR700-TP	AFP	10	2,3	9-12,000shp		
BR 710	AFF	1,10	2,2	20000#	3520	G-V
BR 715	AFF	1,2,10	2,3	18500-21000#	4545	MD-95
CFE Co., Phoenix,AZ						
CFE 738-1-1B	ACFF	1,5,1	2,3	5918#	1325	Falcon 2000
CFM International, Cincinnati, OH						**CFM**
CFM56-2-C1/-C-3-C5/-C6	AFF	1,3,9	1,4	22000#	4635	DC-8
CFM56-2A2	AFF	1,3,9	1,4	24000#	4820	DC-8
CFM56-2B1	AFF	1,3,9	1,4	22000$	4617	E-3, E-6
CFM56-3-B1	AFF	1,3,9	1,4	20000#	4276	KC-135R
CFM56-3-B2	AFF	1,3,9	1,4	22000#	4301	737
CFM56-3C-1	AFF	1,3,9	1,4	23500#	4301	737
CFM56-5-A1	AFF	1,3,9	1,4	25000#	4860	A320
CFM56-5A3	AFF	1,3,9	1,4	26500#	4975	A320
CFM56-5A4	AFF	1,3,9	1,4	22000#	4975	A319

*first one or two letters designate.d compressor: A=Axial, C=Centrifugal; next letter for flow: final letters define output: F=Fan, J=Jet, LJ=Lift Jet, P=Propeller, S=Shaft

GAS TURBINE ENGINES

Model	Type*	Stages Comp. turbine		Power @S.L.	Dry wt.,lb.	Notes, use
CFM cont.						
CFM56-5A5	AFF	1,3,9	1,4	23500#	4975	A321
CFM56-5B1	AFF	1,4,9	1,4	30000#	5250	A321
CFM56-5B2	AFF	1,4,9	1,4	31000#	5250	A321
CFM56-5B3	AFF	1,4,9	1,4	32000#	5250	A321
CFM56-5B4	AFF	1,4,9	1,4	27000#	5250	A320
CFM56-5B5	AFF	1,4,9	1,4	22000#	5250	A319
CFM56-5B6	AFF	1,4,9	1,4	23500#	5250	A319
CFM56-5C-2	AFF	1,4,9	1,5	31200#	5250	A340
CFM56-5C-3	AFF	1,4,9	1,5	32500#	5700	A340
CFM56-5C-4	AFF	1,4,9	1,5	34000#	5700	A340
CFM56-7B18	AFF	1,3,9	1,4	19500#	5700	737-600
CFM56-7B20	AFF	1,3,9	1,4	20600#	5234	737-600
CFM56-7B22	AFF	1,3,9	1,4	22700#	5234	737-600
CFM56-7B24	AFF	1,3,9	1,4	24200#	5234	737-700
CFM56-7B26	AFF	1,3,9	1,4	26300#	5234	838-800
CFM56-7B27	AFF	1,3,9	1,4	27300#	5234	
CFM56-9	AFF	1,2,9	1,3	1800-22000#	4252	

Eurojet Turbo, Munich

Model	Type*	Stages Comp. turbine		Power @S.L.	Dry wt.,lb.	Notes, use
EJ200	AFF	3.5	1.1	26300#	2180	EF2000

General Electric Co., GE Aircraft Engines, Evendale, OH GE

Model	Type*	Stages Comp. turbine		Power @S.L.	Dry wt.,lb.	Notes, use
CF34-3A	AFF	1F,14C	2,4	9220#	1625	CL-601
CF34-3A1	AFF	1F,14C	2,4	9220#	1655	Can. RJ
CF34-3B/3B1	AFF	1,14	2,4	9220#	1670	CL-601, RJ
CF34-8C	AFF	1,10	2,4	13790#	2235	CRJ-X
CF6-50C1, C2	AFF	1, 3,14	2,4	52500#	8731	DC-10,A300
CF6-50C2B	AFF	1,3,14	2,4	54000#	8731	DC-10
CF6-50C2R	AFF	1,3,14	2,4	51000#	8731	DC-10
CF6-50E	AFF	1,3,14	2,4	52500#	8490	747-200
CF6-50E1	AFF	1,3,14	2,4	52500#	8490	747-200
CF6-50E2	AFF	1,3,14	2,4	52500#	8768	747-200
CF6-50E2B	AFF	1,3,14	2,4	54000#	8768	747-200
CF6-D1A, -60	AFF	1,1,16	2,5	41500#	7896	DC-10-10
CF6-6K2, -6K	AFF	1,1,16	2,5	41500#	7937	DC-10-10
CF6-80A	AFF	1, 3,14	2,4	48000#	8496	767
CF6-80A2	AFF	1,3,14	2,4	50000#	8496	767
CF6-80A3	AFF	1,3,14	2,4	50000#	8720	A310
CF6-80C2A1	AFF	1,4,14	2,5	59000#	9135	A300-600
CF6-80C2A2	AFF	1,4,14	2,5	53500#	9135	A310-300
CF6-80C2A3	AFF	1,4,14	2,5	60200#	9135	A300,MD-11
CF6-80C2A5	AFF	1,4,14	2,5	61300#	9135	A300-600R
CF6-80C2A8	AFF	1,4,14	2,5	59000#	9135	A310
CF6-80C2B1	AFF	1,4,14	2,5	56700#	9399	747-300
CF6-80C2B2	AFF	1,4,14	2,5	52500#	9399	747-400
CF6-80C2B2F	AFF	1,4,14	2,5	52700#	9499	767-200
CF6-80C2B4	AFF	1,4,14	2,5	57900#	9399	767-300ER
CF6-80C2B4F	AFF	1,4,14	2,5	58100#	9499	767-300
CF6-80C2B5F	AFF	1,4,14	2,5	60800#	9499	767-300ER

*first one or two letters designate.d compressor: A=Axial, C=Centrifugal; next letter for flow: final letters define output: F=Fan, J=Jet, LJ=Lift Jet, P=Propeller, S=Shaft

Model	Type*	Stages		Power @S.L.	Dry wt.,lb.	Notes, use
		Comp.	turbine			
General Electric cont.						
CF6-80C2B6	AFF	1,4,14	2,5	60800#	9399	767-300Er
CF6-80C2B6F, FA, 7F	AFF	1,4,14	2,5	60800#	9499	767-300ER
CF6-80C2D1F	AFF	1,4,14	2,5	61960#	9634	MD-11
CF6-80E1A2	AFF	1,4,14	2,5	65800#	10726	A330
CF6-80E1A4	AFF	1,4,14	2,5	68100#	10726	A330
CF700-2D-2	AFF	1F,8C	2,1	4500#	767	Sabreliner 75
CT58-140	AFS	10	2,1	1500shp	340	S-61, S-62
CT64-820-4	AFP	14	2,2	3133shp	1145	DHC-5D
CT7-2A	A/CFS	5A,1C	2,2	1725shp	429	214ST
CT7-2D/2D1	A/CFS	5A,1C	2,2	1725shp	442/446	S-70C
CT7-5A2	A/CFP	5A,1C	2,2	1735shp	783	Saab 340A
CT7-6/6A	A/CFS	5A,1C	2,2	2000shp	485	EH101
CT7-6D	A/CFS	5A,1C	2,2	2325shp	504	S-92
CT-6E	A/CFS	5A,1C	2,2	2737shp	531	
CT-7A	A/CFP	5A,1C	2,2	1700shp	783	
CT-8	A/CFS	5A,1C	2,2	2430shp		
CT7-9B/C	A/CFP	5A,1C	2,2	1870shp	805	Saab 340B
CT7-9D	A/CFP	5A,1C	2,2	1940shp	805	
F101-GE-102	AFF	2,9	1,2	30780#	4400	B-1B
F110-GE-100	AFF	3,9	1,2	28000#	3920	F-16
F110-GE-129	AFF	3,9	1,2	29000#	3940	F-16C/D
F110-GE-400	AFF	3,9	1,2	27000#	4400	F-14B/D
F118-GE-100	AFF	3,9	1,2	19000#	3200	B-2
F118-GE-101	AFF	3,9	1,2	17000#	3150	U-2
F404/F1D2	AFF	3,7	1,1	18100#	1730	F-117A
F404/RM12	AFF	3F,7C	1,1	18000#	2325	
F404-GE-100	AFF	3F,7C	1,1	18000#	2310	
F404-GE-100D	AFF	3F,7C	1,1	11000#	1820	
F404-GE-400	AFF	3F,7C	1,1	16000#	2195	F/A-18
F404-GE-400D	AFF	3F,7C	1,1	10800#	1795	A-6F
F404-GE-402	AFF	3F,7C	1,1	17800#	2282	F/A-18C/D
F404-GE-F2J3	AFF	3F,7C	1,1	18300#	2335	
GE90-76B	AFF	1,3,10	3,6	76900#	16644	777-200
GE90-85B/90B	AFF	1,3,10	3,6	90000#	16644	777-200/300
GE90-92B	AFF	1,3,10	3,6	92000#	16644	777200/300
J79-GE-8	AFJ	17	3	17000#	3695	F-4B, N
J79-GE-10, 17	AFJ	17	3	17820#	3855	F-4J,G,S,E
J79-GE-15	AFJ	17	3	17000#	3699	F-4C, D
J85-GE-17A,B	AFJ	8	2	2850#	400	A-37A
J85-GE-4B	AFJ	8	2	2950shp	415	
J85-GE-5J	AFJ	8	2	3850#	584	T-38A
T58-GE-10	AFS	10	2,1	1400shp	350	UH-46
T58-GE-16	AFS	10	2,2	1870shp	443	CH-46E
T58-GE-5/100/402	AFS	10	2,1	1500shp	335	HH-3E
T58-GE-8F	AFS	10	2,1	1350shp	305	CH-46A
T64/P4D	AFP	14	2,2	3400shp	1188	
T64-GE-7A	AFS	14	2,2	3936shp	720	
T64-GE-100	AFS	14	2,2	4330shp	720	S-65A
T64-GE-413A	AFS	14	2,2	3925shp	716	CH-53D
T64-GE-415/416/416A	AFS	14	2,2	4380shp	720	CH-53D/E
T64-GE-419	AFS	14	2,2	4750shp	755	CH-53E
T700-GE-401	A/CFS	2,2	1	1690shp	434	SH-60B

*first one or two letters designate.d compressor: A=Axial, C=Centrifugal; next letter for flow: final letters define output: F=Fan, J=Jet, LJ=Lift Jet, P=Propeller, S=Shaft

GAS TURBINE ENGINES

Model	Type*	Stages Comp. turbine		Power @S.L.	Dry wt.,lb.	Notes, use
General Electric cont.						
T700-GE-401C	A/CFS	2,2	1	1800shp	458	SH-60B
T700-GE-700	A/CFS	2,2	1	1622shp	437	UH-60A
T700-GE-701	A/CFS	2,2	1	1698shp	437	AH-64
T700-GE-701A	A/CFS	2,2	1	1723shp	437	S-70C
T700-GE-701C	A/CFS	2,2	1	1890shp	456	AH-64
T700-GE-T6A	A/CFS	2,2	1	2040shp	485	EH-101
T700-GE-T6E	A/CFS	5A,1C 2,2		2255shp	531	
TF34-GE-100	AFF	1F,14C 2,4		9065#	1440	A-10
TF34-GE-400A	AFF	1F,14C 2,4		9275#	1478	S-3A
International Aero Engines AG, Glastonbury, CT						
V2500-A1	AFF	1,3,10 2,5		25000#	5210	A320
V2522-A5	AFF	1,4,10 2,5		22000#	5230	A319
V2424-A5	AFF	1,4,10 2,5		23500#	5230	A319
V2525-D5	AFF	1,4,10 2,5		25000#	5610	MD-90
V2527-A5	AFF	1,4,10 2,5		26500#	5230	A320
V2528-D5	AFF	1,4,10 2,5		28000#	5610	MD-90
V2530-A5	AFF	1,4,10 2,5		31400#	5230	A321-100
V2530-D5	AFF	1,4,10 2,5		30000#	5610	MD-90-50
V2533-A5	AFF	1,4,10 2,5		33000#	5230	A321-200
Ishikawajima-Harima, Tokyo, Japan						IHI
CT700-IHI-70/A	ACFS	5,,1	2,2	1723shp	437	UH-60J
CT700-IHI-401C	ACFS	5,1	2,2	1940shp	458	SH-60J
F-3-IHI-30	AFF	2,5	1,2	3680#	750	
F100-1H1-100	AFF	13	4	25000#	3021	
F100-IHI-220E	AFF	3,10	2,2	23450#	3227	F-15J
F-100-IHI-229	AFF	3,9	1,2	29000#	3218	
T56-IHI-14	AFP	14	4	4910sh	1885	
T64-IHI-10J	AFP	14	2+2	3490shp	1240	
TF40-IHI-801A	AFF	2+5	1+1	7300#	1633	
Light Helicopter Turbine Engine Co., St. Louis, MO						LHT
T800-LHT-800	CFS	2	2,2	1334shp	310	
T800-LHT-801	CFS	2	2,2	1550shp	330	RAH-66
CTS 800-4N	CFS	2	2,2	1299shp	383	
CTS800-5D/51	CFS	2	2,2	1580shp	330	
Microturbo, Toulouse, France						MICROTURBO
TRI 60-1	AFJ	3	1	787#	108	
TRI 60-2	AFJ	3	1	832#	108	
TRI 60-3	AFJ	3	1	900#	116	
TRI 60-5	AFJ	3	1	990#	116	
TRI 60-20	AFJ	4	1	1200#	142	
TRI 60-30	AFJ	4	1	1200#	135	
TRS 18	CFJ	1	1	225#	81.5	
TRS 18 (RPV)	CFJ	1	1	254#	81.5	
TRS 18-1/18-1 (RPV)	CFJ	1	1	326/337#	85	
Mitsubishi, Tokyo, Japan						MITSUBISHI
CT63-M-5A	ACFS	6+1	4	317shp	139	

*first one or two letters designate.d compressor: A=Axial, C=Centrifugal; next letter for flow: final letters define output: F—Fan, J—Jet, LJ=Lift Jet, P—Propeller, S=Shaft

Model	Type*	Stages Comp.	turbine	Power @S.L.	Dry wt.,lb.	Notes, use
Motoren-umndTurbinen-Union, Munich						MTU
250-MTU-C20B	ACFS	6,1	2,2	420shp	158	License prod.
CF6-80C2-A1	AFF	5,14	2,5	59000#	9135	in share w/GE
CF6-80E1	AFF	4,14	2,4	67500#		in shre w/GE
EJ200	AFF	3+5	2	20000#		MTU 33%
F117-PW-100	AFF	1,4,12	2,5	40900#	7700	PW partner
J79-MTU-17A	AFJ	17	3	17900	3855	License prod.
JT8D-219	AFF	14	4	21700#	4515	PW share
MTR390	CFS	2	3	1285#		
PW305	AFF	1,5	2,3	5920#	992	PWC share
PW500	AFFC	1,2,1	3	4200#	765	PWC share
PW2037	AFF	1,4,12	2+5	38250#		
PW2040	AFF	1,4,12	2,5	41700#		
PW4084	AFF	1,6,11	2,7	84000#		
PW4090	AFF	1,6,11	2,7	90000#		
PW4098	AFF	1,7,11	2,7	98000#		
RB199	AFF	3,9	2	16000#		
T64-MTU-7	AFS	14	2,2	3925shp	712	CH-53G
Tyne-MTUMk 21/22	AFP	6,9	1,3	5665shp	2194	C-160
MTU, Turbomeca, Rolls-Royce GMBH,Munich						
MTR 390	CFS	2	1,2	1556shp	372	Tiger PAH-2
Polskie Zaklady Lotnicze, Rzeszow, Poland						PZL
SO-1,3	AFJ	7	1	2205#	668	
Isotov GTD-350	ACFS	7,1	2	396shp	298	
Pratt & Whitney Canada, Inc., Longueuil, Quebec						PRATT & WHITNEY
JT15D-5,-5A	ACFF	1,1,1	1,2	2900#	632	B-400, Cit. 5
JT15D-5C	ACFF	1,1,1	1,2	3190#	665	SM S-211
JT15D-5D	ACFF	1,1,1,	1,2	3045#	627	Cit. Ultra
PT6A-25/25A	ACFP	3,1	1,1	580shp	328	PC-7
PT6A-25C	ACFP	3,1,	1,1	750shp	346	EMB-312
PT6A-27	ACFP	3,1	1,1	680shp	328	DHC-6
PT6A-34	ACFP	3,1	1,1	750shp	331	EMB-110
PT6A-42	ACFP	3,1	1,2	850shp	403	
PT6A-60A	ACFP	3,1	1,2	1050shp	475	B-300/350
PT6A-61	ACFP	3,1	1,1	850shp	429	Cheyenne 3A
PT6A-65AG	ACFP	4,1	1,2	1300shp	486	T. Thrush
PT6A-65AR	ACFP	4,1	1,2	1424shp	486	Shorts 360
PT6A-65B	ACFP	4,1	1,2	1100shp	481	B-1900
PT6A-54R	ACFP	4,1	1,2	1376shp	481	Shorts 360
PT6A-66	ACFP	4,1	1,2	850shp	456	P.180
PT6A-67	ACPF	4,1	1,2	1200shp	506	RC-12K
PT6A-67AF	ACFP	4,1	1,2	1424shp	532	Shorts 360
PT6A-67B	ACFP	3,1	1,2	1200shp	515	PC-12
PT6A-67D	ACFP	3,1	1,2	1271shp	515	B-1900D
PT6A-67R	ACFP	4,1	1,2	1424shp	515	Shorts 360
PT6A-112	ACFP	3,1	1,1	500shp	334	Conquest
PT6A-114/1124A	ACFP	3,1	1,1	600/675shp	350	Caravan I
PT6A-135B	ACFP	3,1	1,1	750shp	346	
PT6T-3B	ACFS	3,1	1,1	981shp	378	S-76B

*first one or two letters designate.d compressor: A=Axial, C=Centrifugal; next letter for flow: final letters define output: F=Fan, J=Jet, LJ=Lift Jet, P=Propeller, S=Shaft

Model	Type*	Stages		Power	Dry	Notes,
		Comp.	turbine	@S.L.	wt.,lb.	use
Pratt & Whitney cont.						
PW118/118A	CFP	2	1,1,2	1800shp	861/866	EMB-120
PW119	CFP	2	1,1,2	2288shp	877	Do.328
P120	CFP	2	1,1,2	2000shp	921	ATR-42
PW121	CFP	2	1,1,2	2150shp	936	ATR42-320
PW121A	CFP	2	1,1,2	2200shp	957	ATR42-400
PW123/123AF	CFP	2	1,1,2	2380shp	992	DHC-8
PW123D	CFP	2	1,1,2	2150shp	992	DHC-8-200
PW123E	CFP	2	1,1,2	2380shp	992	DHC-8-315
PW124B	CFP	2	1,1,2	2400shp	1060	ATR72
PW125B	CFP	2	1,1,2	2626shp	1060	Fokker 50
PW127, B/C/D/F	CFP	2	1,1,2	2750shp	1060	ATR72-210
PW127E	CFP	2	1,1,2	2400shp	1060	ATR42-500
PW150	ACFP	3A,1C	1,1,2	5071shp	1521	DHC-8-400
PW206A	CFS	1	1,1	640shp	237	
PW206B	CFS	1	1,1	621shp	236	EC 135
PW206C	CFS	1	1,1	640shp		Agusta A109
PW206D	CFS	1	1,1	611shp		Bell 427
PW305A	ACFF	1,5	2,3	4679#	993	Lear 60
PW305B	ACFF	1,5	2.3	5266#	993	Hawker 1000
PW306A	ACFF	1,5	2,3	5700#	1043	Astra Galaxy
PW530A	ACFF	1,3	1,2	2750#		Cit. Bravo
PW545A	ACFF	1,1,3	1,3	3640#		Cit. Excel
Commercial Engine Business, East Hartford, CT						
JT8D-11	AFF	13	4	15000#	3389	DC-9
JT8D-15	AFF	13	4	15500#	3414	727/37,DC-9
JT8D-15A	AFF	13	4	15500#	3474	727/37,DC-9
JT8D-17	AFF	13	4	16000#	3430	727/37/DC-9
JT8D-17A	AFF	13	4	16000#	3475	727/37,DC-9
JT8D-17AR	AFF	13	4	17400#	3500	727
JT8D-209	AFF	14	4	19250#	4410	MD-80
JT8D-217,217A	AFF	14	4	20850#	4430	MD-80
JT8D-217C	AFF	14	4	20850#	4515	MD-80
JT8D-219	AFF	14	4	21700#	4515	MD-80
JT8D-9, 9A	AFF	13	4	14500#	3377	727/37,DC-9
JT9D-20J	AFF	15	6	50000#	8560	DC-10-40
JT9D-59	AFF	16	6	53000#	9140	DC-10,A300
JT9D-7	AFF	15	6	45600#	8850	747
JT9D-7A	AFF	15	6	46250#	8850	747
JT9D-7F	AFF	15	6	48000#	8850	747
JT9D-7Q,Q3	AFF	16	6	53000#	9295	747-200
JT9D-7R4D, D1	AFF	16	6	48000#	8905	767, A310
JT9D-7R4E, E1	AFF	16	6	50000#	8905	767, A310
JT9D-7R4E3/E4	AFF	16	6	50000#	8905	767, A210
JT9D-7R4G2	AFF	16	6	54750#	9100	747-300
JT9D-7R4H1	AFF	16	6	56000#	8885	A300-600
JT9D-70A	AFF	16	6	53000#	9155	747-200
PW2037	AFF	17	7	38250#	7160	757, C-17
PW2040	AFF	17	7	40900#	7196	757
PW4000 Series	AFF	16	6	50-64000#	9400	747,767
PW4052	AFF	16	6	52200#	9400	767

*first one or two letters designate.d compressor: A=Axial, C=Centrifugal; next letter for flow: final letters define output: F=Fan, J=Jet, LJ=Lift Jet, P=Propeller, S=Shaft

Model	Type*	Stages Comp.	turbine	Power @S.L.	Dry wt.,lb.	Notes, use
Pratt & Whitney cont.						
PW4056	AFF	16	6	56750#	9400	767
PW4060	AFF	16	6	60000#	9400	767,747
PW4084	AFF	18	9	84000#	13701	777
PW4090	AFF	18	9	90000#	15585	777
PW4090	AFF	19	9	98000#	16165	777
PW4152	AFF	16	6	52000#	9400	310-300
PW4156	AFF	16	6	56000#	9400	A300-600
PW4158	AFAF	16	6	58000#	9400	A300-600R
PW4168	AFF	17	7	68000#	14350	A330
PW4460	AFF	16	6	60000#	9400	ND-11
Government Engine Business, West Palm Beach, FL						
J52-P-8A, B, C	AFJ	12	2	9300#	2129	A-6
J52-P-6A, B, C	AFJ	12	2	8500#	2056	TA-4J
J52-P-408	AFJ	12	2	11200#	2318	EA-6B, A-6
J57-P-43W	AFJ	16	3	13750#	3870	KC-135A
J57-P-59W	AFJ	16	3	13750#	4320	KC-135A
J58	AFJ	9	2	34500#	6326	SR-71
J75-P-19W/P-13B	AFJ	15	3	26500#	5960	TR-1
TF30-P-100/P-111	AFF	16	4	25100#	4022	F-111F
TF30-P-3/P-103	AFF	16	4	18500#	4079	F-111A
TF30-P-7/P107	AFF	16	4	20350#	4144	F-111
TF30-P-9/P109	AFF	16	4	20840#	4105	F-111
TF30-P-414/414A	AFF	16	4	20900#	4251	F-14A
TF33-P-3	AFF	15	4	17000#	3900	B-52H
TF33-P-7, 7A	AFF	16	4	21000#	4650	C-141
TF33-PW-100A	AFF	16	4	21000#	4790	E-3A
TF33-PW-102,102A	AFF	15	4	18000#	4340	KC-135,C-18
F100-PW-100	AFF	13	4	23830#	3064	F-15
F100-PW-200	AFF	13	4	23830#	3117	F-16
F100-PW-220	AFF	13	4	23830#	3200	F-15, F-16
F100-PW-220E	AFF	13	4	23830#	3180	F-15
F100-PW-220P	AFF	13	4	27000#	3405	F-15, F-16
F100-PW-220LE	AFF	3	10	23700#	3378	F-15, F-16
F100-PW-229	AFF	13	4	29100#	3705	F-15, F-16
F100-PW-229A	AFF	3,10	4	29100	3745	F-15, F-16
F117-PW-100	AFF	17	7	41700#	7100	C-17
F119-PW-100	AFF	3,6	2	35000#		F-22
Rolls-Royce Commercial Aero Engines, Ltd., Derby, England						**ROLLS-ROYCE**
RB183-555, 15P	AFF	4,12	2,2	9900#	2287	F-29
RB211-22B	AFF	1,7,6	1,1,3	42000#	9195	L-1011
RB211-524B2	AFF	1+7+6	1+1+3	50000#	9814	747,L-1011
RB211-524C2	AFF	1+7+6	1+1+3	51500#	9859	747
RB211-524D4/upgrade	AFF	1+7+6	1+1+3	53000#	9874	747
RB211-524G	AFF	1+7+6	1+1+3	58000#	9499	747
RB211-524H	AFF	1+7+6	1+1+3	60600#	9499	747, 767
RB211-535C	AFF	1,6,6	1,1,3	37400#	7294	757
RB211-535E4B	AFF	1,6,6	1,1,3	43100#	7264	757, Tu-204
RB211-535E4	AFF	1,6,6	1,1,3	40100	7264	757, Tu-204

*first one or two letters designate.d compressor: A=Axial, C=Centrifugal; next letter for flow: final letters define output: F=Fan, J=Jet, LJ=Lift Jet, P=Propeller, S=Shaft

Model	Type*	Stages Comp. turbine		Power @S.L.	Dry wt.,lb.	Notes, use
Rolls-Royce cont.						
RB211-Trent 768	AFF	1,8,6	1,1,4	67500#	10550	A330
RB211-Trent 772	AFF	1,8,6	1,1,4	71000#	10550	A330
B211-Trent 875	AFF	1,8,6	1,1,5	74900#	13100	777
RB211 Trent 877	AFF	1,8,6	1,1,5	77200#	13100	777
RB211 Trent 884	AFF	1,8,6	1,15	84900#	13100	777
RB211 Trent 892	AFF	1,8,6,	1,1,5	91300#	13100	777
RB211 Trent 895	AFF	1,8,6	1,1,5	95500#	13100	777
RB211 Trent 900		1,8,6	1,1,5	82250#	12400	747, A3XX
Spey 506-14W	AFF	4,12	2,2	10400#	2297	BAC 111
Spey Mk.512-14	AFF	5,12	2,2	12000#	2540	BAC 111
Spey Mr.512-14 DW	AFF	5,12	2,2	12550#	2609	BAC 111
Tay 611	AFF	1,3,12	2,3	13850#	2951	G-IV
Tay 620	AFF	1,3,12	2,3	13850#	3185	Fokker 100
Tay 650	AFF	1,3,12	2,3	15100#	3340	Fokker 100
Tay 651	AFF	1,3,12	2,3	15400#	3380	727
Military Aero Engines Ltd., Bristol						
GEM-2	AFCS	4,1	1,1,2	900shp	330	Lynx
GEM-41-1/-42	AFCS	4,1	1,1,2	1120shp	346	Lynx
Gnome H1400	AFS	10	2,1	1480shp	326	Sea King
Gnome H1400-1	AFS	10	2,1	1660shp	326	Boeing 107
Pegasus 11	AFF	3,8	2,2	21300#	3226	Harrier
Pegasus 11-21 (F402-RR-406)	AFF	3,8	2,2	22000#	3235	AV-8B
Pegasus 11-61	AFF	3,8	2,2	23800#	3560	Harrier 2
RR 1004	AFCS	4,1	1,1,2	1018shp	313	Agusta 129
Spey Mk 101	AFF	4,12	2,2	11030#	2471	Buccaneer
Spey Mk 202	AFF	5,12	2,2	20515#	4093	Phantom
Spey Mk 250/251	AFF	5,12	2,2	11995#	2740	Nimrod
Spey RB. 168-807	AFF	4,12	2,2	11030#	2417	
Tyne R Ty. Mk 101, 515	AFP	6,9	1,3	5505shp	2219	CL-44
Tyne R Ty. Mk 21, 22	AFP	6,9	1,3	6100shp	2203	
Tyne R. Ty. 20 Mk 801	AFP	6,9	1,3	4380shp	2187	G.222
Viper 11	AFJ	7	1	2500#	625	
Viper 522	AFJ	8	1	3360#	760	HS.125
Viper 531	AFJ	8	1	3120#	790	
Viper 535,540	AFJ	8	1	3360#	790	
Viper 601	AFJ	8	2	3750#	790	HS.125-600
Viper 632	AFJ	8	2	4000#	825	MB326. 339
Viper 633	AFJ	8	2	5000#	928	IAR 93
Viper 680	AFJ	8	2	4370#	836	MB339K
Rolls-Royce, SNECMA						
Olympus 593	AFJ	7,7	1,1	38000#	6780	Concorde
Rolls-Royce, SNECMA, MTU, FN						
Tyne R Ty, 20 Mk 21, 22	AFP	6,9	1,3	6100shpp	2203	Atlantic
Rolls-Royce Turbomeca Ltd.						
Adour-871	AFF	2,5	1,1	5990#	1306	Hawk
Adour Mk 811, 815	AFF	2,5	1,1	8400#	1633	Jaguar Intl
Adour Mk 861	AFF	2,5	1,1	5710#	1282	Hawk
RTM 322-01	ACFS	3,1	2,2	2400shp	529	EH-101

*first one or two letters designate.d compressor: A=Axial, C=Centrifugal; next letter for flow: final letters define output: F=Fan, J=Jet, LJ−Lift Jct, P−Propeller, S−Shaft

Model	Type*	Stages Comp.	turbine	Power @S.L.	Dry wt.,lb.	Notes, use
Snecma, Paris, France						SNECMA
Atar 9K50	AFJ	9	2	15870#	3490	Mirage F1, 50
M53-5	AFF	3,5	2	19850#	3240	Mirage 2000
M53-P2	AFF	3,5	2	21360#	3300	Mirage 2000
M88-2	AFF	3,6	1,1	16500#	1920	Rafale C,D,M
M88-3	AFF	3,6	1,1	19000#	2013	
Teledyne CAE, Toledo, OH						TELEDYNE
J69-T-25A	CFJ	1	1	1025#	358	T-37B
J69-T-29	ACFJ	1,1	1	1700#	340	BQM-34A
J69-T-41A	ACFJ	1,1	1	1920#	350	BQM-34A
J69-T-406	ACFJ	1,1	1	1920#	360	BQM-34E,F
J402-CA-400	ACFJ	1,1	1	660#	102	Harpoon
J402-CA-700	ACFJ	1,1	1	640#	113	RPVs
J402-CA-701	ACFJ	1,1	1	725#	113	RPVs
J402-CA-702	ACFJ	2,1	1	960#	138	MQM-107
305-7E	CFJ	1	1	90#	19	Missles
305-10	CFJ	1	1	179#	27	Missles
312-1	CFJ	1	1	177#	34	Missles
320-1	CFJ	1	1	240#	45	Missles
320-2	ACFJ	1,1	1	350#	60	Missles
383-8B	ACFJ	2,1	1	960#	138	Missles
382-10	ACFF	1,1	1	990#	145	UAV-MR
Societe Turbomeca, Bordes, France						TURBOMECA
Arbizon 3B	ACFJ	1,1	1	907#	253	Missle
Arriel 1B	ACFS	1,1	2,1	641shp	254	AS 350B
Arriel 1C, 1C1	ACFS	1,1	2;1	700shp	262	SA365N
Ariel 1C2	ACFS	1,1	2,1	771shp	262	AS 565
Arriel 1D/1D1	ACFS	1,1	2,1	684/712shp		AS350B1/B2
Ariel 1E2	ACFS	1,1	2,1	770shp	271	BK118C1
Arriel 1K/1K1	ACFS	1,1	2,1	724/772shp	266	Agusta A109
Ariel 1M/1M1	ACFS	1,1	2,1		271	AS565
Arriel 1S/1S1	ACFS	1,1	2,1	778/802shp	271	S-76A
Arrius 1A, 1M	ACFS	1	1,1	547shp	203	AS 355, 555
Arrius 2B	ACFS	1	1,1	696shp	228	EC 135
Arrius 2C	ACFS	1	1,1	696shp	236	MD Explor.
Arrius 2D	ACFP	1	1,1	497shp	265	AS Omega
Artouste 3B	ACFS	1,1	3	562shp	401	SA 315, 316
Astazou 3	ACFS	1,1	3	590shp	324	SA 341
Astazou 14H	ACFS	2,1	3	590shp	348	SA 342
Astazou 16D	ACFP	2,1	3	968shp	452	Jetstream
Astazou 16G	ACFP	2,1	3	1021shp	503	Pucara
Makila 1A1	ACFS	3,1	2,2	1878shp	518	SA 332
Mikila 1A2	ACFS	3,1	2,2	2104shp	518	AS332/532
TM 333 A/B	ACFS	2,1	1,1	912/1056shp	297	SA365M
Turmo 3C7	ACFS	1,1	2,2	1610shp	660	SA 321
Turmo 4C	ACFS	1,1	2,1	1558shp	515	SA330
Turbo-Union Ltd., Bristol						
RB199 Mk 103/104	AFF	3,3,6	1,1,2	16000#	1549	Tornado IDS
RB199 Mk105	AFF	3,3,6	1,1,2	16800#	1611	Tornado ECR

*first one or two letters designate.d compressor: A=Axial, C=Centrifugal; next letter for flow: final letters define output: F=Fan, J=Jet, LJ=Lift Jet, P=Propeller, S=Shaft

Model	Type*	Stages Comp. turbine		Power @S.L.	Dry wt.,lb.	Notes, use
Volvo Flygmotor ab, Trollhaettan, Sweden						**VOLVO**
RM8A	AFF	2,4,7	1,3	25990#	4630	JA37
RM8B	AFF	3,3,7	1,3	28110#	4895	JA37
RM12	AFF	3,7	1	18000#	2315	JAS39
Walter, Prague, Czech Republic						
M601T	ACFP	2,1	1,1	798shp	445	PZL Orlik
M601Z	ACFP	2,1	1,1	543shp	434	Z-13T
M502A	ACFP	2	1,2,2	1824shp	1256	L-610
M701	CFJ	1	1	1962#	738	L-29
Williams International Corp., Walled Lake, MI						**WILLIAMS**
F107-WR-101	ACFF	4,1	1,2	600#	146	
F107-WR-402	ACFF	4,1	1,2	700#		
F112-WR-100	ACFF	4,1	1,2			
F121-WR-100	AFJ	1,6	2	150#	42	
P8300	AFJ	2,5	2	1000#		
WJ24-8	ACFJ	1,1	1	240#	50	
WJ119-2	AFJ	6	1	105#		
WR2-6	CFJ	1	1	125#	28	CL-89
WR24-7	ACFJ	1,1	1	176#	44	
WTS34-16	CFS	1	1	50shp	53	
Williams-Rolls, Inc.						
FJ44-1	ACFF	1,1,1	1,2	1900#	445	CitationJet
JF44-2	ACFF	1,3,1	1,2	2300#		SJ30-2

*first one or two letters designate.d compressor: A=Axial, C=Centrifugal; next letter for flow: final letters define output: F=Fan, J=Jet, LJ=Lift Jet, P=Propeller, S=Shaft

TOP 20 AIRLINES
by revenue passenger miles
Source: *Aviation Week & Space Technology Aerospace Source Book*, 1997

Airline	1995 RPM (million)	Airline	1995 RPM (million)
United Airlines	111,539	Air France	39,773
American Airlines	102,668	Singapore Airways	30,075
Delta Air Lines	85,108	KLM Royal Dutch	27,625
Northwest Airlines	62,502	All Nippon Airways	26,629
British Airways	58,323	Trans World Airlines	24,905
Japan Airlines	43,357	Southwest Airlines	23,330
Lufthansa	38,278	Cathay Pacific	21,949
USAir	37,619	Korean Air	20,991
Continental Airlines	35,512	Alitalia	19,727
Qantas Airways	32,231	Thai Airways International	16,810

TOP TEN REGIONAL AIRLINES
by passengers (x million)

Airline	pass.	'95 rank	Airline	pass.	'95 rank
Simmons Airlines (AA)	6,010	1	Horizon Air (AS/NW)	3,557	4
Comair (DL)	4,651	2	Flagship Airlines (AA)	2,939	6
Mesa Airlines (HP/UA/US)	4,135	5	Piedmont Airlines (US)	2,744	8
Continental Express (CO)	4,090	3	SkyWest Airlines (DL/CO)	2,593	9
Atlantic Southeast Airlines (DL)	3,632	7	Trans States(AS/NW/TW/US/US)	2,121	13

TOP 10 ALL-CARGO AIRLINES
by freight ton miles (FTM)

Airline	1995 FTM (million)	Airline	1995 FTM (million)
Federal Express	5,070	Emery Worldwide	935
United Parcel Service	3,377	American International	767
Cargolux	1,329	Southern Air Transport	578
Nippon Cargo Airlines	1,111	Airborne Express	485
Polar Air Cargo	977	Evergreen International	405

V. GOVERNMENT

FAA/Government

FEDERAL AVIATION REGULATIONS

137	Agricultural Aircraft Operations
139	Certification and Operation of Land Airports serving CAB-Certificated Air Carriers
141	Pilot Schools
143	Ground Instructors
145	Repair Stations
147	Aviation Maintenance Technician Schools
149	Parachute Lofts

Airports and Navigation Facilities

150	Airport Noise Compatability Planning
151	Federal Aid to Airports
152	Airport Aid Programs
153	Acquisition of U.S. land for Public Airports
154	Acquisition of U.S. Land for Public Airports under the Airport and Airway Development Act of 1970
155	Release of Airport Property from Surplus Property Disposal Restrictions
157	Notice of Construction, Alternation, Activation, and Deactiviation of Airports
159	National Capital Airports
169	Expenditure of Federal Funds for Nonmilitary Airports or Air Navigation Facilities Thereon
171	Non-Federal Navigation Facilities

Agency Procedures

183	Representatives of the Administrator
185	Testimony by Employees and Production of Records in Legal Proceedings and Service of Legal Process and Pleadings
187	Fees
189	Use of Federal Aviation Administration Communication System

FAA FUNDING 1990-97
($millions)

Account	1990	1991	1992	1993	1994	1995	1996	1997*
Operations	3824	4036	4360	4530	4580	4572	4643	4955
(General Fund)	(3017)	(2033)	(2250)	(2251)	(2285)	(2122)	(2420)	(3255)**
(Trust Fund)	(807)	(2003)	(2110)	(2279)	(2295)	(2450)	(2223)	(1700)
Percentage from Trust Fund	21%	50%	48%	50%	50%	54%	48%	34%
Airport Improvement Program								
(Obligation Limitation)	1425	1800	1900	1800	1690	1450	1450	1460
Facilities & Equipment	1721	2095	2409	2302	2055	1960	1866	1938
Research Eng.& Development	170	105	218	230	254	252	186	208
Total	7140	8136	8887	8862	8579	8234	8145	8561
(General Fund)	(3017)	(2033)	(2250)	(2251)	(2285)	(2122)	(2420)	(3255)**
(Trust Fund)	4123)	(6103)	(6637)	(6611)	(6294)	(6112)	(5725)	(5306)
Percentage from Trust Fund	58%	75%	75%	75%	73%	74%	70%	62%

*FAA's fiscal year 1997 budget as enacted
**Under Public Law 104-205, the FAA must use up to $75 million in user fees charged for ari traffic control and related services to non-governmental aircraft that fly over but do not take off or land in the US in lieu of General Fund financing.

U.S. DEPARTMENT OF TRANSPORTATION

400 7th St., SW
Washington, DC 20590
(202) 366-4000

FEDERAL AVIATION ADMINISTRATION

800 Independence Avenue, Washington, DC 20591, (202) 267-4000

Administrator
Deputy Administrator
Executive Director for System Operations
Executive Director for Acqjuisition and Safety Oversight
Assistant Administrator for Civil Aviation Security
Assistant Administrator for Policy, Planning, and International Aviation
Assistant Administrator for Airports
Associate Administrator for Airway Facilities
Associate Administrator for Air Traffic
Associate Administrator for Aviation Safety
Associate Administrator for Regulation and Certification
Associate Administrator for Aviation Standards
Director of Airport Planning and Programming
Director of Airport Safety and Standards
Director of Aviation Policy, Plans, and Management Analysis
Federal Air Surgeon

FIELD and REGIONAL OFFICES

Aviation Standards National Field Office
John M. Howard, Director
P.O. Box 25082
Oklahoma City, OK 73125
(405) 868-2305

FAA Technical Center
Edward Harris, Director
Atlantic City Airport, NJ 08405
(609) 484-4000

Mike Monroney Aeronautical Center
Homer McClure, Director
P.O. Box 25082
Oklahoma City, OK 73125

Alaskan Regional Office
Franklin L. Cunningham, Director
P.O. Box 14
Anchorage, AK 99513
(907) 271-5645

Central Regional Office
Paul Bohr, Director
Federal Office Bldg.
601 E. 12th St.
Kansas City, MO 64106

Eastern Regional Office
Fitzgerald Federal Bldg.
JFK International Airport
Jamaica, NY 11430
(718) 917-1023

East Central Regional Office
Timothy Forte, Director
2300 E. Devon Ave.
Des Plaines, IL 60018
(312) 694-7410

Northeast Regional Office
Arlene Feldman, Director
12 New England Executive Park
Burlington, MA 01803
(617) 273-7244

Northwest Mountain Regional Office
Fredrick Isaac, Director
1601 Lind Ave. SW
Renton, WA 98055
(206) 227-1913

South Central Regional Office
Donald Watson, Director
P.O. Box 1689
Fort Worth, TX 76101
(817) 624-5001

GOVERNMENT AGENCIES

Southeastrn Regional Office
Garland Castleberry, Director
P.O. Box 30320
Atlanta, GA 30320
(404) 763-7222

Southwest Regional Office
P.O. Box 92007
Los Angeles, CA 90009
(213) 297-1427

Europe, Africa & Middle East Office
Ben Demps, Director
c/o American Embassy
Brussels, APO New York 09667-1011
2 513-3830, Ext. 2700

OTHER GOVERNMENT AGENCIES, COMMISSIONS and OFFICES

Aviation Safety Commission
John Albertine, Chairman
Suite 1115
1725 I St., NW
Washington, DC 20006
(202) 634-4860

U.S. Coast Guard
Adm. P.A. Yost, Commandant
Department of Transportation
2100 2nd St., SW
Washington, DC 20593-0001
(202) 267-0952 (Av. Div.)

Customs Service
Carroll Hallett, Commissioner
1301 Constitution Ave., NW
Washington, DC 20229
(202) 566-2101

Environmental Protection Agency
William K. Reilly, Administrator
401 M St., SW
Washington, DC 20460
(202) 382-4700

Executive Office of the President
The White House
1600 Pennsylvania Ave., NW
Washington, DC 20500
(202) 395-3000

Federal Communications Commission (FCC)
1919 M St., NW
Washington, DC 20554
(202) 632-7000

National Aeronautics & Space Administration (NASA) Vice Administrator
6th & Independence Ave., SW
Washington, DC 20546
(202) 453-1000

National Air & Space Museum
6th & Independence Ave., SW
Washington, DC 20560
(202) 357-2700

National Oceanic & Atmospheric Administration (NOAA)
Dept. of Commerce
Commerce Bldg., Rm. 5128
Washington, DC 20230
(301) 443-8910

National Transportation Safety Board (NTSB)
Carl Voght, Chairman
800 Independence Ave., SW
Washington, DC 20594
(202) 382-6600

Passport Services
Dept. of State
22nd & C Sts., NW
Washington, DC 20520
(202) 632-2422

STANDING SENATE AND HOUSE COMMITTEES & SUBCOMMITTEES, CHAIRS AND RANKING MINORITY MEMBERS:

U.S. SENATE
The Capitol
Washington, DC 20510
(202) 224-3121

Committee on Appropriations
Hon. Ted Stevens (R-AK),
Chairman
S-128 Capitol Bldg.
Washington, DC 20510
(202) 224-3471
Hon. Robert C. Byrd (D-WV)
Ranking Minority Member
S-206 Capital Bldg.
Washington, DC 20510
(202) 224-7200

**Subcommittee on Transportation
and Related Agencies**
Hon. Richard C. Shelby (R-OR),
Chairman
SD-133 Dirksen Sen. Office Bldg.
Washington, DC 20510
(202) 224-7281
Hon. Frank R. Lautenberg (D-NJ)
Ranking Minority Member
SD-196 Sen. Office Bldg.
Washington, DC 20510
(202) 224-6280

Committee on the Budget
Hon. Pete V. Domenici (R-NM),
Chairman
SD-621 Dirksen Sen. Office Bldg.
Washington, DC 20510
(202) 224-6988; fax (202) 224-
4835
Hon. Frank R. Lautenberg (D-NJ)
Ranking Minority Member
SD-621 Dirksen Sen. Office Bldg.
Washington, DC 20510
(202) 224-0642; fax (202) 224-
1447

**Committee on Commerce,
Science
and Transpotation**
Hon. John McCain (R-AZ),
Chairman
SR-254 Russell Sen. Office Bldg.
Washington, DC 20510
(202) 224-1251; fax (202) 224-
1259
Hon. Ernest F. Hollings (D-SC)
Ranking Minority Member

SD-558 Dirksen Sen. Office Bldg.
Washington, DC 20510
(202) 224-0427

Subcommittee on Aviation
Hon. Slade Gorton (R-WA),
Chairman
SH-427 Hart Sen. Office Bldg.
Washington, DC 20510
(202) 224-4852; fax (202) 228-
0326
Hon. Wendell H. Ford (D-KY)
Ranking Minority Member
SD-558 Dirksen Sen. Office Bldg.
Washington, DC 20510
(202) 224-0411

Committee on Finance
Hon. William V. Roth, Jr. (R-DE),
Chairman
SD-219 Dirksen Sen. Office Bldg.
Washington, DC 21510
(202) 224-4515; fax (202) 224-
5920
Hon. Daniel P. Moynihan (D-NY)
Ranking Minority Member
SH-203 Hart Sen. Office Bldg.
Washington, DC 20510
224-5315

**Committee on Governmental
Affairs**
Hon. Fred Thompson (R-TN),
Chairman
SD-340 Dirksen Sen. Office Bldg.
Washington, DC 20510
(202) 224-4751
Hon. John Glenn (D-OH)
Ranking Minority Member
SD-326 Dirksen Sen. Office Bldg.
Washington, DC 20510
(202) 224-2627; fax (202) 224-
9682

**U.S. HOUSE OF
REPRESENTATIVES**
The Capitol
Washington, DC 20515
(202) 224-3121

Committee on Appropriations
Hon. Bob Livingston (R-LA),
Chairman
H-218 Capitol Bldg.
Washington, DC 20515
(202) 225-2771

SENATE AND HOUSE COMMITTEES

(Committee on Appropriations)
Hon. David Obey (D-WI)
Ranking Minority Member
1016 Longworth House Office
Bldg.
Washington, DC 20515
(202) 225-3481

**Subcommittee on Transportation
and Related Services**
Hon. Frank R. Wolf (R-VA),
Chairman
2358 Rayburn House Office Bldg.
Washington, DC 20515
(202) 225-2141
Hon. Martin Olav Sabo (D-MN)
Ranking Minority Member
1016 Longworth House Office
Bldg.
Washington, DC 20515
(202) 225- 3481

Committee on the Budget
Hon. John R. Klasich (R-OH),
Chairman
309 Canoon House Office Bldg.
Washington, DC 20515
(202) 226-7270; fax (202) 226-
7174
Hon. John Spratt (D-SC)
Ranking Minority Member
214 O'Neil House Office Bldg.
Washington, DC 20515
(202) 226-7200; fax (202) 226-
7233

**Committee on Government
Reform and Oversight**
Hon. Dan Burton (R-IN),
Chairman
2157 Rayburn House Office Bldg.
Washington, DC 20515
(202) 225-5074; fax (202) 225-
3974
E-mail:
http://www.house.gov/reform
Hon. Henry Waxman (D-CA)
Ranking Minority Member
2153 Rayburn House Office Bldg.
Washington, DC 20515
(202) 225-5051

**Subcommittee on Government
Management, Information &
Technology**
Hon. Steve Horn (R-CA),
Chairman
B-373 Rayburn House Office
Bldg.
Washington, DC 20515
(202) 225-5147; fax (202) 225-
2373
Hon. Carolyn B. Maloney (D-NY)
Ranking Minority Member
B-250ARayburn House Office
Bldg.
Washington, DC 20515
(202) 225-5051; fax (202) 225-
8185

**Committee on Transporation
and Infrastructure**
Hon. Bud Shuster (R-PA),
Chairman
2165 Rayburn House Office Bldg.
Washington, DC 20515
(202) 225-9446; fax (202) 225-
6782
E-mail:
http://www.house.gov/transportation
Hon. James L. Oberstar (D-MN)
Ranking Minority Member
2163 Rayburn House Office Bldg.
Washington, DC 20515
(202) 225-4472; fax (202) 226-
1270

Subcommittee on Aviation
Hon. John J. Duncan, Jr. (R-TN),
Chairman
2251 Rayburn House Office Bldg.
Washington, DC 20515
(202) 226-3220; fax (202) 225-
4629
Hon. William O. Lipinski (D-IL)
Ranking Minority Member
2251 Rayburn House Office Bldg.
Washington, DC 20515
(202) 225-9161

Committee on Ways and Means
Hon. Bill Archer (R-TX),
Chairman
1102 Longworth House Office
Bldg.
Washington, DC 20515
(202) 225-3625
Hon. Charles B. Rangel (D-NY)
Ranking Minority Member
1106 Longworth House Office
Bldg.
Washington, DC 20515
(202) 225-4021

Subcommittee on Oversight
Hon. Nancy L. Johnson (R-CT),
Chair
1136 Longworth House Office
Bldg.
Washington, DC 20515
(202) 225-7061

Hon. William Coyne (D-PA)
Ranking Minority Member
1106 Longworth House Office
Bldg.
Washington, DC 20515
(202) 225-4021

Subcommittee on Trade
Hon. Phillip M. Crane (R-IL),
Chairman
1104 Longworth House Office
Bldg.
Washington, DC 20515
(202) 225-6649
Hon. Robert T. Matsui (D-CA)
Ranking Minority Member
1106 Longworth House Office
Bldg.
Washington, DC 20515
(202) 225-4021

STATE & TERRITORY AERONAUTICS AGENCIES

Alabama Dept. of Aeronautics
817 S. Court St.
Birmingham, AL 36130
(205) 261-4480

Alaska DOT & Public Facilities
P.O. Box 190649
Anchorage, AK 99519-0649
(907) 243-4396

American Samoa Dept. of Port
Administration
P.O. Box 1539
Pago Pago, American Samoa
(684) 633-4251, (684) 639-9101

Arizona DOT, Aeronautics Div.
1801 W. Jefferson, 426M
Phoenix, AZ 85007
(602) 255-7691

Arkansas Dept. of Aeronautics
3rd Floor, Regional Airport
Terminal
1 Airport Dr.
Little Rock, AR 72202
(501) 376-6781

California Dept. of Aeronautics
P.O. Box 1499, 1130 K St.
Sacramento, CA 95807
(916) 322-3090

Colorado Airport Planning Staff
Dept. of Local Affairs, Div. Local
Govt.
Room 520, 1313 Sherman St.
Denver, CO 80293
(303) 866-3004

Connecticut DOT
Bureau of Aeronautics, Drawer A
24 Wolcott Hill R.
Wethersfield, CT 06109
(203) 566-4417

Delaware Transportation
Authority, Aeronautics Division
Box 778
Dover, DE 19903
(302) 736-3264

Florida DOT, Aviation Bureau
605 Suwannee St., MS 46
Tallahassee, FL 32399-0450
(904) 488-8444

Georgia DOT, Bureau of
Aeronautics
2017 Flightway Dr.
Chamblee, GA 30341
(404) 986-1350

Guam Airport Authority
Guam International Air Terminal
Tamuning, Guam 96911
(671) 646-0300

Hawaii State DOT
Honolulu International Airport
Honolulu, HI 96819 (808) 548-
3205

Idaho Bureau of Aeronautics and
Public Transportation
3483 Rickenbacker St.
Boise, ID 83705
(208) 334-3183

Illinois Div. of Aeronautics
Capital Airport
1 Langhorne Bond Rd.
Springfield, IL 62707
(217) 785-8515

Indiana DOT
Division of Aeronautics
143 W. Market St., Suite 300
Indianapolis, IN 46204
(317) 232-1470

Iowa DOT, Air & Transit Div.
State House
Des Moines, IA 50319
(515) 281-4280

Kansas DOT, Aviation Div.
Docking State Office Bldg.
Topeka, KS 66612-1568
(913) 296-2553

Kentucky Transportation Cabinet,
Office of Aeronautics
U.S. 127 Annex
Frankfort, KY 40622
(502) 564-4480

Louisiana DOT, Office of Aviation
& Public Transport
P.O. Box 94245
Baton Rouge, LA 70804-94245
(504) 379-1235

Maine DOT, Div. of Aeronautics
Transportation Bldg., Child St.
Augusta, ME 04333
(207) 289-3185

Maryland State Aviation
Administration
P.O. Box 8766
Baltimore/Washington Int'l
Airport, MD 21240
(301) 859-7100

Massachusetts Aeronautics
Commission
State Transportation Bldg.
10 Park Plaza, Room 6620
Boston, MA 02116-3966
(617) 973-7350

Michigan Aeronautics Commission
DOT, Capital City Airport
Lansing, MI 48906
(517) 373-1834

Minnesota DOT, Aeronautics
Division
Room 417, Transportation Bldg.
St. Paul, MN 55155
(612) 296-8202

Missouri Highway &
Transportation Dept., Aviation
Division
Box 720
Jefferson City, MO 65102
(314) 751-2589

Montana Aeronautics Board
Box 5178
Helena, MT 59604
(406) 444-2506

Nebraska Dept. of Aeronautics
Box 82008, Municipal Airport
Lincoln, NE 68501
(402) 471-2371

New Hampshire DOT, Division of
Aeronautics
65 Airport Rd.
Concord, NH 03301-5298

New Jersey DOT, Aeronautics
Division
1035 Parkway Ave., CN 600
Trenton, NJ 08625
(609) 530-2900

New Mexico Highway &
Transportation Dept., Aviation
Division
Box 579
Santa Fe, NM 87504-0579
(505) 827-0332

New York State DOT, Aviation
Division
1220 Washington Ave.
Albany, NY 12232
(518) 457-2820

North Carolina DOT
Aviation Division
Box 25201
Raleigh, NC 27611
(919) 787-9618

North Dakota Aeronautics Comm.
Box 5010
Bismarck, ND 58502
(701) 224-2748

Ohio DOT, Aviation Division
University Airport
2829 W. Granville Ave.
Worthington, OH 43085
(614) 889-2533

Oklahoma Aeronautics Comm.
DOT Bldg., Room B-7
200 NE 21st St.
Oklahoma City, OK 73105
(405) 521-2377

Oregon Public Utility Comm.
Labor & Industries Bldg.
Salem, OR 97310
(503) 378-6664

Pennsylvania DOT, Bureau of
Aviation
Transportation & Safety Bldg.
Harrisburg, PA 17120
(717) 783-2282

Puerto Rico Ports Authority
Luis Munoz Marin International
Airport
G.P.O. Box 2829
San Juan, PR 00936
(809) 723-2260

Rhode Island DOT, Div. of
Airports
T.F. Green State Airport
Warwick, RI 02886
(401) 737-4000

South Carolina Aeronautics
Comm.
Drawer 1987
Columbia, SC 29202
(803) 739-5400

South Dakota DOT, Office of
Aviation Services
700 Broadway Ave., E.
Pierre, SD 57501
(605) 773-3574

Tennessee DOT, Aeronautics Div.
Box 17326
Nashville, TN 37217
(615) 741-3208

Texas Dept. of Aviation
Box 12607, Capitol Station
410 E. 5th St.
Austin, TX 78711
(512) 476-9262

Utah DOT, Div. of Aeronautics
135 N. 2400 W.
Salt Lake City, UT 84116
(801) 328-2066

Vermont Agency of
Transportation, Division of
Aeronautics
State Adm. Bldg., 133 State St.
Montpelier, VT 05602
(802) 828-2828

Virgin Islands Port Authority
P.O. Box 1707
St. Thomas, VI
(809) 774-1629

Virginia Dept. of Aviation
4508 S. Labumum Ave., Box
7716
Richmond, VA 23231
(804) 786-1364

Washington State DOT,
Aeronautics Division
King County International Airport
8600 Perimeter Rd.
Seattle, WA 98108
(206) 764-4131

Wisconsin DOT, Aeronautics
Bureau
Box 7914
Madison, WI 53707
(608) 266-3351

Wyoming Aeronautics
Commission
State of Wyoming
Cheyenne, WY 82002-0090
(307) 777-7481

TECHNICAL DATA

GREENWICH MEAN TIME AND TIME ZONES

Time is calculated from the Greenwich (Zero degree) Meridian; its time zone extending 7½ degrees east and west of the meridian. 24 fifteen-degree arcs define the basic time zones around the world. Zones are established so that when the sun is directly above a meridian it is noon at that location.

Greenwich Mean Time (GMT) is used in aviation to provide one standard time that eliminates confusion when crossing several time zones during flight, and the 24-hour clock is used to avoid confusion concerning a.m. or p.m. Hours on the 24-hour clock are specified as four-digit numbers beginning with 0100 (1:00 a.m. GMT) and continuing through 12:00 (Noon GMT), 1300 (1:00 p.m. GMT), and so on until 2400 (Midnight GMT).

Conversion to Greenwich Mean Time from:
Atlantic Standard or Eastern Daylight...add 4 hrs.
Eastern Standard or Central Daylight...add 5 hrs.
Central Standard or Mountain Daylight..add 6 hrs.
Mountain Standard or Pacific Daylight...add 7 hrs.
Pacific Standard..add 8 hrs.
Hawaii/Alaska Standard..add 10 hrs.
Western Alaska/Aleutian (Bering Standard)...add 11 hrs.

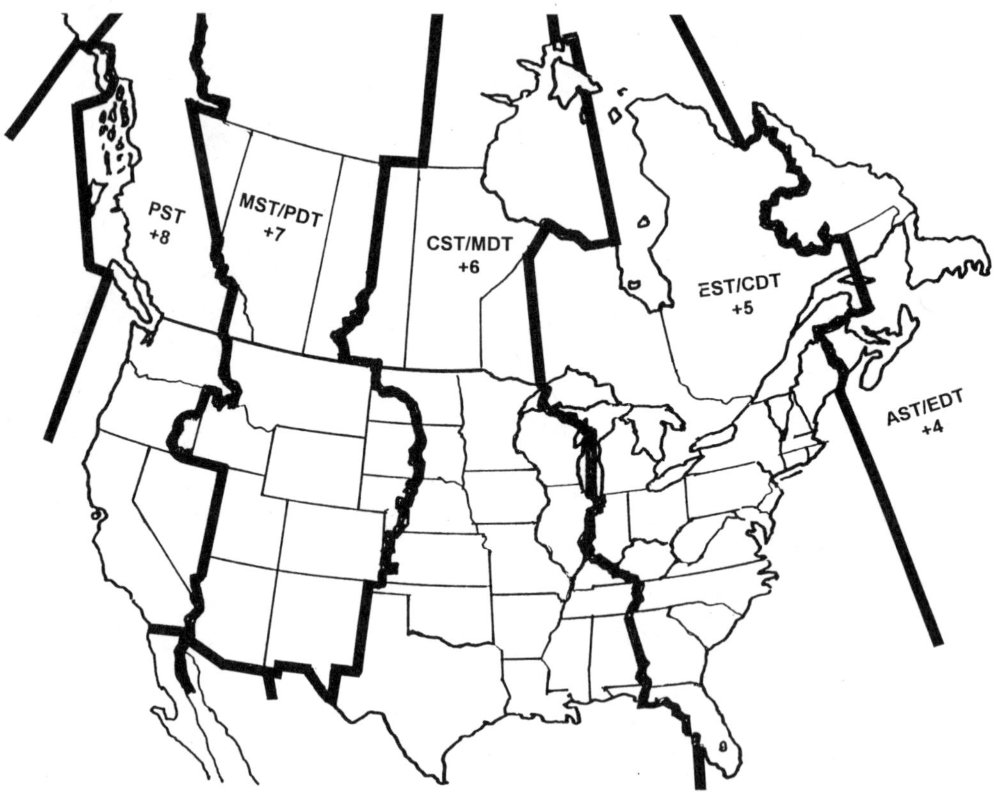

Computations and Conversions
Mileage

1 nautical mile=6,076.11549 feet

1 nautical mile=1.15 statute miles

1 nautical mile=1.852 kilometers

1 statute mile=5,280 feet

1 statute mile=0.87 nautical miles

1 statute mile=1.6 kilometers

1 kilometer=3,280.8 feet

1 kilometer=0.62137 statute miles

1 kilometer=.54 nautical miles

To convert	to	Multiply
Statute miles (sm)	Nautical miles (nm)	sm by 0.87
Nautical miles (nm)	Statute miles (sm)	nm by 1.15
Kilometers (km)	Statute miles (sm)	km by 0.62
Statute miles (sm)	Kilometers (km)	sm by 1.6

Speed

1 foot per second=60 feet per minute or 3,600 feet per hour

1 mile per hour=1.46 feet per second or 88 feet per minute

1 kt=1.688 feet per second or 101.288 feet per minute

To figure speed, use the formula d/t x 60=V, in which **d** is distance, **t** is time (in minutes and hundredths of minutes), and **V** is velocity, or speed. Example: if an aircraft is timed over a known distance of five statute miles in three minutes, five seconds, the distance--5 mi.--is divided by the time--3.08 minutes (185 seconds ÷ 60) and then multiplied by 60 to arrive at the speed (97.4 mph).
5÷3.08 x 60=97.4 mph. If known distance is nautical miles, speed will be in kts., if known distance is in kilometers, speed will be in km/hr.

Ground Speed Estimates
Seconds per mile, km or kt=Speed*

sec/mi (kt, km)	*speed*	*sec/mi (kt, km)*	*speed*	*sec/mi (kt, km)*	*speed*
60	60	32	112	22	164
55	65	31	116	21	171
45	80	30	120	20	180
42	86	29	124	19	189
40	90	28	128	18	200
38	95	27	133	17	212
36	100	26	138	16	225
35	103	25	144	15	240
34	106	24	150	10	360
33	109	23	156	8	450

*Known distance in statute miles will produce speeds in mph; known distance in kilometers will produce km/hr; known distance in nautical miles will produce kts.

Speed of sound--1,088 feet per second (742.8 mph) in air at 32° F. at sea level, varying at other temperatures and altitudes. The speed of sound in feet per second through any given temperature of air can be calculated:

$$V=1088 \quad \frac{273 +t}{16.52}$$

in which V is speed, t is the temperature C.

Temperature Conversion

To get degrees F.--multiply degrees C. by 9/5 (1.8) and add 32.F.=9/5 C. + 32
To get degrees C.--multiply degrees F. minus 32 by 5/9 (.555)C.=5/9(F. -32)

Celsius	Fahrenheit	Celsius	Fahrenheit	Celsius	Fahrenheit
-30	-22	10	50	50	122
-20	-4	20	68	60	140
-10	14	30	86	70	158
0	32	40	104	80	167

Weight of Liquids

One gallon of	*=lbs.*
Gasoline	6.0
Oil	7.5
Kerosene (jet fuel)	7.5
Methanol	6.62
Water	8.33

Liquid Equivalents

1 U.S. gallon=0.83268 Imperial Gal.
1 U.S. gallon=231 cubic inches
1 U.S. gallon=0.134 cubic feet
1 U.S. gallon=3.785 liters
1 cu. ft.=7.5 U.S. gallons

Decibels (dB)

A decibel is a measure of the relative loudness or intensity of sound. A 20-decibel sound is 10 times louder than a 10-decibel sound; 30 decibels is 100 times louder; 40 decibels is 1,000 times louder, etc. One decibel is the smallest difference between sounds detectable by the human ear. A 120-decibel sound is painful.

10 decibels=a light whisper	40 decibels=light traffic	80 decibels=rock music, subway
20 decibels=quiet conversation	50 decibels=loud conversation	90 decibels=heavy traffic, thunder
30 decibels=normal conversation	60 decibels=noisy office	100 decibels=jet taking off
	70 decibels=normal traffic	

Btu (British Thermal Unit)=Amount of heat required to increase the temperature of one pound of water by one degree Fahrenheit (252 calories). One Therm equals 100,000 Btu's.

Inches as decimal equivalents of feet

Inches	1"	2"	3"	4"	5"	6"	7"	8"	9"	10"	11"
Feet	0.083	0.166	0.250	0.333	0.416	0.500	0.583	0.666	0.750	0.833	0.916

Mathematical Formulas

To find the CIRCUMFERENCE of a circle, multiply the diameter by 3.1416.
To find the AREA of a:
> **Circle**--multiply the square of the diameter by 0.7854.
> **Rectangle**--multiply the length of the base be the height.
> **Sphere (surface)**--multiply the square of the radius by 3.1416 and multiply by 4.
> **Square**--square the length of one side.
> **Trapezoid**--Add the two parallel sides, multiply by the height, and divide by 2.
> **Triangle**--multiply the base by the height and divide by 2.

To find the VOLUME of a:
> **Cone**--multiply the square of the radius of the base by 3.1416, multiply by the height, divide by 3.
> **Cube**--cube the length of one edge.
> **Cylinder**--Multiply the square of the radius of the base by 3.1416 and multiply by the height.
> **Pyramid**--multiply the area of the base by the height and divide by 3.
> **Rectangular prism**--multiply the length by the width by the height.
> **Sphere**--multiply the cube of the radius by 3.1416, multiply by 4 and divide by 3.

Prefixes and Multiples

Milli (m)=thousandth part	Deci (d)=Tenth part	Kilo (k)=thousandfold
Centi (c)=hundredth part	Deka (da)=tenfold	Mega (M)=millionfold
	Hecto (h)=hundredfold	

Metric Conversions

To get	from	multiply by
Inches (in)	millimeters (mm)	0.04
Inches (in)	centimeters (cm)	0.40
Feet (ft)	meters (m)	3.3
Yards (yd)	meters (m)	1.1
Statute miles (sm)	kilometers (km)	0.6
Statute miles (sm)	nautical miles (nm)	1.15
Cubic feet (cu ft)	cubic meters	35.3145
Ounces (oz)	grams (g)	0.035
Pounds (lbs)	kilograms (kg)	2.2
Pints (pt)	liters (l)	2.1
Quarts (qt)	liters (l)	1.06
Gallons (gal)	liters (l)	0.26
Centimeters (cm)	inches (in)	2.5
Centimeters (cm)	feet (ft)	30
Meters (m)	feet (ft)	0.3
Meters (m)	yards (yd)	0.9
Kilometers (km)	statute miles (sm)	1.6
Nautical miles(nm)	statute miles (sm)	0.87
Cubic meters	cubic feet (cu ft)	.0283
Grams (g)	ounces (oz)	28
Kilograms (kg)	pounds (lb)	0.45
Liters (l)	pints (pt)	0.47
Liters (l)	quarts (qt)	0.95
Liters (l)	gallons (gal)	3.8

Wind Chill Table

Source: National Weather Service

Both temperature and wind cause heat loss from body surfaces. A combination of cold and wind makes a body feel colder than the actual temperature. The table shows, for example, that a temperature of 20° Fahrenheit, plus a wind of 20 mph, causes a body heat loss equal to that at -10° with no wind. In other words, the wind makes 20° feel like -10°.

ACTUAL TEMPERATURE (°F.)

Wind, mph	35	30	25	20	15	10	5	0	-5	-10	-15	-20	-25	-30	-35	-40
5	33	27	21	16	12	7	0	-5	-0	-15	-21	-26	-31	-36	-42	-47
10	22	16	10	3	-3	-9	-15	-22	-27	-34	-40	-46	-52	-58	-64	-71
15	16	9	2	-5	-11	-18	-25	-31	-38	-45	-51	-58	-65	-72	-78	-85
20	12	4	-3	-10	-17	-24	-31	-39	-46	-53	-60	-67	-74	-81	-88	-95
25	8	-1	-7	-15	-22	-29	-36	-44	-51	-59	-66	-74	-81	-86	-90	-103
30	5	-2	-10	-18	-25	-33	-41	-49	-56	-64	-71	-79	-86	-93	-101	-109
35	4	-4	-12	-20	-27	-35	-43	-52	-58	-67	-74	-82	-89	-97	-105	-113
40	3	-5	-13	-21	-29	-37	-45	-53	-60	-69	-76	-84	-92	-100	-107	-115
45	2	-6	-14	-22	-30	-38	-46	-54	-62	-70	-76	-85	-93	-102	-109	-117

Wind speeds above 45 mph have little additional chilling effect

The National Weather Service and Weather

TORNADO--a violent rotating column of air pendant from a thundercloud, usually recognized as a funnel-shaped vortex. With rotating winds of up to 300 mph, it is the most destructive type of storm, varying in length from a few feet to several hundred miles (average 5 miles) and in width to over a mile (average 220 yards). It covers the ground at an average of 25-40 mph. A CYCLONE is an atmospheric circulation of winds covering a broader area (see Hurricane).

A TORNADO WATCH is issued for a specific area where it is reasonably possible that tornadoes may occur during the valid time of a watch. A watch alerts to look for tornado activity and listen for a Tornado Warning.

A TORNADO WARNING means that a tornado has been sighted or is indicated by radar, and that safety precautions should be taken at once.

HURRICANE--a severe cyclone originating over tropical ocean waters and having winds over 74 mph (in the western Pacific, it is known as a typhoon). the area of strong winds takes the form of a circle or oval up to 500 miles in diameter. In lower latitudes, the storm usually moves toward the west or northwest at 10-15 mph. When the center approaches 25-30 degrees North Latitude, direction of the motion often changes to northeast, with increased forward speed.

A HURRICANE WATCH means that an existing hurricane poses a threat to coastal and inland communities in the area specified.

A HURRICANE WARNING means hurricane force winds and/or dangerously high water and exceptionally high waves are expected in a specified coastal area within 24 hours.

Beaufort Wind Scale

Originated in 1806 on the effect of wind on a full-rigged warship at sea, the scale was adopted for international use in 1874 and since 1946 has been determined by anemometer measurement 10 m above the ground.

Force	Description	speed, kts	Characteristics
0	Calm	1	Smooth water, smoke rises vertically
1	Light Air	1-3	Direction shown by smoke drift, but not by weather vane or windsock
2	Light Breeze	4-6	Felt on face, vane movement.
3	Gentle Breeze	7-10	Leaves in motion, extends flag or windsock, large wavelets, crests begin to break
4	Moderate Breeze	11-16	Dust raised, small waves become longer
5	Fresh Breeze	17-21	Small trees begin to sway, crested wavelets form on inland waters
6	Strong Breeze	22-27	Large branches in motion, wires sing, large waves form, white foam crests
7	Moderate Gale, Near Gale	28-33	Whole trees in motion, sea heaps up and white foam from breaking waves streaks in direction of wind
8	Fresh Gale, Gale	34-40	Twigs break off trees, progress impeded, moderately high waves
9	Strong Gale	41-47	Slight structural damage possible, high waves
10	Whole Gale, Storm	48-55	Trees uprooted, considerable structural damage, very high waves
11	Storm, Violent Storm	56-63	Widespread damage, extremely high waves
12	Hurricane	64+	

NWS Definitions

BLIZZARD--temperature at or below 20° F., winds of 35 mph or higher, ¼" accumulation of snow, visibility of less than 1/4 mile.
SEVERE BLIZZARD--temperature at or below 10° F., winds of 45 mph or higher, visibility reduced by snow to at or near zero.

Extremes of Weather in the U.S. and the World

Highest Temperature
Death Valley, CA, 7/10/13--134° F.
El Azisia, Libya, 9/13/22--136° F.

Lowest Temperature
Prospect Creek, AK, 1/23/71-- -80° F.
Rogers Pass, MT, 1/20/54-- -70° F.
Vostok, Antarctica, 7/23/83-- -129° F.

Greatest Rainfall in 24 hours
Alvin, TX, 7/25-26/79--43"
Foc-Foc, La Reunion, 1/7-8/66--72"

Greatest Snowfall in 24 hours
Silver Lake, CO, 4/14-15/21--76"

Highest Average Mean Temperature
Key West, FL (30 years)--78.2° F.
Dallol, Ethopia (1960-66)-- -94° F.

Lowest Average Mean Temperature
Barrow, AK (30 years)-- -9.3° F.
Plateau Station, Antarctica-- -70° F.

Greatest Average Annual Rainfall
Mt. Walaleale, Kauai, HI (32 years)--460"
Cherrapunji, India (74 years--450"

Minimum Average Annual Rainfall
Death Valley, CA (42 years, including 767 consecutive days without rain)--1.63"
Arica, Chile (59 years, including 14 consecutive years without rain)--0.03"

Speeds Of Winds in the U.S.
Annual averages above 10 mph

Station	Avg. (mph)	High
Mt. Washington, NH	35.3	231
Casper, WY	12.9	81
Boston, MA	12.5	61
Buffalo, NY	12.0	91
Milwaukee, WI	11.6	54
Honolulu, HI	11.4	67
Cape Hatteras, NC	11.2	72
Galveston, TX	11.0	100*
Dallas, TX	10.9	73
Kansas City, MO	10.7	70
Minneapolis, MN	10.6	92
Omaha, NE	10.6	109
Cleveland, OH	10.6	74
Chicago, IL	10.3	58
Detroit, MI	10.3	48
Bismarck, ND	10.2	72

* Recorded before anemometer blew away at estimated 120 mph

A GLOSSARY OF AVIATION TERMS AND ACRONYMS

AAS--Airport Advisory Service
AD--Airworthiness Directive
ADF--Automatic Direction Finder
ADIZ--Air Defense Identification Zone
AGL--Above Ground Level
ADI--Attitude Deviation Indicator
AHRS--Attitude-heading reference system
AIM--Airman's Information Manual
ALS--Approach Light System
Angle of Attack--the angle at which a wing meets
  oncoming airstream
AOA--Angle Of Attack
AOPA--Aircraft Owners & Pilots Assn.
AP--autopilot
APU--Auxiliary Power Unit
ARSA--Airport Radar Surveillance Area
ARSR--Air Route Surveillance Radar
ARTCC--Air Route Traffic Control Center
ASL--Above Sea Level
ASR--Airport Surveillance Radar
ATC--Approved Type Certificate--also Air
  Traffic Control
ATIS--Automatic Terminal Information Service
AWOS--Automated Weather Observing System

BCM--Back Course Marker
BFR--Biennial Flight Review
BOW--Basic Operating Weight
Brg--Bearing

C--Centigrade, or Celsius
CAS--Calibrated Airspeed
CAT--Clear Air Turbulence
CAVU--Ceiling and Visibility Unlimited
Cerfificated--FAA-approved
CDI--Course Deviation Indicator
CDU--Control Display Unit
CFI--Certified Flight Instructor
CFII--Certified Flight Instructor--Instrument
CFR--Crash, Fire, and Rescue
CG--Center of Gravity
CH--Compass Heading
CHT--Cylinder Head Temperature
Com or Comm--communication
CRT--Cathode Ray Tube
CVR--Cockpit Voice Recorder

dB--Decibel, measure of noise
DF--Direction Finder
DG--Directional Gyro
DH--Decision Height
DME--Distance Measuring Equipment
DOT--Department of Transportation
DR--Dead Reckoning

EAA--Experimental Aircraft Assn.
ECDI--Electronic Course Deviation Indicator
EFAS--Enroute Flight Advisory Service
EFIS--Electronic Flight Instrumentation System
EGT--Exhaust Gas Temperature
ELT--Emergency Locater Transmitter
EOW--Empty Operating Weight
ETA--Estimated Time of Arrival

F--Farenheit
FAR--Federal Air Regulation
FBO--Fixed Base Operator
FD--Flight Director
FDR--Flight Data Recorder
FE--Fire Extinguisher
FF--Fuel Flow
FL--Flight Level
FSS--Flight Service Station

G--force of gravity
GAMA--General Aviation Manufacturers Assn.
GCA--Ground Controlled Approach
GMT--Greenwich Mean Time
GPS--Global Positioning System
GS--Glide Slope
GWT--Gross Weight

Hg--symbol for Mercury as used in barometers
HF--High Frequency
HIGE--Hover In Ground Effect
HIRL--High Intensity Runway Lights
Hobbs--type of engine-hour gauge
HOGE--Hover Out of Ground Effect
HP--horsepower
HPSE--High-Performance Single Engine
HSI--Horizontal Situation Indicator

IAS--Indicated Airspeed
IFR--Instrument Flight Rules
IFSS--International Flight Service Station
ILS--Instrument Landing System
INS--Inertial Navigation System
IRS--Inertial Reference System

KCAS--Knots Calibrated Airspeed
KIAS--Knots Indicated Airspeed
Known Ice--certified for flight into known icing conditions
Kt--knot, or nautical mile. The length of one minute
  of the arc of a great circle; 1.15 statute miles, 6,076 ft.

LCD--Liquid Crystal Display
LE--Left Engine
LF--Low Frequency
LMM--Middle Marker
LOC--Localizer

LOM--Compass Locater at Outer Marker
Loran--Hyperbolic grid navigation system
LSE--Light Single Engine

M--Mach number, airplane speed divided by the
  speed of sound at a given altitude.
MB--Marker Beacon
MCP--Maximum Continuous Power
ME--Multi-engine
MEA--Minimum Enroute Altitude
MEL--Minimum Equipment List
MFD--Multifunction Display in an EFIS
MLS--Microwave Landing System
MM--Middle Marker
Mmo--Mach limit, maximum operating speed
MS or Milspec--Military Specification
MSL--Mean Sea Level

NAS--National Airspace System
NATA--National Air Transportation Assn.
Nav--Navigation
Nav/Com--combined Navigation/Communication
NBAA--National Business Aircraft Assn.
NDB--Non-Directional Beacon
NDH--No Damage History
NDT--Non-Destructive Testing
NM--Nautical Mile
NMS--Navigation Management System
Notam--Notice to Airmen

OBI--Omni-Bearing Indicator
OEI--One Engine Inoperative
OM--Outer Marker

PCA--Positive Contol Area
PAPI--Precision Approach Path Indicator
Part--Refers to class of FAR
Pireps--Pilot weather report
PMA--Parts Manufacturers Assn.
PTT--Push To Talk

RAM--Random Access Memory
RE--Right Engine
RG--Retractable Gear
RMI--Radio Magnetic Indicator
ROC--Rate Of Climb
RNAV--Random Navigation

SAR--Search And Rescue
SM--Statute Miles
STC--Supplemental Type Certificate
SCMOH--time Since Chrome Major Overhaul
SFAR--Special FAR
SFRM--time Since Factory Remanufacture
SHP--Shaft Horsepower
SID--Standard Instrument Departure
Sigmet--Significant Meteorological information

SM--Statute Mile
SMOH--time Since Major Overhaul
SOH--time Since Overhaul
SPOH--time Since Partial Overhaul
SR--Specific Range--miles/lb.
SSB--Single Sideband
STOH--time Since Top Overhaul
STOL--Short Take Off and Landing
SVFR--Special VFR

TAC--Terminal Aera Chart
TACAN--Tactical Air Navigation
TAS--True Airspeed
TBO--Time Between Overhauls
TC--Type Certificate
TCA--Terminal Control Area
TO--Take Off
Tracon--Terminal radar approach control
Transceiver--combination transmitter-receiver
TRSA--Terminal Radar Service Area
TSO--Technical Standard Order
TT--Total Time
TTA--Total Time, Airframe
TTAE--Total Time, Airframe and Engine
TTE--Total Time, Engine
TTSN--Total Time Since New

Va--Maneuvering speed
VASI--Visual Approach Slope Indicator
Vfe--Flap Extension speed
VFR--Visual Flight Rules
VHF--Very High Frequency
Vle--maximum Landing gear Extension speed
VLF--Very Low Frequency
Vlo--maximum Landing gear Operating speed
Vmc--Minimum Control speed
Vmo--airspeed limit
Vne--Never Exceed speed
Vno--maximum structural cruising speed
VOR--Very high frequency Omni-Range
VORTAC--VHF Omni-Range/Tactical Air navigation
Vs--Stall speed
VSI--Vertical Speed Indicator
Vso--Minimum controllable landing speed
VTOL--Vertical Take Off and Landing
Vx--Best angle of climb speed
Vy--Best rate of climb speed

WS--Windshield, also Weather Service

X-Band--radar frequency
Xcvr--transceiver
Xmtr--transmitter
Xpndr--transponder

Z or Zulu--Zero Meridian--Greenwich Mean Time

THE INTERNATIONAL PHONETIC ALPHABET

A	Alpha	H	Hotel	O	Oscar	V	Victor
B	Bravo	I	India	P	Papa	W	Whiskey
C	Charlie	J	Juliet	Q	Quebec	X	X-Ray
D	Delta	K	Kilo	R	Romeo	Y	Yankee
E	Echo	L	Lima	S	Sierra	Z	Zulu
F	Foxtrot	M	Mike	T	Tango		
G	Golf	N	November	U	Uniform		

INTERNATIONAL AIRCRAFT MARKINGS

A-2	Botswana	HS	Thailand	TI	Costa Rica
A3	Tonga	HZ	Saudi Arabia	TJ	Cameroon
A5	Bhutan	H4	Solomon Islands	TL	Central African Rep.
A6	United Arab Emirates	I	Italy	TN	Congo
A7	Qatar	J2	Djibouti	TR	Gabon
A9C	Bahrain	J3	Grenada	TS	Tunisia
A40	Oman	J5	Guinea Bissau	TT	Chad
AP	Pakistan	J6	St. Lucia	TU	Côte d'Ivoire
B	China	J7	Dominica	TY	Benin
C2	Nauru	J8	St. Vincent, Grenadines	TZ	Mali
C5	Gambia	JA	Japan	T9	Bosnia/Herzegovina
C6	Bahamas	JY	Jordan	VH	Australia
C9	Mozambique	LN	Norway	VP, VQ,	U.K. Colonies
C, CF	Canada	LV, LQ	Argentina	VR	& Protectorates
CC	Chile	LX	Luxembourg	VT	India
CN	Morocco	LY	Lithuania	XA, XB,	
CP	Bolivia	LZ	Bulgaria	XC	Mexico
CR, CS	Portugal	N	U.S.A.	XT	Burkina Faso
CU	Cuba	OB	Peru	XU	Dem. Kampucha
CX	Uruguay	OD	Lebanon	XV	Vietnam
D	Germany	OE	Austria	XY, XZ	Myanmar
D2	Angola	OH	Finland	YA	Afghanastan
D4	Cape Verde	OK	Czech Rep.	YI	Iraq
DQ	Fiji	OO	Belgium	YK	Syrian Arab Rep.
EC	Spain	OY	Denmark	YL	Latvia
EI, EJ	Ireland	P	N. Korea	YR	Romania
EK	Armenia	P2	Papua New Guinea	YS	El Salvador
EL	Liberia	P4	Aruba	YV	Venezuela
EP	Iran	PH	Netherlands	Z	Zimbabwe
ER	Rep. of Modova	PI	Netherlands Antilles	ZK, ZL,	
ES	Estonia	PK	Indonesia	ZM	New Zealand
ET	Ethiopia	PP, PT	Brazil	ZP	Paraguay
EW	Belarus	PZ	Surinam	ZS, ZT	
EX	Kyrgyzstan	RA	Russian Fed.	ZU	South Africa
EY	Tajikistan	RP	Philippines	3A	Monaco
EZ	Turkmenistan	RDPL	Lao	3B	Mauritius
F	France	S2	Bangladesh	3C	Eq. Guinea
G	United Kingdom	S5	Slovenia	3D	Swaziland
HA	Hungary	S7	Seychelles	3X	Guinea
HB (+nat'l emblem) Switzerland		S9	San Tome & Principe	4K	Azerbaijan
HB (+ nat'l. emblem) Liechtenstein		SE	Sweden	4L	Georgia
HC	Ecuador	SP	Poland	4R	Sri Lanka
HH	Haiti	ST	Sudan	4X	Israel
HI	Dominican Republic	SU	Egypt	5A	Libya
HK	Columbia	SX	Greece	5B	Cyprus
HL	Republic of Korea	TC	Turkey	5H	Tanzania
HP	Panama	TF	Iceland	5N	Nigeria
HR	Honduras	TG	Guatemala	5R	Madagascar

5T	Mauritania
5U	Niger
5V	Togo
5W	Western Samoa
5X	Uganda
5Y	Kenya
60	Somalia
6V, 6W	Senegal
6Y	Jamaica
70	Yemen
7P	Lesotho
7QY	Malawi
7T	Algeria
8P	Barbados
8Q	Maldives
8R	Guyana
9A	Coatia
9G	Ghana
9H	Malta
9J	Zambia
9K	Kuwait
9L	Sierra Leone
9M	Malaysia
9N	Nepal
9Q	Zaire
9U	Burundi
9V	Singapore
9XE	Rwanda
9Y	Trinidad & Tobago

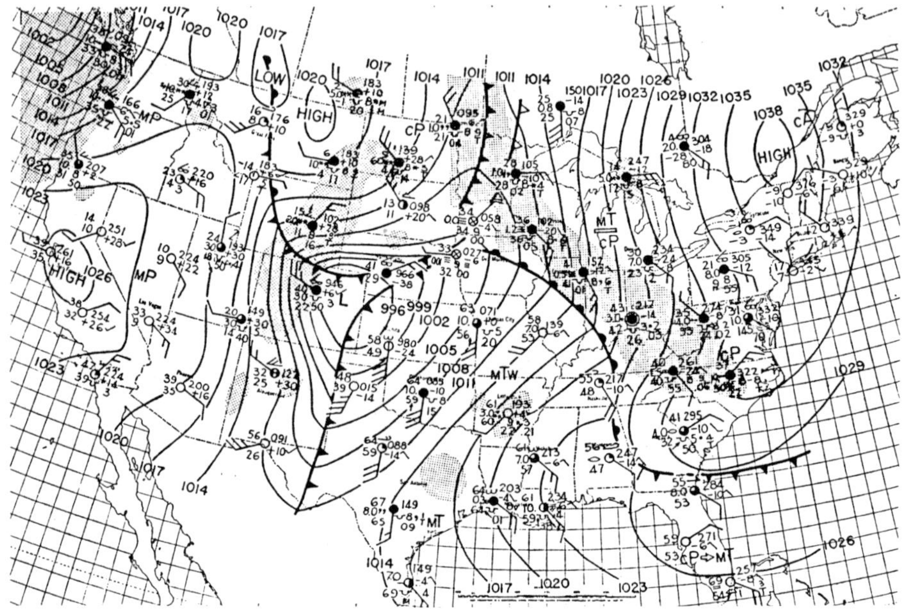

HIGH
LOW
HIGH
MP
CP
MT
CP
MT
CP
MTW
MT
HIGH
CA
CP
CP⇔MT

VII. AVIATION ORGANIZATIONS

Aviation Organizations

*denotes information not verified as current

-A-

Academy of Model Aeronautics--AMA
5151E. Memorial Dr.
Muncie, IN 47302
(317) 287-1256

Aerial Firefighting Industry Assn.--AFIA
4700 Flightline Dr.
Kingman, AZ 86401
(602) 757-1002

Aerobatic Club of America
P.O. Box 397
Springfield, TN 37172
(615) 255-9299

Aero Club of New England
P.O. Box 183
E. Boston, MA 02128
(617) 631-6680

Aero Club of Pennsylvania
P.O. Box 748
Blue Bell, PA 19422
(610) 534-2425

Aero Club of Pittsburgh
207 Dormont Village, 2961 W. Liberty Ave.
Pittsburgh, PA 15216
(414) 341-5090

Aero Medicos
P.O. Box 538
Goleta, CA 93116
(805) 964-2771

***Aero Club of Washington**
1133 15th St. NW, Suite 1000
Washington, DC 20005
(202) 293-0177

***Aeronaut Society**
1241 High St.
Oakland, CA 94601
(415) 261-4222

Aeronautical Repair Station Assn.--ARSA
121 N. Henry St.
Alexandria, VA 22314-3524
(703) 739-9543

Aeronca Aviators Club
55 Oakley Ave.
Lawrenceburg, IN 47025
(812) 537-9354

***Aeronca Lover's Club**
P.O. Box 3
Clark, SD 51225
(605) 532-3862

Aeronca Sedan Club
115 Wendy Ct.
Union City, CA 94587
(510) 489-5642

Aerospace Industries Assn. of America--AIA
1250 Eye St. NW
Washington, DC 20005
(202) 371-8400

Aerospace Medical Association
320 S. Henry St.
Alexandria, VA 22314-3524
(703) 739-2240

Aerostar Owners Association
341 Albion St.
Denver, CO 80220
(303) 322-AERO

Airborne Law Inforcement Association
14268 Linda Vista Dr.
Whittier, CA 90602
(213) 485-2011

Air Cadets, USA
HC2 Box 359
Blue Eye, MO 65611
(417) 779-4274

***Aircraft Apprasial Assn. of America**
P.O. Box 599207
Oklahoma City, OK 73159
(405) 350-1635

Aircraft Electronics Association--AEA
P.O. Box 1963
Independence, MO 64055
(816) 373-6565

***Aircraft Exterior Refinishers & Outfitters**
P.O. Box 80945
Lincoln, NE 68501
(402) 474-5074

Aircraft Owners and Pilots Assn.--AOPA
421 Aviation Way
Frederick, MD 21709
Member services: (301) 695-2000 or
(800) 872-2672

AOPA cont
Aircraft Title & Escrow Service:
 (800) 682-2511
News media and PR: (301) 695-2160
AOPA Speakers Bureau: (301) 695-2155
AOPA Air Safety Foundation
Inquiries: (301) 695-2170
Course information: (800) 638-3101

Airpark Owners & Operators Assn.
11470 W. 800N
Colfax, IN 46035
(317) 324-2424

***Airport Consultants Council--ACC**
421 King St., Suite 220
Alexandria, VA 22314
(703) 683-5900

Airports Council International, North America
1220 19th St. NW
Washington, DC 20036
(202) 293-8500

Air Force Association--AFA
1501 Lee Highway
Arlington, VA 22209-1198
(703) 247-5800

Air Force Historical Foundation
1535 Command Dr., Suite A-1222
Andrews AFB, MD 20331
(301) 736-1959

***Air Freight Association--AFA**
1710 Rhode Island Ave. NW
Washington, DC 20036
(202) 293-1030

Air Line Pilots Association--ALPA
P.O. Box 1169
Herndon, VA 22070
(703) 689-2270

Air Rescue Association
8124 E. Gail Rd.
Scottsdale, AZ 85260
(602) 948-6660

Air Traffic Control Association--ATCA
2020 N. 14th St., Suite 410 Arlington, VA 22201
(703) 522-5717

Air Transport Association of America-- ATA
1301 Pennssylvania Ave. NW, Suite 1100
Washington, DC 20004-1707
(202) 626-4000

***AirLifeLine**
1116 24th St.
Sacremento, CA 95816
(916) 446-0995

***AirLifeLine of Texas**
P.O. Box 79366
Houston, TX 77279-9366
(214) 977-8539; (214) 644-9523

Airport Operators Council International
1220 19th St. NW, Suite 200
Washington, DC 20036
(202) 293-8500

***Airport Owners Assn.**
P.O. Box 100
Hebron, OH 43025
(614) 928-5300

***Alaska Air Carriers Association**
4134 Ingra St.
Anchorage, AK 99503
(907) 562-1272

***Alpha Eta Rho**
Room 220, 4627 Ocean Blvd.
San Diego, CA 92109
(619) 270-7114

American Air Racing Society--AARS
4060 W. 158th St.
Cleveland, OH 44135
(216) 941-0089

American Assn. of Airport Executives-- AAAE
4212 King St.
Alexandria, VA 23302
(703) 824-0500

***American Astronautical Society**
6212-B Old Keene Mill Court
SPringfield, VA 22152
(703) 866-0020

American Aviation Historical Society
2333 Otis St. Santa Ana, CA 92704
(714) 549-4818

***American Bar Association Forum Committee on Air & Space Law**
750 N. Lake Shore Dr.
Chicago, IL 60611
(312) 988-5579

American Bonanza Society
P.O. Box 12888
Wichita, KS 67277
(316) 945-6913

*American Electronics Association
P.O. Box 10045
Palo Alto, CA 94303
(415) 857-9300

American Hartz Association
16225 143rd Ave. SE
Yelm, WA 98597
(206) 894-2862

American Helicopter Society--AHS
217 N. Washington St.
Alexandria, VA 22314
(703) 684-6777

American Institute of
Aeronautics & Astronautics
370 l'Enfant Promenade
Washington, DC 20024
(202) 646-7504

*American Medical Support Flight Team
5524 Meander Lane
Lake Wales, FL 33853
(813) 439-3780; (813) 294-7361
3237 Donald Douglas Loop S.
Santa Monica, CA 90405
(213) 390-2958
2452 Brookhurst Dr.
Atlanta, GA 30339
(404) 458-0674

American Maule Society
P.O. Box 720
Mirror Lake, NH 03853
(603) 569-1338

American Navion Society
P.O. Box 1810
Lodi, CA 95241-1810
(209) 339-4213

*American Society of Aviation Artists
1805 Meadowbrook Heights Rd.
Charlottesville, VA 22901
(804) 296-9771

*American Yankee Association
P.O. Box 1531
Cameron Park, CA 95682
(916) 676-4292

American Tiger Club
Rt. 1
Moody, TX 76557
(817) 853-2008

*Angel Planes
2595 Chandler, Suite 12
Las Vegas, NV 89120
(702) 736-1604

*Animal Transportation Association
P.O. Box 797095
Dallas, TX 75379-7095
(214) 713-9954

Antique Airplane Association--AAA
Route 2, Box 172
Ottumwa, IA 52501
(515) 938-2773

Arizona Pilots Association
P.O. Box 61242
Phoenix, AZ 85083

Army Association of America
49 Richmondville Ave.
Westport, CT 06880-2000
(203) 226-8184

*Associated Airtanker Pilots
311 E. Perkins
Ukiah, CA 95482
(707) 462-7528

Association of Air Medical Services
35 S. Raymond Ave., Suite 205
Pasadena, CA 91105
(818) 793-1232

Association of Aviation Psychologists
Behavior Sciences Dept.
USAF Academy
Colorado Springs, CO 80840
(614) 459-4299

Association of Naval Aviation
5205 Leesburg Pike, Suite 200
Falls Church, VA 22041
(703) 998-7733

*Association of Professional Flight
Attendants
1004 W. Euless Blvd.
Euless, TX 76040
(800) 395-2732

*Aviation and Computer Enthusiasts
2009 Camelot Dr.
Las Cruces, NM 88005
(505) 526-5645

Aviation Crime Prevention Institute
P.O. Box 30
Hagerstown, MD 21741
(800) 694-5444

Aviation Distributors & Manufacturers
Assn.--ADMA 1900 Arch St.
Philadelphia, PA 19103
(215) 564-3484

Aviation Exploring Division,
Boy Scouts of America
1325 Walnut Hill Lane
Irving, TX 75038
(214) 580-2427

Aviation Maintenance Foundation Intl.
P.O. Box 2826
Redmond, WA 98073
(206) 828-3917

Aviation Speakers Bureau
P.O. Box 6030
San Clemente, CA 92674
(800) 247-1215

Aviator/Aviatrix Single Pilot Association
11111 W. Jefferson Blvd., Suite 3505
Culver City, CA 90231

*Av-Nurse International
P.O. Box 1247
Ballwin, MO 63022-1947

-B-

B-17 Combat Crewmen & Wingmen
P.O. Box 482
South Gate, CA 90280
(213) 803-9406

B-24 Liberator Club
P.O. Box 15-2424
San Diego, CA 92195
(619) 582-5445

*B-26 Marauder Historical Society
P.O. Box 1051
Columbia, MO 65205
(314) 445-3040

*Baja Bush Pilots
P.O. Box 34280
San Diego, CA 92163
(619) 297-5587

Balloon Federation of America
P.O. Box 400
Indianolat, IA 50125
(515) 961-8809

Bede-Micro Aviation
1484 Carmel Dr.
San Jose, CA 95125
(408) 293-3023

Beechcraft Duke Association
P.O. Box 819
Mesa, AZ 85201
(602) 969-2291

Bellanca-Champion Club
P.O. Box 708
Brookfield, WI 53008-0708
(414) 784-4544

Bird Airplane Club
P.O. Box 328
Harvard, IL 60033
(815) 943-7205

Bird Dog Association
3939 C-8 San Pedro NE
Albuquerque, NM 87110
(505) 884-4822

Bücker Club
6438 W. Millbrook
Remus, MI 49340
(517) 561-2393

Buckeye Pietenpol Association
43 Ridge Rd.
Placitas, NM 87043
(505) 867-0713

Bush Pilots Association
1291 E. Vista Way
Vista, CA 92084
(800) 544-2252

-C-

California Pilots Associaiton
P.O. Box 6868
San Carlos, CA 94070
(415) 594-9300

Cardinal Club
1701 Saint Andrews Drive
Lawrence, KS 66047
(916) 842-7016

*Caterpillar Club
P.O. Box 1325
Trenton, NJ 08607
(609) 587-3300

Central States Association
9283 Lindbergh Blvd.
Olmsted Falls, OH 44138

Cessna 150-152 Club
P.O. Box 15388
Durham, NC 27704
(919) 471-9492

Cessna 172/182 Club
P.O. Box 22631
Oklahoma City, OK 73123
(405) 495-8664

*Cessna 190-195 Association
25575 Butternut Ridge Rd.
North Olmstead, OH 44070
(216) 777-4025

Cessna Airmaster Club
9 S. 135 Aero Dr.
Napierville, IL 60564
(708) 904-8416

Cessna Owner Organization
P.O. Box 337
Iola, WI 54945
(715) 445-5000

Cessna Pilots Association
P.O. Box 12948
Wichita, KS 67277
(316) 946-4777

Cessna T-50 Flying Bobcats
3821 53rd St. SE
Auburn, WA 98002
(206) 833-1068

Cherokee Pilots Association
P.O. Box 7927
Tampa, FL 33673
(813) 935-7492

*Children's Transplant Association
P.O. Box 53699
Dallas, TX 75253
(214) 287-8484

China, Burma, India Hump Pilots
P.O. Box 309
Broken Bow,k NE 68822
(308) 872-2896

*Christian Air Ministry
P.O. Box 1245
Griffin, GA 30224
(404) 228-2884

Christian Pilots Association
P.O. Box 603
West Covina, CA 91793-0603
(818) 962-7591

Civil Air Patrol--CAP
Bldg. 714
Maxwell AFB, AL 36112-5572
(205) 293-6019

Civil Aviation Medical Association
P.O. Box 23864
Oklahoma City, OK 73116
(405) 840-0199

Classic Jet Aircraft Association
P.O. Box 3086
Oshkosh, WI 54903
(414) 426-4800

Collegiate Soaring Association
909 Logan No. 11H
Denver, CO 80203
(303) 860-0485

Commander Alliance
P.O. Box 3103
Antioch, CA 94531
(415) 754-6033

Commander Flying Association
899 W. Foothill Blvd. No. E
Monrovia, CA 91016-1938
(818) 359-1040

Confederate Air Force--CAF
Midland International Airport
Midland, TX 79711
(915) 563-1000

Continental Luscombe Association
5736 Esmar Rd.
Ceres, CA 95307
(209) 537-9943

Corben Club
P.O. Box 127
Blakesburg, IA 52536
(515) 938-2733

Corporate Angel Network
Westchester Co. Airport Bldg. 1
White Plains, NY 10604
(914) 328-1313

Cub Club
P.O. Box 2002
Mt. Pleasant, MI 48858
(517) 561-2393

Culver Club
60 Skywood Way
Woodside, CA 94062
(415) 851-0204

Culver PQ-14 Association
29621 Kensington Drive
Laguna Niguel, CA 92677
(714) 831-0173

Curtiss Robin Preservation Society
21 Sunset Lane
Bushnell, IL 61422
(309) 772-2067

-D-

Dart Club
3958 Washburn Dr.
Port Clinton, OH 43452
(419) 797-2434

de Havilland Moth Club 1021 Serpentine
Lane
Wyancote, PA 19095
(215) 635-7000

Dragonfly Group
1112 Layton Dr.
Olathe, KS 66061
(913) 764-5118

-E-

EAA Warbirds of America
Wittman Airfield
Oshkosh, WI 54903-3086
(414) 426-4800

***Early Birds of America**
4072 Cadle Creek Rd.
Edgewater, MD 21037

Early Birds of Aviation
8725 Oland Ave.
Sun Valley, CA 91352
(818) 767-7042

East Central Ohio Pilots Association
630 Deerfield Dr. SW
N. Canton, OH 44720

Eastern 190/195 Association
25575 Butternut
N. Olmsted, OH 44070
(2160) 777-4025

Eastern Region Helicopter Councijl
1188 Raymond Blvd.
Newark, NJ 07102
(215) 321-3425

Eastern Region/US Air Racing Association
26726 Henry Rd.
Bay Village, OH 44140
(216) 961-9010

***Emergency Volunteer Air Corps**
3200 Airport Ave., Suite 3
Santa Monica, CA 90405
(213) 837-0762

Ercoupe Owners Club
P.O. Box 15388
Durham, NC 27704
(919) 471-9492

Experimental Aircraft Association--EAA
P.O. Box 3086
Oshkosh, WI 54903-3086
(414) 426-4800

-F-

Fairchild Club
7645 Echo Point Rd.
Cannon Falls, MN 55009
(507) 263-2414

Fairchild Fan Club
P.O. Box 127
Blakesburg, IA 52536
(515) 938-2773

***First Flight Society**
P.O. Box 1903
Kitty Hawk, NC 27949

Fleet Club
4880 Duquid Rd.
Manlius, NY 13104
(315) 682-6380

Flight Safety Foundation
2200 Wilson Blvd., Suite 500
Arlington, VA 22201
(703) 522-8300

Flights for Life
5644 W. Greenbriar Dr.
Glendale, AZ 85308
(602) 843-7220

Florida Aero Club
8550 NW 24th Ct.
Pembroke Pines, FL 33024
(305) 432-2069

Flying Apache Association
6778 Skyline Dr.
Delray Beach, FL 33446
(407) 499-1115

*Flying Architects Association
571 E. Hazelwood Ave.
Rahway, NJ 07065
(201) 388-5298

Flying Chiroprators Association
7301 Hasbrook Ave.
Philadelphia, PA 19111
(215) 722-7200

Flying Dentists Association
4700 Chamblee-Dunwoody Rd.
Dunwoody, GA 30338
404) 457-1351

Flying Doctors of America
1951 Airport Rd.
Atlanta, GA 30341
(404) 451-3068

*Flying Educators Association
2922 Alleghany Dr. NE
Cedar Rapids, IA 52402
(319) 366-4545

*Flying Engineers International
8506 Louis Drive
Huntsville, AL 35802

Flying Funeral Directors of America
256 Spring Cove Drive
Elgin, IL 60123
(312) 931-4232

Flying Judges Association
262 Old Country Rd.
Mineola, NY 11501
(516) 571-2631

Flying Physicians Association
P.O. Box 677427
Orlando, FL 32867
(407) 359-1423

Flying Veterinarians Association
330 Waukegan Rd.
Glenview, IL 60025
(708) 729-5200

Funk Aircraft Owners Association
933 Dennstedt Pl.
El Cajon, CA 92020
(619) 466-1461

-G-

General Aviation Manufacturers
Association--GAMA
1400 K St. NW, Wuite 801
Washington, DC 20005-2485
(202) 393-1500

Great Lakes Club
P.O. Box 127
Blakesburg, IA 52536
(515) 938-2773

-H-

Hatz Club
P.O. Box 127
Blakesburg, IA 52536
(515) 938-2773

Heath Parasol Club
6431 Paulson Rd.
Winneconne, WI 54986
(414) 582-4454

Helicopter Association International--HAI
and Helicopter Foundation International
1635 Prince St.
Alexandria, VA 22314-3406
(703) 683-4646

Helicopter Club of America
16003 Edgewood Dr.
Dumfries, VA 22026
(703) 670-8882

-I-

Illinois Pilots Association
P.O. Box 934
South Holland, IL 60473
(708) 331-2117

International 180/185 Club
3958 Cambridge Rd, Suite 185
Cameron Park, CA 95682
(916) 672-2620

International 195 Club
P.O. Box 737
Merced, CA 95340
(209) 722-6283

International Aerobatic Club--IAC
Wittman Airfield
Oshkosh, WI 54903-3086
(414) 426-4800

International Aeronca Association
P.O. Box 3, East Clark, SD 57225
(605) 532-3862

International Airline Passengers
Association
P.O. Box 870188
Dallas, TX 75287-0188

International Association of
Natural Resource Pilots
200 Patrick St. SW
Vienna, VA 22180
(703) 560-1271

International Bird Dog Association
3939 San Pedro NE, Suite C-8
Albequerque, NM 87110
(505) 884-4822

International Cessna 120-140 Association
6425 Hazelwood Ave.
Northfield, MN 55057

International Cessna 170 Association
P.O. Box 1667
Lebanon, MO 65536
(417) 532=4847

International Challenger Owners
Association
817 Lovers Lane
Wichita Falls, TX 76310
(817) 855-6977

International Civil Aviation Organization
1000 Sherbrooke St. W
Montreal, Quebec, Canada H3A 2R2
(514) 285-8219

International Comanche Society
Hangar 3, Wiley Post Airport
Oklahoma City, OK 73008
(405) 491-0325

International Council of Air Shows--ICAS
1910 Horton Rd.
Jackson, MI 49203
(517) 782-2424

International Fellowship of Flying
Rotarians
03 Nadine Place
Westerville, OH 430812
(614) 891-7721

International Flying Farmers
P.O. Box 9124
Wichita, KS 67277
(316) 943-4234

International Flying Nurses Association
P.O. Box 837
Dadeville, AL 36853

Int'l Liaison Pilot and Aircraft Assn.
16518 Ledgestone
San Antonio, TX 78232
(512) 490-LIPA

*International Nude Pilots Organization
P.O. Box 461
Grayson, GA 30221

International Pietenpol Association
P.O. Box 127
Blakesburg, IA 52536
(515) 938-2773

*International Shrine Aviation Assn.
P.O. Box 94922
Las Vegas, NV 89193
(702) 565-7273

International Taylorcraft Owners Club
12809 Greenbower Rd.
Alliance, OH 44601
(216) 823-9748

International Wheelchair Aviators
1117 Rising Hill Way
Escondido, CA 92029
9619) 746-5018

Interstate Club
P.O. Box 127
Blakesburg, IA 52536
(515) 938-2773

-J-

*Jodel Club
1852 Ironwood Dr.
Fairborn, OH 45324
(513) 879-3184

*Jungle Aviation and Radio Service--
JARS
P.O. Box 248
Waxhaw, NC 28173
(704) 843-6000

-L-

Lake Amphibian Flyers Club
815 N. Lake Reedy Blvd.
Frostproof, FL 33843
(813) 635-3381

Lake County Aviation Association
8807 Airport Blvd.
Leesburg, FL 34748

Lawyer-Pilots Bar Association
500 E St. NW, Suite 930
Washington, DC 20024
(202) 863-1000

*LifeLine
913 Harrington
Champaign, IL 61821
(217) 373-4195

Light Aircraft Manufacturers Association
22 Deer Oaks Court
Pleasanton, CA 94588
(415) 462-9122

Lighter-Than-Air Society
1800 Triplett Blvd.
Akron, OH 44306
(216) 868-4508

Living With Your Airplane Association
P.O. Box 39099
Tacoma, WA 98439
(206) 471-9888

Long Island Early Fliers Club
P.O. Box 962
Mellville, NY 11747

Luscombe Association
6438 W. Millbrook
Remus, MI 49340
(517) 561-2393

-M-

**Malibu/Mirage Owners and Pilots
Association**
341 Albion St.
Denver, CO 80220
(303) 399-1440

Marine Corps Aviation Association
P.O. Box 296
Quantico, VA 22134
(800) 336-0291

Martin B-26 Historical Society
9211 SW 192nd Dr.
Miami, FL 33157
(305) 233-8557

Maule Aircraft Association
5630 SA. Washington Rd.
Lansing, MI 48911
(800) 594-4634

*Mercy Airlift
P.O. Box 603
West Covina, CA 91793
(818) 962-7591

*Mercy Med-Flights
6883 Emerald Shores Dr.
Troy, MI 48098
(313) 879-2146

*Mercy Medical Airlift of Mid-America
P.O. Box 197
Lawrenceville, IL 62439
(618) 943-1019

*Mercy Medical Airlift of Washington,
DC
P.O. Box 1628
Arlington, VA 22210
(703) 361-1191

Meyers Aircraft Pilots Association
5852 Bogue Rd.
Yuba City, CA 95993
(916) 673-2724

Michigan Helicopter Association
P.O. Box 2613
Southfield, MI 48037
(313) 669-3080

*Model 18 Association
P.O. Box 4427
Allentown, PA 18105

**Mooney Aircraft Pilots Association--
MAPA**
10715 Gulfdale, No. 285
San Antonio, TX 78216
(210) 525-8008

-N-

National Aeronautic Association--NAA
1815 N. Fort Myer Dr., Suite 700
Arlington, VA 22209
(703) 527-0226

National Aeronca Association
806 Lockport Rd.
Terre Haute, IN 47802
(812) 232-1491

National Agricultural Aviation Assn.--
NAAA
1005 E St. SE
Washington, DC 20003
(202) 546-5722

*National Air Carrier Association
1730 M St. NW, Suite 710
Washington, DC 20036
(202) 833-8200

National Air Racing Group
5063 Brett Court
Fremonte, CA 94538
(410) 657-5015

National Air Transportation Association--
NATA
4226 King St.
Alexandria, VA 22302
(703) 845-9000

National Aircraft Resale Association--
NARA
4226 King St.
Alexandria, VA 22302
(703) 671-8273

National Association of Air Traffic
Specialists
11303 Amherst Ave, Suite 4
Wheaton,, MD 20902
(301) 933-6228

National Association of Flight Instructors
Ohio State U. Aiport Box 793
Dublin, OH 43017
(703) 323-8763; (614) 889-6148

National Association of Priest Pilots (RC)
921 12th St.
Boone, IA 5006
(515) 432-1971

*National Association for Search and
Rescue
P.O. Box 3709
Fairfax, VA 22038
(703) 352-1349

National Association of State
 Aviation Officials--NASAO
8401 Colesville, No. 505
Silver Spring, MD 20910
(301) 588-0587

National Biplane Association
Hangar 5, Jones-Riverside Airport
Tulsa, OK 74132
(918) 299-2532

National Büker Club
Rt. 1
Moody, TX 76557
(817) 853-2008

National Business Aircraft Association--
NBAA
1200 18th St. NW, Suite 200
Washington, DC 20036
(202) 783-9000

National Business Travel Association
1650 King St.
Alexandria, VA 22314-2747
(703) 684-0836

*National EMS Pilots Association
P.O. Box 8272
Rapid City, SD 57709
(605) 341-0273

National Flight Nurses Association
6900 Grove Rd.
Thorofare, NJ 08086
(609) 384-6725

National Flight Paramedics Association
35 S. Raymond Ave., Suite 205
Pasadena, CA 91105
(818) 405-9851

National Intercollegiate Flying
Association
P.O. Box 3204
Cleveland, MS 38733
(601) 846-4205

*National Recreational Fliers Association
P.O. Box 423
Huntington, PA 16652
(814) 643-5027

National Ryan Club
4421 W. 112th Terrace
Leawood, KS 66211
(800) 373-6202

National Stinson Club
14418 Skinner Rd.
Cypress, TX 77429
(713) 373-0418

National 210 Owners Association
P.O. Box 1065
La Canada Flintridge, CA 91011
(818) 952-6212

National Waco Club
700 Hill Ave.
Hamilton, OH 45015
(513) 868-0084

National WWII Glider Pilots Asoociation
136 W. Main St.
Freehold, NJ 07728
(201) 462-1838

Naval Helicopter Associaiton
P.O. Box 180460
Coronado, CA 92178
(619) 435-7139

Naval Aircraft Restorers Association
3320 Northridge Dr.
Grand Junction, CO 81506
(303) 245-7899

Negro Airmen International
P.O. Box 1340
Tuskegee, AL 36088
(205) 727-0721

**New England Helicopter Pilots
Association**
P.O. Box 88
Belford, MA 01730

Ninety-Nines
P.O. Box 59965
Oklahoma City, OK 73159
(405) 685-7969

North American Trainer Association
25801 NE Hinness Rd.
Brush Prairie, WA 98606
(206) 256-0066

Northwest Balloon Association
3790 N. Cole Rd.
Boise, ID 83704
(208) 322-8521

-O-

Order of Daedalians
P.O. Box 249
Randdolph AFB, TX 78148
(210) 945-2111

OV-1 Mohawk Association
1360 Keenan Way
San Jose, CA 95125
(408) 371-9156

OX5 Aviation Pioneers
207 Dormont Village, 2961 W. Liberty Ave.
Pittsburgh, PA 15216
(412) 341-5650

-P-

PBY Catalina International Association
1510 Kabel Dr.
New Orleans, LA 701131
(504) 392-1227

***ilots for Christ International**
P.O. Box 9 Parkesburg, PA 19365
(215) 857-1234

Pilots International Association
4000 Olson Memorial Hwy.
Minneapolis, MN 55422
(612) 588-5175

Piper L-4 Grasshopper Wing
P.O. Box 2002
Mt. Pleasant, MI 48858
(517) 561-2393

Piper Owner Society
P.O. Box 337
Iola, WI 54945
(800) 331-0038

Popular Rotorcraft Association
P.O. Box 8756
Clinton, LA 70722
(504) 683-3545

Porterfield Airplane Club
1019 Hickory Rd.
Ocala, FL 32672
(904) 687-4859

**Professional Aviation Maintenance Assn.--
PAMA**
500 NW Plaza, Suite 809
St. Ann, MO 63074
(314) 739-2580

*Professional Race Pilots Association
220 Greenstone Dr.
Reno, NV 89512
(702) 322-1421

-R-

Real Estate Aviation Chapter
5440 St. Charles Rd.
Berkeley, IL 60163-1287
(708) 547-7100

Rearwin Club
P.O. Box 127
Blakesburg, IA 52536
(515) 938-2773

Regional Airline Association--RAA
1200 19th St. NW, Suite 300
Washington, DC 20036
(202) 857-1170

Replica Fighters Association
2409 Cosmic Dr.
Joliet, IL 60435
(815) 436-6948

Retired Airline Pilots Association
10 Los Alamitos
Wylie, TX 75098
(214) 442-3313

Rocky Mountain Aviation and Aerospace
Association
12579 E. Cedar Ave.
Aurora, CO 80012
(303) 367-0670

-S-

Sailplane Homebuilders Association
13555 El Camino Real
Atascadero, CV 93422

Seabee Club International
6761 NW 32nd Ave.
Ft. Lauderdale, FL 33309
(305) 979-5470

Seaplane Pilots Association
421 Aviation Way
Frederick, MD 21701
(301) 695-2375

Short-Wing Piper Club
P.O. Box 66.
Collinsville, OK 74201
(918) 371-9665

Silver Wings Fraternity
P.O. Box 11579
Burke, VA 22009

*Skymaster Club
P.O. Box 1950
Liberal, KS 67905
(316) 624-2281

Soaring Society of America
P.O. Box E
Hobbs, NM 88241-1308
(505) 392-1177

*Society of Air Racing Historians
36250 Lakeshore Blvd. 518
Eastlake, OH 44095-1442
(216) 946-9069

Society of Experimental Test Pilots
P.O. Box 986
Lancaster, CA 93534-0986
(805) 942-9574

Society of Flight Test Engineers
P.O. Box 4047
Lancaster, CA 93539
(805) 948-3067

South Dakota Aviation Association
P.O. Box 52
Lantry, SD 576-36
(605) 964-3450

Southwest Stinson Club
812 Shady Glen
Martinez, CA 95070
(408) 867-7892

Spartan School of Aeronautics Alumni
Assn.
8820 E. Pine St.
Tulsa, OK 74115
(918) 836-6886

Sport Aircraft Manufacturers Association
1330 E. Fremont Dr.
Tempe, AZ 85282
(602) 730-0421

*Sport Balloon Society of the USA
P.O. Box 2247
Menlo Park, CAd 94026
(415) 326-7679

Staggerwing Club
1885 Millsboro Rd.
Mansfield, OH 44906
(419) 755-1011

Stampe Club International
2940 Falcon Way
Midlothian, TX 76065
(214) 723-1504

Stearman Restorers Association
P.O. Box 10663
Rockville, MD 20859
(301) 990-8788

Straight Tail Cessna Club
2 Forest Land
Gales Ferry, CT 06335
(230) 464-1995

Super Cub Pilots Association
P.O. Box 9823
Yakima, WA 98909
(509) 248-9491

*Suppliers of Advanced Composite
Materials Assn.
1600 Wilson Blvd., Suite 1008
Arlington, VA 22209
(703) 841-1556

Swift Museum Foundation, Inc.
P.O. Box 644
Athens, TN 37303
(615) 745-9547; (615) 744-9696

-T-

T-34 Association
P.O. Box 925
Champaign, IL 61824
(217) 356-3063

Tailhook Association
9696 Businesspark Ave.
San Diego, CA 92131
(619) 689-9223

Taylorcraft Owners Club
12809 Greenbower Rd.
Alliance, OH 44601
(216) 823-9748

The International Group for Historic
 Aircraft Recovery--TIGHAR
1121 Arundel Dr.
Wilmington, DE 19808
(302) 994-4410

Tomahawk Society
Mountaintop Airstrip
Wolfeboro Falls, NH 03896

Travel Air Club
P.O. Box 127
Blakesburg, IA 52536
(515) 938-2773

Travel Air Div. of Staggerwing Museum
P.O. Box 550
Tullahoma, TN 37388
(615) 455-1974

Twin Beech Association
P.O. Box 8186
Fountain Valley, CA 92728-8186
(714) 964-4864

Twin Bonanza Association
19684 Lakeshore Dr.
Three Rivers, MI 49093
(616) 279-2540

Twin Cessna Flyers
8531 Wealthwood Dr.
New Haven, IN 46774
(800) 825-5310

Twirly Birds International
P.O. Box 18029
Oxon Hill, MD 20745
(301) 567-4407

*310 Owners of America
8531 Wealthwood
New Haven, IN 46774
(219) 749-2520; (800) 825-5310

-U-

United Flying Octogenarians--UFO
3531 Larga CIrcle
San Diego, CA 92110
(619) 224-6758

United States Air Racing Association
26726 Henry Rd.
Bay Village, OH 44140
(216) 871-3781; (216) 961-9010

United States Parachute Association
1440 Duke St.
Alexandria, VA 22314
(703) 836-3495

United States Pilots Association
483 S. Kirkwood Rd., Suite 10
St. Louis, MO 63122
(314) 843-2766

University Aviation Association
3410 Skyway Dr.
Auburn, AL 36801
(205) 844-2434

-V-

Vietnam Helicopter Pilots Association
7 W. 7th St., Suite 1940
Cincinnati, OH 45202
(513) 721-8472

Vintage Sailplane Association
Rt. 1, Box 239
Lovettsvillre, VA 22080
(703) 822-5504

-W-

Waco Historical Society
1013 Westgate Rd.
Troy, OH 45373
(513) 335-2621

Warbird Restoration Society
20 Emerald Court
Airdrie, Alta. Canada T4B 1B8
(403) 948-7432

Warbirds Worldwide, Ltd.
5 White Hart Chambers, 16 White Hart St.
Mansfield, Notts NG18 1DG, England
(0623) 24288

Whirly-Girls
P.O. Box 7446
Menlo Park, CA 94026
(415) 462-1441

The Wings Club
52 Vanderbilt Ave.
New York, NY 10017
(212) 867-1770

World Beechcraft Society
1436 Muirlands Dr.
La Jolla, CA 92037
(619) 459-2745

World War I Aeroplanes
15 Crescent Rd.
Poughkeepsie, NY 12601
(914) 473-3679

World War II Glider Pilots Association
136 W. Main St.
Freehold, NJ 07728
(201) 462-1838

-Z-

Zenair Associtioan
6438 W. Millbrook Rd.
Remus, MI 49340
(517) 561-2393

AIRCRAFT OWNERS AND PILOTS ASSOCIATION

AOPA GUIDE TO MEMBER SERVICES

TOLL-FREE MEMBER ASSISTANCE HOTLINE   800/USA-AOPA (872-2672)

Experts standing by to offer advice and answer your questions--as close as your telephone and free of charge to all members in the United States and Canada.

From a touch-tone telephone: please listen to AOPA's automated voice directory, then press the number of the department you wish to reach. From a rotary-dial telephone: please remain on the line after AOPA's short automated voice directory is finished. A representative will answer your call in turn. Member assistance hours: 8:30 a.m. to 6 p.m. Eastern time, Monday through Friday.

CHART ORDERS
- Individual charts or subscriptions. NOS or Jeppesen

AVIATION INFORMATION
- Advice on aviation matters, including questions related to aircraft operations, flight training, aircraft insurance
- FAA regulations, pilot certification, enforcement, attorney referrals
- Other technical questions
- Assistance with flight planning, Customs or immigration procedures, transoceanic flights

MEDICAL QUESTIONS
- For all matters related to FAA medical certification

MEMBER SERVICES
- Address changes, renewals, credential requests
- Membership cards, decals, lapel wings
- AOPA Pilot or AOPA's Airport Directory inquiries
- Project Pilot
- AOPA Certified programs
- AOPA FlightPlus Aviation AD&D Insurance
- AOPA Legal Services Plan

AOPA AIR SAFETY FOUNDATION
Your authority on aviation education and air safety.

Aviation Safety Database	301/695-2171
Course information/registration	800/638-3101
Audiovisual and aviation book sales	800/543-8633
Flight Instructor Department	301/695-2237
Foundation Development	301/695-2027
Inquiries	301/695-2170
Program Development	301/695-2172
Special Courses Development	301/695-2179

COMMUNICATIONS
Building a positive public image of general aviation.

AOPA Speakers Bureau	301/695-2154
News Media and Public Relations	301/695-2162

FAX ON DEMAND

A wide variety of documents, forms, and other information available by facsimile. A touch-tone telephone is required. Fax on Demand 800/GO-AVFAX (462-8329)

AOPA ONLINE
AOPA Online, offered through the Internet's World Wide Web and on CompuServe, features a wide variety of aviation information available to members around the clock and around the world.

AOPA Online   http://www.aopa.org   CompuServe   800/GO2-AOPA (462-2672)

GOVERNMENT AND TECHNICAL AFFAIRS
Working hard to protect your right to affordable, accessible, and enjoyable flying.

Air Traffic Control	301/695-2200
Regional Affairs	
Airports	301/695-2201
Regional representatives	301/695-2201
Regulatory Policy	301/695-2203

Government and Technical Affairs cont.
    Airspace standards; pilot and aircraft certification
      State Legislative Affairs   301/695-2200

AOPA LEGISLATIVE ACTION-PAC
Your voice in Congress and state legislatures across the nation.
Congressional relations, state legislation, Political Action Committee
(AOPALA-PAC)   202/479-4050

AOPA EXPO CONVENTION
Presenting you with educational seminars, exhibits, and static aircraft displays.
      AOPA EXPO   800/942-4269

NEW ADDRESS?
Please send your correct address and AOPA membership number to: AOPA, 421 Aviation Way, Frederick, Maryland 21701-4798; or fax to: 301/695-2375. On CompuServe, E-mail address changes to 76702,2724; via the Internet, use 76702.2724@compuserve.com.

FOR SERVICES NOT LISTED
    Dial   301/695-2000, Fax 301/695-2375
Office hours: 8:30 a.m. to 5 p.m. Eastern time, Monday through Friday. Mailing address: 421 Aviation Way, Frederick, Maryland 21701.
E-mail AOPA via CompuServe 76702,2721   Internet 76702.2721@compuserve.com
Internet home page   http://www.aopa.org

AOPA Certified Services
AIRCRAFT SERVICES
    AOPA Aircraft Insurance Program   800/622-AOPA (622-2672)
    AOPA Aircraft Financing Program   800/62-PLANE (627-5263)
    AOPA Title And Escrow Service   800/654-4700

PILOT SERVICES
    AOPA Legal Services Plan   800/USA-AOPA (872-2672)
    AOPA Non-Owned Aircraft Insurance Program 800/622-AOPA (622-2672)
    AOPA Aviation Charts Discounts   800/4-CHARTS (424-2787)
    AOPA Insignia Merchandise   800/LIFTOFF (543-8633)

CREDIT AND FINANCIAL

Available through MBNAAmerica®

    AOPA Gold MasterCard® & Custom Visa® Program 800/847-7378 Ext. 2500
            http://www.webapply.com/aopa/
    AOPA Line Of Credit, Money Market Account & CD 800/345-0397
    AOPA Aircraft Financing Program   800/62-PLANE (627-5263)
    AOPA Installment Loan Program   800/626-2760

TRAVEL
    AOPA Travelers Club   800/888-AOPA (888-2672)
    AOPA Rental Car Discounts
    When calling, mention code in parentheses
          Alamo: 800/354-2322 (JE #434861)
          Avis: 800/331-1800 (AWD #A451300)
          Hertz: 800/654-2200 (CDP #10232)

INSURANCE
    AOPA FlightPlus AD&D Insurance   800/USA-AOPA (872-2672)
    AOPA Group Term Life Plan   800/328-9343
    AOPA Individual Term Life Plan   800/328-9343 Ext. 3336
    AOPA Health and Disability Programs   800/882-5583

EXPERIMENTAL AIRCRAFT ASSOCIATION
EAA Aviation Center, P.O. Box 3086, Oshkosh, WI 54903-3086
(414) 426-4800, fax (414) 426-6560   Web Page at http://www.eaa.org
Homebuilt/Ultralight Safety Internet Site: www.safetydata.com

EAA and Division Membership Services (800) JOIN-EAA (564-6322), (414) 426-6761
Office hours 8:15 a.m.-5:00 p.m., CST Monday-Friday

PROGRAMS AND ACTIVITIES
Fax numbers for programs listed below: (414) 426-6560

Aircraft general questions (414) 426-4821
AME Pilot Advocates (414) 426-6522
Auto Fuel STCs (414) 426-4843
B-17 Tour (800) FLY-NB17
Build/restore information (414) 426-4821
Chapters: locating/organizing (414) 426-4876
Chapters: insurance (414) 426-4822
Education--EAA Air Academy, EAA Scholarships,
  EAA Young Eagles camps (414) 426-6815

Flight Advisors information (414) 426-6522
Flight Instructor information (414) 426-6801
Government issues (414) 426-6522
Library Services/Research (414) 426-4848
Medical questions (414) 426-4821
Museum information (414) 426-4818
Technical Counselors (414) 426-4821
Ultralight Programs (414) 426-6522
Young Eagles (414) 426-4831

BENEFITS
Aircraft Financing (800) 999-4515   Aircraft Insurance (AVEMCO)
(800) 638-8440
Rental cars: Avis (mention code AWD #A948700), US: (800) 331-1212; International: (800) 331-1084
   Budget (mention code #V041300), US: (800) 527-0700; International: (800) 472-3325
Member Insurance plans (Wohlers) (800) 323-2106

EAA AVIATION FOUNDATION
Artifact donations (414) 426-4877   Financial support (800) 236-1025

EAA OFFICERS AND DIRECTORS

Paul H. Poberezny, Oshkosh, WI
S.H. Schmid, Milwaukee, WI
John Baugh, Jr., Lebanon, TN
Bill Eickhoff, St. Petersburg, FL
Dr. Dean M. Hall, Ft. Collins, CO
Espie Joyce, Greensboro, NC
Robert Reece, San Angelo, TX
Harry Zeisloft, Mesa, AZ

Tom Poberezny, Oshkosh, WI
John Beetham, Indianapolis, IN
Susan Dusenbury, Stoneville, NC
Malvern Gross, Jr., Eastsound, WA
Jack Harrington, Chicago, IL
Leonard McGinty, Jr., Tampa, FL
Richard Rihn, Plesant Hill, CA
R.J. Lickteig, Albert Lea, MN

Ray Scholler, Random Lake, WI
Louis Andrew, Jr., Fond du Lac, WI
Jack Eggspuehler, Dublin, OH
Robert Gyyllenswan, Rockford, IL
Verne Jobst, McHenry, IL
Vern Raburn, Scottsdale, AZ
Rem Walker, Regina, Sask., Canada

EAA AVIATION FOUNDATION, INC. OFFICERS AND DIRECTORS

Paul H. Poberezny, Oshkosh, WI
Carl Bury, Savannah, GA
Dr. Margaret Baty, W. Alexandria, OH
Thomas Davis, Winston-Salem, NC
William Harrison, Jr., Tulsa, OK
Morton Lester, Martinsville, VA
J. Dawson Ransome, New Hope, PA
Tom Warner, Pala, CA

Tom Poberezny, Oshkosh, WI
James Gorman, Mansfield, OH

Marion Gregory, Climax Springs, MO
E.E. Hilbert, Union, IL
Lynda Moerschbaecher, San Rafael, CA
Ray Scholler, Random Lake, WI

John Parrish, Tullahoma, TN
Jim Barton, Oshkosh, WI
James Brown, Piqua, OH
Richard Hansen, Batavia, IL
Jimmy Leeward, Ocala, FL
David Pasahow, Toronto, Ont.
Don Taylor, Hampshire, IL
Kermit Weeks, Polk City, FL

ANTIQUE/CLASSIC OFFICERS AND DIRECTORS

Espie Joyce, Greensboro, NC
Charles Harris, Tulsa, OK`
Phil Coulson, Lawton, MI
John Copeland, Shrewbury, MA
Bob Lickteig, Albert Lea, MN
Bob Lumley, Brookfield, WI
George York, Mansfield, OH

George Daubner, Hartford, WI
John Berendy, Cannon Falls, MN
Bob Brauer, Chicago, IL
Dale Gustafson, Indianapolis, IN
Jeannie Hill, Harvard, IL
Wes Schmid, Wauwatosa, WI
Buck Hilbert, Union, IL

Steve Nesse, Albert Lea, MN
Gene Morris, Roanoke, TX
Joe Dickey, Lawrenceburg, IN
Stan Gomoll, Minneapolis, MN
Dean Richardson, Madison, WI
Geoff Robison, New Haven, IN

INTERNATIONAL AEROBATIC CLUB OFFICERS AND DIRECTORS

Dr. Richard Rihn, Pleasant Hill, CA
Doug McConnell, Olympia Fields, IL
Bob Davis, Woodstock, IL
William Weaver, Enterprise, AL

Howard Stock, Woodstock, IL
Brian Becker, Pompano Beach, FL
Tom Meyers, Palo Alto, CA
Louis Andrew, Jr., Fond du Lac, WI

Betty Steward, Spring Grove, IL
Mike Heuer, Cordova, TN
Gerry Molidor, Jr., McHenry, IL
Gen. Merrill McPeak, Arlington, VA

WARBIRDS OF AMERICA, INC. OFFICERS AND DIRECTORS

Jack Harrington, Chicago, IL
Richard Ervin, Indianapolis, IN
Dave Clinton, Leucadia, CA
Rick Fernalld, Yelm, WA
Joe Howard, McMinnville, TN
Carl Schmieder, Phoenix, AZ
Vic Barrett, Sevierville, TN
Vern Raburn, Belleville, WA

Rick Hegenberger, Charleston, SC
John Baugh, Jr., Lebanon, TN
Richard Dieter, South Bend, IN
John Harrison, Sacramento, CA
Charlie Nogle, Champaign, IL
Jerry Walburn, Ocala, FL
John Galt Bowman, Edmonds, WA
John Russman, Lincoln Park, MI
Walter Ohlrich, Jr., Virginia Beach, VA

Michael Schloss, NY, NY
Connie Bowlin, Griffin, GA
John Ellis, Battle Creek, MI
William Harrison, Jr., Tulsa, OK
Dave Schlingman, Tampa, FL
Tom Wise, Morris, IL
Michael Pangia, Washington, DC
John Seidel, Plantation FL

HALLS OF FAME

Antique/Classic: George York, Buck Hilbert, Elrey Jeppesen
Homebuilders: Sam Burgess, Nicholas D'Apuzzo, Edward Heath, Volmer Jensen
Warbirds: John Baugh, Jr., William Harrison, Jr., Walter Olrich, Jr., Paul Poberezny, Richard Dieter, Charles Nogle

---

EAA OSHKOSH FLY-IN CONVENTION
Wednesday, July 30-Tuesday, August 5, 1997
Wednesday, July 29-Tuesday, August 4, 1998

General information (800) JOIN-EAA
Fax-On-Demand (May 1-August 1) (908) 885-6711
Exhibit information (414) 426-6543
Housing information (Oshkosh Convention and Visitors Bureau) (414) 235-3007
EAA Public Relations (414) 426-6514

State, city	Chap.	Meeting
ALABAMA		
Auburn	998	Call (334) 887-3871
Bessemer	25	1 Sa, 9 a., Bessemer AP
Birmingham	152	3 Fr, 7 p., So. Museum Flt.
Cullman	1163	2 Su, 2 p, Cullman AP
Enterprise	351	Call (205) 588-2422
Evergreen	909	3 Su, 2 p., Middleton Fld.
Fort Payne	890	2 Sa, 10 a., Isbell Field
Gadsden	1048	1 Tu, 7 p., Gadsden AP
Guntersville	683	2 Th, Guntersville AP
Huntsville	190	3 Tu, 7 p. (205) 498-3454
Jasper	1061	2 Sa, Bevil Fld.
Mobile	416	3 Tu, 7 p.,(334) 344-6465
Tuscaloosa	557	2 Th, Tuscaloosa AP
Wetumpka	822	2 Th, Wetumpka AP
ALASKA		
Anchorage	42	Last Th, 7 p., (907) 243-5151
Fairbanks	1129	2 W, 6:30 p., Hutchinson Career Center.
Soldotna	975	2 M, 7 p. (907) 283-3501
ARIZONA		
Buckeye	1053	Call (602) 935-6396
Camp Verde	952	2 W, 7:30 p., (602) 567- 2744
Cochise Co.	776	3 Th, 6:30 p., (602) 378-2517
Flagstaff	856	2 W, Pulliam Airport
Goodyear	128	2 W, 7:30 p.,Glendale AP
L.Havasu City	681	3 Sa, 10 a., L.H. AP
Mesa	228	4 Tu, 7:30 p., (602) 962-7624
Overgaard	1044	2 Sa, 9 a., Mogollon AP
Payson	810	2 M, 7 p., Payson AP
Phoenix	538	2 Sa, 6:30 p., (602) 439-1923
Prescott	658	2 Tu, 7 p., Embry Riddle
Salome	1144	Last Tu, 7 p., In. Hills AP
Show Low	586	3 Su, 2 p.,(520) 537-7405
Tucson	81	3 Tu, 7 p., Pima Cm. Coll.
Tucson	1153	1 W, 7 p., Catalina H.S.
Wickenburg	883	3 W, 7:30 p., Comm. Ctr.
Yuma	590	1st Tu, 7 p., 1463 E. 10th
ARKANSAS		
Batesville	1001	1 Su, 2 p., Batesville AP
El Dorado	626	2 Tu, 6:30 p., Chap. hgr. or Good Times Grill
Harrison	669	4 Su, 2 p., Boone Co. AP
Jonesboro	437	2 Su, 2 p.,(501) 236-7393
Little Rock	165	3 S, 2 p., Grady's Pizza
Mena	1142	Last M, Mena Air Ctr.
Monticello	1120	2 Sa, 10 a., Ellis Field

State, city	Chap.	Meeting
Mtn. Home	775	3 F, 7 p., Baxter Co. AP
NW Ark.	732	3 Su, 2 p., Ark. Air Mus'm.
Paris	1156	1 Su, 2 p., Paris AP
Russellville	356	3 Su., 2 p., R'ville AP
Springfield	1152	3 Su., 2 p., Heifer Crk. AP
CALIFORNIA		
Alhambra	224	Last Sa, 7:30 p., FAST, Inc., El Monte AP
Apple Valley	768	1 Sa, 10 a., A.V. AP Term.
Bakersfield	71	2 W, 7 p., Bakersfield AP
Berm. Dunes	1116	1 W, 6:30 p., B.D. AP
Camarillo	723	2 Sa, 10 a., Camarillo AP
Chico	427	3 Th, 7:30 p.,Ranchero AP
Concord	393	4 W.,7:30 p.,Buchanan AP
Corning	1148	1 Sa, 12:00, Corning AP
Corona	494	3 Su, 2 p., Corona AP
Dunsmuir	654	1 Th, 6:30 p., (916) 238-8255
Edwards AFB	1000	3 Tu, 5 p., Test Pilot Sch.
Exeter/Lindsey	1124	1 Sa, 9:30 a. T'hawk Fld.
Fresno	97	2 Sa, 9 a., Chandler Coffee Shop
Fresno	376	Last Tu, 7:30 p., 4443 W. Vandergrift
Georgetown	1074	Call (916) 642-1097
Half Moon B.	639	2 W, 7 p., HMB AP
Hanford	1138	2 E, 7 p., (209) 583-0352
Hayford	824	4 M, 7 p., Weverville AP
Lancaster	49	1 W, 7:30 p., S'dale Sch.
Livermore	663	1 Th, 7:30 p., L'rmore AP
Lompoc	275	4 W, 7 p., Lompoc AP
Long Beach	7	2 Th, 7:30 p., Airflite
Marina	204	2 Th, 7 p., Marina AP
Modesto	90	1 Tu, 7:30 p., 4118 N. Tegner Rd., Turlock
Napa	167	1 Tu, 7:30 p., Bakers Sq.
N. San Diego	286	1 Th, 7:30 p., Joslyn Ctr., 111 Richmar, San Marcos
Orange Co.	92	1 W, 7:30 p., Los Castillos
Oroville	1112	3 Sa, 8 a., 129 C. Yeager
Paradise	735	1 W, 7 p., (916) 876-1131
Placerville	512	3 W, 7:30 p., P'Ville AP
Redding	157	3 Tu, 7:30 p., Benton AP
Redlands	845	2 Th, 7 p., Redlands AP
Redwood City	20	2 Tu, 7:30 p.,S. Carlos AP
Riverside	1	2 Su, 2 p., Flabob AP
Roseville	526	2 W, 7:30 p., Placer S&L
Sacramento	52	Last Tu.,7:30 p.,S. Exec. AP
Salinas	1150	3 Tu, 7 p., Salinas AP
San Andreas	484	2 F, 7 p., Calaveras AP
San Diego	14	3 Tu., 7:30 p., Joyce Beers Comm. Center

State, city	Chap.	Meeting
California cont.		
San Jose	62	1 Th, 7:30 p., V. Miller Av., Reid Hillview AP
San Jose	338	2 Tu, 7:30 p.,Ewerts Photo
San Luis Ob.	170	3 Th, 7 p., 203 Granada
Santa Barbara	527	2 Th, 7:30 p., Pilot Center
Santa Maria	499	2 Tu, 7:30 p., Mus. of Flt.
Santa Monica	11	2 Th, 7:30 p., S.M. AP
Santa Rosa	124	1 W, 7:30 p., Sonoma AP, W. end Taxiway Alpha
Santa Ynez	491	3 Tu, 7 p., S.Y. AP Hgr.J6
Susanville	794	1 M, 6 p., (916) 257-0334
Torrance	96	3 Th, 7:30 p., Torrance AP
Truckee	1073	3 W, 7 p., Truckee AP
Tulare	262	4 Sa, 11 a., Visalla AP
Upland	448	2 F, 8 p., Cable AP N. side
Van Nuys	40	3 F, 8 p., (818) 725-4247
Watsonville/ Santa Cruz	119	1 Tu, 7:30 p., Harbor HS
Willits	1027	1 Th, 7 p., Willits AP

COLORADO

State, city	Chap.	Meeting
Aspen	835	Call (303) 927-4425
Colo. Springs	72	3 Tu, 7:30 p., (719) 596-3295
Delta	800	2 Sa,10 a.,(970) 921-3715
Denver	43	2 Sa, 7 p., Jeffco AP
Denver	301	3 F, 7 p., Old Lowery AFB
Denver	660	3 Sa, Call (303) 973-8395
Ft. Collins/ Loveland	515	1 Tu, 7:30 p., Mt. Flyers
Greeley	720	2 Th, 7 p., Weld Co. AP
Greeley	1117	3 M, 8 p., Weld Co. AP
Longmont/ Boulder	648	2 M, 7 p., Flatiorns Av.
Pueblo	808	2 W, 7:30 p., Pueblo AP
Salida	1161	3 Th, 7 p., AP Terminal
Steamboat Sp.	649	2 Tu, 7:30 p., Yampa Valley Electric Assn.

CONNECTICUT

State, city	Chap.	Meeting
Bethlehem	1097	Call (203) 266-0293
Danbury	130	2 F, 7:30 p., Danbury AP
Daneilson	1035	2 Tu, 7 p., Danielson AP
Hartford	166	Last Su, 7:30 p., P&W Cust. Training Center
Meriden	27	3 Su, 10 a., Meriden AP
Simsbury	324	1 W, 7:30 p., Simsbury AP
Waterford	334	2 W, 7:30 p., NL/Groton AP

DELAWARE

State, city	Chap.	Meeting
Wilmington	240	1 M, 7:30 p., Wm.Penn HS

DISTRICT OF COLUMBIA

State, city	Chap.	Meeting
Washington	4	2 Th, 8 p., Coll. Park AP

FLORIDA

State, city	Chap.	Meeting
Clearwater	282	1 M, 7 p., Clearwater AP
Daytona Beach	288	3 Th, 7:30 p., Embry-Riddle
De Land	635	2M, 7:30 p., Bob Lee AP
Fernandina B.	943	1 W, 7 p., Check at AP
Ft. Lauderdale	133	3 Th, 7:30 p., Am. Exec. AP
Ft. Meyers	66	2 Tu, 7:30 p., Page Fld.
Ft. Pierce	908	3 M, 7:30 p., 3100 Indust.
Homestead	620	3 M, 7:30 p., Tamiami #501
Jacksonville	193	2 Th, 7:30 p., (904) 259-5914
Kissimmee	813	1 M, 7 p., Kissimmee AP
Lake City	977	1 Sa, 7 p., Cannon Creek AP
Lakeland	454	1 M, 7 p., Sun 'N Fun
Leesburg	534	4 W, 7:30 p., Leesburg AP
Live Oak	797	Last Th,7:30 p.,Suwanee AP
Merritt Island	724	2 W, 7:30 p., Merritt Is. AP
Miami	37	3 F, 7:30 p., Baker School
Naples	1067	1 W, 7:30 p. Naples AP
New Port Richey	791	2 Th, 7:30 p., Hidden L. AP
Ocala	812	3 Sa, 9 a., Ocala AP
Okeechlobee	1102	3 Su, 1 p., 19515 NW 80th
Orlando	74	3 Tu, 7:30 p., Orlando Exec.
Pal Beach Gar.	203	2 Th, 7:30 p.. (561) 774-1432
Panama City	202	3 Sa,10:30 a.,Sandy Cr. AP
Pensacola	485	1 Tu, 7 p., Tryon Br. Libr.
Plymouth	1137	1 M, 7 p., Plymouth Woodshed
Punta Gorda	565	2 M, 7:30 p., Charl. Co. AP
Sarasota	180	4 W, 7:30 p., Sarasota AP
Sebring	803	2 Th, 7:30 p., Sebring AP
St. Petersburg	47	1 F, 7:30 p., Whitted AP
Tallahassee	445	2 Tu, 6:30 p.. Flightline
Tampa	175	4 Su, 2 p., Vandenburg hgr.
Titusville	866	1 W, 7:30 p., A. Dunn AP
Winter Haven	229	2 M, 7:30 p., W.H. AP

GEORGIA

State, city	Chap.	Meeting
Atlanta	1062	1 M, 7:30 p., Peachtree AP
Blythe	172	2 Th, 7:30 p., Pea Patch AP
Carrollton	976	2 Sa, 4 p., W. Ga. Reg. AP
Columbus	677	2 Th, 7:30 p., (706) 568-3250
Gainesville	611	1 Tu, 7:30 p., Beef Corral
Gwinnett Co.	690	2 F, 8 p., Gwinnett Co. AP
Marietta/ Cartersville	268	3 Th, 7 p., McCullom AP
Moultrie	1082	2 Sa, 1:30 p., Maule Air
Newnan	6	2 Sa, 10 a., Coweta AP
Rome	709	1 Sa, 10 a., Eagles Nest, Russell AP
Savannah	330	3 Sa, 11 a., Briar Patch AP
St. Simons Is.	905	2 Su, 2 p., McKinnon AP
Statesboro	489	1 Su, 4 p., Brooklet AP
Stockbridge	468	4 Sa, 3 p., Berry Hill AP
Stone Mt.	1025	3 Th, 7:30 p., (770) 972-6589
Toccoa	1011	1 M, 7:30 p., Toccoa AP
Warner Robb.	38	1 Th, 7:30 p., W.R. AP

State, city	Chap.	Meeting
HAWAII		
Paia Maui	882	1 Th, 7 p., (808) 879-7638
IDAHO		
Idaho Falls	407	Last Sa, 6 p., (208) 237-4143
Nampa	103	1 Th, 7:30 p., Nampa AP
Sandpoint	757	3 Th, 7 p., Sandpoint AP
ILLINOIS		
Addison	101	3 Tu, 7:30 p., Sch'berg AP
Aurora	579	4 Th, 7:30 p., Aurora AP
Barrington	790	4 Tu, 7:30 p., 30 S.Barr.
Bloomington	129	3 Th, 7 p., Hgr. F5 BN AP
Cahokia	1003	1 W, 6:30 p., Parks Coll.
Canton	832	Last Tu, 7:30 p., (309) 647-7557
Champaign	29	1 Th, 7 p., Willard AP
Chicago Hts.	260	3 F, 8 p., Lansing AP
Countryside	86	1 F, 8 p., 6406 Joliet Rd.
Danville	622	1 Tu, 7 p., V'milion Co. AP
De Kalb	241	2 Tu, 7:30 p., 8242 S. Malta
Decatur	274	2 Tu, 7:30 p., Decatur AP
East Alton	864	3 Th, 7 p., STL Reg. AP
Flora	16	3 Th, 7 p., Flora AP
Freeport	475	3 Th, 7:30 p., Albertus AP
Grayslake	414	3 Th, 7:30 p., Lake Co. H.S.
Harrisburg	750	1 Th, 7 p., (618) 993-8689
Jerseyville	213	Last Tu, 7:30 p., J'ville AP
Kankakee	990	1 Tu, 7:30 p., Gr. Kank. AP
Lostant	948	2 Tu,7:30 p., Hartenb'r AP
Madison Co.	1162	2 Tu, 7 p., Shafer Metro AP
Mendota	263	3 Tu, 7:30 p., Old LaSalle AP
Monmouth	350	3Sa, 7:30 p., Monmouth AP
Morris	95	4 Tu, 7:30 p., Morris AP
Mt. Prospect	89	2 Tu, 7:30 p., St. Marks Lutheran
Mt. Verson	1155	2 Tu, 7 p., Outland AP
O'Fallon	64	1 Tu, 7:30 p., (618) 233-4297
Peoria	563	2 W,7:30 p.,Mt. Hawley AP
Peotone	1159	2 W, 7:30 p., Howell AP
Plainfield	461	2 Tu, 7:30 p., Clow AP
Quad Cities	75	2 Sa, 7 p., Deere Wiman Carr. House, Moline
Rochester	770	2 Th, 7 p., (217) 483-5713
Rock Falls	410	1 Sa, 7:30 p., Whiteside AP
Rockford	22	1 Tu, 7:30 p., Cottonwood AP
Romeoville	15	2 F, 7:30 p., Lewis U. AP
Sandwich	872	3 Tu, 7:30 p., Sandwich AP
Schaumburg	153	2 F, 7:30 p., Sch'burg AP
Springfield	137	2 E, 7:30 p., (217) 744-9059
Urbana	1069	1 Th, 7 p., Frasca Air Mus.
Wonder Lake	932	2 M, 7:30 p., Galt AP
INDIANA		
Anderson	226	3 Su, 6:30 p., Anderson AP
Bloomington	650	2 M, 7 p., Wings of Indiana
Columbus	729	2 Th, 7 p., (812) 376-0530
Elkhart	132	4 Th, 7 p., Elkhart AP
Evansville	21	2 W, 7 p., Evansville AP
Ft. Wayne	2	2 F, 7 p., Smith Field
French Lick	888	3 M, 7:30 p., Fr. Lick AP
Indianapolis	900	1 Tu, 7:30 p., 1125 S. Franklin Rd.
Kokomo	235	Last Th, (317) 883-5939
Michigan City	966	1 F, 7 p.s, (219) 77*-4117
Nappanee	938	1 Su, 6 p., Naappanee AP
Richmond	373	3 Th, 7 p., (317) 489-4292
Sellersburg	1042	Last Tu, 7:30 p., Purdue AP
Terre Haute	83	2 F, 7:30 p., (812) 234-2834
Valaparaiso	104	2 F, 7 p., Porter Co. AP
Vincennes	625	2 Th, 7:30 p., (812) 882-3332
Warsaw	1109	1 M, 7 p., Warsaw AP
W. Lafayette	256	3 Th, 7:30 p., Purdue AP
Westfield	67	1 M, 7:30 p., Noblesville AP
IOWA		
Burlington	700	1 W, 7 p., (319) 392-4178
Cedar Rapids	33	1 F, 7:30 p., (319) 926-2238
Des Moines	135	2 Sa, 7 p., Ankeny AP
Dubuque	327	Call (319) 852-7155
Ft. Dodge	214	3 Su, 2 p., (712) 335-4497
Marshalltown	675	2 F, 7:30 p., (515) 752-4729
Mason City	94	3 Su, 2 p., (515) 357-2547
Muscatine	111	1 Sa, 7:30 p., (319) 262-8538
Osceola	1143	Last F, 7 p., Osceola Av.
Ottumwa	409	2 Su, 2 p., (515) 655-7831
Postville	368	2 Tu, 8 p., (319) 536-8649
Sioux City	291	2 Su, 2 p., Martin AP
Spencer	999	2 W, 7:30 p., Spencer AP
Waterloo	227	3 F, 7 p., (319) 266-9880
KANSAS		
Atchison	967	1 Su, 3 p., Earhart Field
Ellsworth	1127	2 Su, 2:30 p., Ellsworth AP
Great Bend	1134	1 W, 7 p., Great Bend AP
Hays	1133	1 W, 7 p., Hays AP
Independence	980	Last M, 7 p., Ind. AP
Olathe	868	1 M, 7:30 p., Marley Cooling Tower
Overland Park	200	2 Th, 7:30 p., 9595 W. 95th
Topeka	313	1 W, 7:30 p., Forbes Av.
Wichita	88	3 Sa, 8 p., Lewis St. Glass
KENTUCKY		
Bowl'g Green	1050	3 M, 7 p., B.G. Warren AP
Georgetown	169	3 Su, 2:30 p., Scott Co. AP

State, city	Chap.	Meeting
Kenturcky cont.		
London	942	2 Tu, 7 p., Chesnut's Hgr.
Louisville	110	2 W, 7:30 p., Bowman Fld.
Murray	920	2 Tu, 7 p., Murray AP
Paducah	807	2 Tu, 7 p., 2705 Olivet Church Rd.
Richmond	892	Last Su, 1:30 p., Madison AP
Russellville	1165	3 Tu, 7 p., Logan Co. AP
Vine Grove	657	1 F, 7:30 p., Vine Grove AP

LOUISIANA

State, city	Chap.	Meeting
Alexandria/ Pineville	614	2 Th, 6:30 p., Eng. Jet Cent.
Baton Rouge	244	1 Th, 7 p., (504) 774-4208
Ferriday	912	4 Su, 2 p.,Concordia Par.AP
Hammond	405	2 Sa, 2 p., Hammond AP
Houma	513	3 Th, 7 p., CAP, Houma AP
Lake Charles	541	1 Sa, 8 a., (318) 433-6057
New Orleans	261	2 Th, 7:30 p., (504) 889-1260
Reserve	971	2 W, 7 p., St. John AP
Shreveport	343	3 Th, 7:30 p., D'ntown AP
Slidell	697	2 Tu, 7:30 p., Slidell AP

MAINE

State, city	Chap.	Meeting
Bangort	827	2 Th, 7 p.,(207) 990-4490
Bowdoinham	87	2 Tu, 7:30 p., Merry Meeting AP
Norridgewock	736	3 Tu, 7 p., (207) 623-4830
Owls Head	696	1 W, 7 p., Knox Co. Fl.Club

MARYLAND

State, city	Chap.	Meeting
Annapolis	571	2 Th, 7:30 p., Lee AP
Cumberland	426	Last Th, 7:30 p., Mex.Farms AP
Easton	532	2 Tu, 7 p., Easton AP
Frederick	524	1 Th, 7:30 p., Frederick AP
Hagerstown	36	1 Tu, 7:30 p., Wash. Co. AP
Lexington Park	478	3 Tu, 7:30 p., Naval Mus.
Pasadena	1002	3 Su, 4 p., Mt Road AP

MASSACHUSETTS

State, city	Chap.	Meeting
Boston	'06	1 F, 7:30 p., Bedford VA
Framingham	196	Last F, 7:30 p., Fr. St. Coll.
Hanson	279	2 F, 7:30 p., Cranland AP
Lawrence	136	2 Th, 7:30 p., (508) 256-8612
Mansfield	701	3 F, 7:30 p., Mansfield AP
Marlboro	673	2 Tu, 7 p., Marlboro AP
Marstons Mills	498	1 F, 7 p., Cape Cod AP
Newburyport	502	1 Su, 2 p., (508) 363-5424

MICHIGAN

State, city	Chap.	Meeting
Alma	134	2 Th, 7:30 p., Alma AP
Alpena	1021	2 M, 7 p., Alpena AP
Ann Arbor	333	w Th, 7:30 p., A.A. AP
Brighton	384	2 W, 7:30 p., Livingston AP
Cadillac	678	3 Th, 7:30 p., Cadillac AP
Canton	113	3 Th, 8 p., Mettetal AP
Central U.P.	439	1 Sa, 9 a., (906) 774-5550
Chesaning	597	1 W, 7:30 p., Nixon AP
Cheboygan	560	2 Su, 2:30 p., (616) 526-2227
Dowagiac	1028	1 Su, 6 p.,Dowagiac AP
Flushing	77	2 Th, 7:30 p., Dalton AP
Fowlerville	1056	1 Th, 6:30 p., 4825 N. Hogback Rd.
Fremont	578	2 Tu, 7:30 p., Fremont AP
Gaylord	1095	2 W, 7 p., Otsego Co. AP
Grand Ledge	1060	4 Tu, 7 p., Abrams AP
Grand Rapids	145	2 Sa, 7 p., Kent Co. AP
Howell	546	1 Sa, 10 a., Livingston AP
Ironwood	1119	3 Th, 7 p., Gogebic Co. AP
Jackson	304	1 W, 7:30 p., Reynolds Field
Kalamazoo	221	1 W, 7 p., WMN Hangar, Kalamazoo AP
Ludington	772	2 Th, 7 p., Mason Co. AP
Marquette	850	3 Sa, 10 a., Marquette AP
Mason	55	2 Sa, 9 a., Mason AP
Midland	1093	1 Th, 7 p., Jack Barstow AP
Mt. Pleasant	907	1 Th, 7 p., Mt Pleasant AP
Niles	865	2 M, 7:30 p., Trinity U.M.C.
Oscoda	901	3 M, 7:30 p., Iosco Co. AP
Pontiac	104	2 Tu (every other mo.), 7 p., (810) 673-5739
Saginaw	159	2 M, 8 p., (517) 684-6982
Smiths Creek	979	3 M, 7:30 p., Orzel Av.
Sparta	704	2 M, 7:30 p., Sparta AP
St. Joseph/ Benton Har.	585	2 W, 8 p., Watervilet Club
Traverse City	234	1 Su, 7 p., (616) 946-4082
Warren	13	1 Th, 7:30 p.,1600 E.Evelyn

MINNESOTA

State, city	Chap.	Meeting
Aitkin	965	2 W, 6:30 p., Aitkin AP
Alexandria	702	Call (612) 846-3600
Austin	386	4 Tu, 7:30, (507) 433-6794
Blaine	237	4 M, 7:30 p., Anoka Co. AP
Duluth	272	3 F, 7:30 p., Duluth AP
Grand Rapids	412	1 Th,. 7:30 p., G.R. AP
Granite Falls	688	Last W, 7:30 p., G.F. SW Tech College
Henning	1065	3 Su, 2:30 p., Henning AP
Hutchinson	951	2 W, 7:30 p., Hutch. AP
Int'l Falls	1012	2 Tu, 7:30 p.,Country Kitch.
Lake Elmo	54	2 M, 7:30 p., Hobo Hangar
Mankato	642	2 Su, 4 p., Mankato AP
Maple Lake	878	2 W, 7:30 p., M.L. AP
Minneapolis	25	3 W, 7:30 p.,ANG Ops Bldg. MSP AP
Minn/St. Paul	587	3 M,7:30 p.,(612) 929-8725
Minnetonka	178	2 Tu, 7:30 p., (612) 939-0489
Owatonna	300	3 Th, 7:30 p., Owat.AP
Richfield	1164	2 Tu, 12 N, Bloomington
Rochester	100	2 F, 7:30p., (507) 281-3356

State, city	Chap.	Meeting
Minnesota cont.		
St. Cloud	551	2 M, 8 p., Leaders AP
Staples	1009	1 Tu, 7 p., Staples AP
Two Harbors	1128	1 Tu, 7 p., Two Harbors AP
White Bear L.	745	2 Tu, 7:30 p., Benson's AP
Willmar	1172	2 Sa, 8:30 a., Willmar AP
Winona	919	3 F, 7:30 p., Sauer Aviation
Worthington	1151	1 M, 6:30 p., Worth. AP
MISSISSIPPI		
Bay St. Louis	479	1 Tu, 7 p., Stennis AP
Jackson	276	2 F, 7 p., (601) 924-4383
Meridian	986	Call (601) 483-9741
Ocean Springs	743	3 Tu, 7 p., Gautier Library
Pontotoc	987	1 Su, 2 p., Pontotoc Co. AP
Tylertown	1157	2 Th, 7 p., Tylertown AP
MISSOURI		
Chillicothe	944	3 Th, 7 p., Hickey Hangar
Jefferson City	429	1 M, 7 p., Jeff City AP
Joplin	463	1 Su, 2 p., Joplin AP Term.
Kansas City	91	3 M, 7 p., Lee's Summit AP
Liberty	612	2 Tu, 7:30 p., Liberty AP
Osceola	1032	3 Th, 7:30 p., Clinton AP
Painton	453	3 Su, 2 p., Painton AP
S. St. Louis	331	2 M, 7:30 p., Sackman Field
Springfield	821	3 Th, 7 p., 235 W. Kearny
St. Louis	32	4 Su, 6 p., (314) 286-9932
Taylor	488	1 Tu, 7:30 p., Taylor AP
MONTANA		
Billings	57	2 Sa, 7 p., (406) 259-7010
Great Falls	1141	3 Tu, 7 p., Gt. Falls AP
Helena	344	2 M, 7 p., (406) 475-3378
Kalispell	102	1 Tu,7:30 p.,(406)837-3455
Missoula	517	3 M,7:30 p.,(406) 728-5428
Polson	1122	2 Th, 7 p., Polson AP
NEBRASKA		
Broken Bow	1170	2 Th, 7 p., B.B. AP
Columbus	876	3 Su (odd mo.), 2 p., (402) 395-6079
Fairbury	680	2 Su, 1:30 p., (913) 243-4537
Hastings	544	3 Su, 2 p., Hastings AP
Kearney	1091	2 Tu, 7 p., 408 W. 39th St.
Lincoln	569	1 Tu, 7:30 p., Shoemaker Truck Stop
Norfolk	918	3 M, 7:30 p., Norfolk AP
North Platte	562	2 Su, 2 p., (308) 532-9518
O'Niell	804	1 M, 8 p., O'Neill AP
Omaha	80	2 M,7:30 p.,(402) 498-3474
Scottsbluff	608	3 Th,7:30 p.,(308)436-7276
York	1055	1 Tu, 7:30 p., York AP
NEVADA		
Carson City	403	1 W, 7:30 p., C. City AP
Elko	1136	3 Tu, 7 p., 2503 Mt. City
Las Vegas	163	1 Tu, 8 p., N. Las Vegas AP
Pahrump	1160	1 Sa, 7 p., Caluda Meadows

State, city	Chap.	Meeting
NEW HAMPSHIRE		
Manchester	336	Last Tu,7:30 p., Manch. AP
Rochester	225	1 W, 7:30 p., 265 Roch. Hill
West Lebanon	740	. 2 Tu, 7 p., W.L. Fire Station
West Ossipee	1085	2 Tu, 7:30 p., 3 Apache Ln.
NEW JERSEY		
Cross Keys	216	3 W, 8 p., C.K. AP
Lakewood	315	1 M, 8 p., Lakewood AP
Lincoln Park	501	2 W, 8 p., L.P. AP
Marmora	287	2 W, 7:30 p., 401 Park Ave.
Millburn	238	4 M, 8 p., (201) 765-0105
Nutley	73	4 Th, 8 p.,641 Franklin Ave.
Ocean County	898	1 Tu, 8 p., Lakhurst NAS
Pittstown	643	2 Su, 10 a., Alexandria Fld.
Trenton	76	2 F, 7:30 p., Pole Barn
NEW MEXICO		
Alamagordo	251	3 Tu, 7 p., White Sands AP
Albuquerque	179	3 Tu, 7:30 p., #9 Fire Sta.
Carlsbad	833	1 W, 7 p., (505) 885-5133
Clovis	969	2 Th, 7:30 p., Clovis AP
Los Alamos/ Santa Fe	691	3 Th, 7:30 p., (505) 672-0148
Los Lunas	530	2 W, 7:30 p.,Mid Valley AP
NEW YORK		
Albion	674	2 Th, 8:30 p., (716) 589-4969
Batavia	1169	4 Tu, 6:30 p., Genesee AP
Bath	953	1 W, 7 p., Lake Country
Bethel	1140	1 Su, 10 a., Sullivan Co. AP
Bethpage	594	2 F, 8 p., (516) 261-3307
Binghamton	53	2 Sa,7:30 p.,(607)642-8184
Buffalo	46	2 W, 7 p., (716) 649-6137
Clarence	656	2 Tu, 7:30 p., Clarence AP
Claverack	1006	2 Sa, 8 p., (518) 828-6149
Cooperstown	1070	1 Tu, 7 p., Westville AP
Edinburg	602	Last M, 7:30 p., Plateau Sky
Elmira	533	4 W, 7:30 p., Elmira AP
Ghent	146	3 M, 7:30 p., Klinekill AP
Glens Falls	353	3 Tu, 7:30 p., Boces Hudson
Ithica	811	3 F, 7 p., (607) 257-2016
Massena	500	2 Tu, 7:30 p., Massena AP
Middlesex	504	Last Tu, 7 p., Mid. Vall. AP
Oswego	362	3 Th, 7:30 p., (315) 668-3794
Rochester	44	3 Tu, 7:30 p., Ledgedale AP
Schenectady	852	2 M, 7 p., Schenctady AP
Suffolk Co.	528	2 F, 8 p., (516) 878-9306
Utica	294	1 W,7:30 p.,(315) 866-2110
Wallkill	474	1 F, 8 p., Kobelt AP
Wappingers F.	246	2 Th, 7 p., Dutchess Co. AP
Warsaw	621	2 Tu, 7:30 p., Perry W. AP
Weedsport	1017	2 Sa, 2 p., Weedsport AP
Westbury	3	4 F, 8:30 p., NY Inst. Tech.
White Plains	1081	3 Sa, 9 a., 1 Blauvelt Pl., Scarsdale

State, city	Chap.	Meeting
NORTH CAROLINA		
Asheville	1016	2 Tu, 7:30 p., (704) 684-6120
Beaufort	1075	1 Th, 7 p., M.J. Smith AP
Burgaw	297	1 Sa, 9:30 a, Stag AP
Cary	1114	Alt. 3 Sa 9 a/3 Tu 7 p., Cox Field
Charlotte	309	Every Tu 5 p., Chap. Hgr.
Furham	879	2 Sa, 9:30 a., Lakeridge AP
Farmville	960	1 Tu, 6:30 p., Flanagan Fld.
Greensboro	8	2 Sa, 12 N, (910) 292-3294
Hickory	731	3 W,7:30 p., Cat. Valley Av.
Long Beach	939	3 Sa, 10 a., Brunswick AP
Maiden	1145	Alt. 1 M 7 p/1 Sa 2 p., Laney's AP
New Bern	1171	4 Sa, 10 a., Craven AP
Raleigh	506	2 Su, 2:30 p., Johnson AP
Rocky Mount/ Wilson	1047	1 Su, 2:30p., Wilson Ind.
Salisbury	1083	2 Tu, 7 p., R. Cabarrus CC
NORTH DAKOTA		
Bismarck	1008	3 Sa, 10 a., (701) 843-7696
Fargo	317	2 M,7:30 p.,(218) 232-6574
Grand Forks	380	3 Tu, 8 p., U. North Dakota
Minot	265	1 W, 7 p., Dak. Terr. Mus.
OHIO		
Akron	147	3 Su, 2 p., Sunset Strip, Marlboro
Athens	737	2 Sa,7:30 p.,(304)773-5406
Bryan	816	1 E, 7 p., Williams Co. AP
Cambridge	928	3 Su, 3 p., Cambridge AP
Canton	82	2 M, 7:30 p., 1922 26th NE
Cincinnati	174	3 Su, 2 p., Clermont Co. AP
Cleveland	127	2 F, 8 p., Hopkins AP 5K
Cleveland	325	3 F, 8 p., Cuyahoga Co. AP
Columbus	9	Call (614) 885-4259
Columbus	443	3 M,7:30 p.,(614) 927-7767
Columbus	1024	1 Sa, call (937) 642-3812
Dayton	48	1 Th, 7:30 p., Moraine AP
Findlay	636	3 Su, 2 p., (419) 422-5594
Geneva	332	3 Th, 8 p., (216) 466-2988
Hamilton	974	1 Su, 2 p., Hamilton AP
Huron	50	1 Su, 1:30 p., Huron AP
La Grange	255	3 Tu, 7:30 p., Harlan AP
Leetonia	507	2 Th, 7 p., York AP
Mansfield/ Wooster/Ash.	148	2 Su, 1:30 p., (419) 529-2332
Marion	516	3 Tu, 7:30 p., Marion AP
Marysville	1063	2 Sa, 9 a., Marysville AP
Middlefield	5	1 Tu, 7:30 p., Geauga AP
New Carlisle	610	4 Th, 7:30 p., Barnhart AP
New Knoxville	922	2 Su, 1 p., N. Armstrong AP
New Philadel.	1077	4 Tu, 7:30 odd mo., N.P.AP
Newark	402	3 Tu, 7:30 p., Newark AP
Niles	117	3 Sa, 7 p., (216) 256-9936
N. Bloomfield	860	Call (330) 889-2337
Portage Co./ Ravenna	1104	1 Sa, 6 p., 6188 St. Rt. 303
St. Clairsville	798	1 F, 7:30 p.,(614) 695-9806
Steubenville	859	1 Tu, 7 p., Jefferson Co. AP
Toledo	582	4 Th, 7 p., (419) 666-9029
Urbana	995	1 Tu, 7 p., Mad River AP
Versailles	109	2 Th, 7:30 p., Vers. AP
Wadsworth	846	2 Th, 7:30 p., 900 AP Rd.
Washington Court House	1033	3 W, 7 p., Fayette Co. AP
Wauseon	149	3 Su, 4 p., (419) 826-4056
Waynesville	284	2 Su, 2 p., Waynesville AP
Youngstown	531	3 Sa, (216) 799-2951
Zanesville	425	1 Su, 7:30 p., Riverside AP
OKLAHOMA		
Ada	1005	2 Th, 7 p., Ada AP Term.
Altus	1014	2 Su, 2 p., Altus AP
Cookson	1004	3 F, 7:30 p., L. Tenkiller AP
Cushing	1139	4 Tu, 6:30 p., Cushing AP
Durant	1146	1 Th, 7 p., (405) 924-6901
Grove	893	Last Su, 1:30 p., Grove AP
Oklahoma City	24	2 Tu, 7:30 p., W.R. AP
Owasso	10	2 M, 7:30, Gundy's AP
Ponca City	1046	2 Th, 7 p., P.C. AP Hgr. 14
Shawnee	1098	2 Su, 2:30 p., Shaw. Vo-Tec
Stillwater	539	1 Su, 2 p., Mulberry H. AP
Woodward	1115	2 M, 7:30 p., Woodward AP
OREGON		
Eugene	31	1 W, 7 p., Creswell AP
Grants Pass	725	1 Th, 7 p.,G.P. AP
Hermiston	219	3 Tu, 7:30 p., (509) 522-1743
Independence	292	2 Th, 7 p., Ind. AP Ch. Hgr.
Klamath Falls	411	2 Tu, 7:30 p., K.F. AP
Medford	319	2 Tu, 7:30 p., Delzell Aero.
Ontario	837	3 W, 7:30 p., Ont. AP EAA
Portland	105	3 Th, 7 p., Twin Oaks AP
Prineville	617	1 Th, 7:30 p., Prineville AP
Roseburg	495	2 Tu, 7 p., Douglas Forest Protective Assn.
The Dalles	1089	1 Tu, 7 p., The Dalles AP
PENNSYLVANIA		
Beaver Valley	68	3 W, 8 p., (412) 843-4063
Bradford	467	1 Su. Mar-June-Sept-Dec., Bradford Reg. AP
Cambridge Sp.	310	Last Sa, 1:30 p., (412) 748-3570
Clarion Co.	994	1 F, 7:30 p., Clarion Co. AP
Erie	160	1 W, 7:30 p., 1607 Asbury
Franklin	988	2 M, 7 p., Venango AP
Hanover	1041	1 M, 7:30 p., Hanover AP
Indiana	993	1 Th, 7:30 p., Ind. Co. AP
Johnstown	633	2 M, 7:30 Johnstown AP
Jonestown	390	2 F, 8 p., Dee Jay AP
Lehighton	1019	1 Tu, 7 p., Arner AP

State, city	Chap.	Meeting
Pennsylvania cont.		
Leigh Valley	70	2 Tu, 7:30 p., Kessler Fld.
Martinsburg	400	3 F, 7:30 p., Altoona AP
Mercer	161	2 Su, 2 p., Grove City AP
Montoursville	401	2 W, 7 p., Cent. Penn Air
New C'land	122	3 M, 7:30 p., Cap. City AP
Philadelphia	78	Last Tu, 8 p., W. Grove NAS
Pittsburgh	45	3 F, 7:30 p., Rostraver AP
Reading	321	3 W, 7:30 p., 311 Penn Ave.
Reedsville	518	1 M, 7 p., Mifflin Co. AP
Riverside	769	3 Th, 7:30 p., Danville AP
Smoketown	540	1 W, 7:30 p., Smoket'n AP
Stroudsburg	839	3 W, 7:30 p., (717) 421-8292
Waynesburg	978	2 Tu, 7:30 p., Green Co. AP

RHODE ISLAND

State, city	Chap.	Meeting
Middletown	51	3 Su, 2 p., Fall River AP
Warwick	381	3 W, 7:30 p., Hangar 1

SOUTH CAROLINA

State, city	Chap.	Meeting
Aiken	1079	2 M, 7 p., Aiken AP
Camden	1132	2 M, 7 p., Woodward Fld.
Greenville	249	1 M, 7:30 p., Greenville AP
Greenwood	1023	1 Th, 7:30 p., Gr. Co. AP
Hartsville	1058	Last Tu, 7 p., (803) 383-2049
Myrtle Beach	1167	3 Sa, 10 a., Conway AP
Timmonsville	629	3 M, 7 p., 713 Pilots Lane

SOUTH DAKOTA

State, city	Chap.	Meeting
Huron	537	2 Su, 2 p., Huron FSS
Rapid City	39	2 Th, 7 p., R.C. AP
Sioux Falls	289	3 Su, 2 p., Gt. Plains AP
Spearfish	806	1 M, 7:30 p., Black H. AP
Yankton	1029	3 Th, 7:30 p., Yankton AP

TENNESSEE

State, city	Chap.	Meeting
Blountville	442	Last F, 7:30 p., Tri-Cities AP
Chattanooga	150	3 Sa, 7:30 p., Collegedale Flight School
Knoxville	17	Last F, 7:30 p., Sky Ranch
Lebanon	863	2 Th, 7 p., Lebanon AP
Memphis	182	4 Sa, 1 p., D. Spain AP
Morristown	779	2 M, 7 p., Morristown AP
Mt. City	1136	Last Su, 2 p., Johnson Co. AP
Murfreesboro	419	1 Th, 6:30 p., M'boro AP
Nashville	162	2 Tu, 7 p., Signature Av. #9
Paris	734	1 Su, 2 p., Paris AP
Sevierville	957	3 Th, 7 p., G'ville/P.F. AP
Winchester	699	1 Tu, 7 p., Winchester AP

TEXAS

State, city	Chap.	Meeting
Abilene	471	3 F, 7 p., Tex. St. T. Coll.
Arlington	34	2 Tu, 7:30 p., UTA Eng.
Austin	187	1 F, 7:30 p., UT Woolrich
Brazoria Co.	347	2 Su odd mo. 2 p., (409) 798-2908
Brookshire	774	3 Th, 7:30 p., Sportflyers
Burleson	280	2 Th, 7:30 p., 150 Shaffstall
Canton	1078	2 Tu, 7 p., First Mon. cabin
Conroe	302	2 Sa, 9 a., Mont. Co. AP
Corpus Christi	191	2 Th, 7:30 p., C.C. AP
Dallas	168	1 Tu, 6:30 p., F.B. Library
Denton	661	1 Tue, 7 p., Denton AP
El Paso	125	Last F, 6:30 p., W. Tex. AP
Fredericksb'g	1088	3 Su, 1 p., (210) 997-7697
Galveston	1149	2 Sa, 12 N, (409) 765-1453
Graham	945	3 M, 7 p., AP Workshop
Granbury	983	2 Th, 7:30 p., Pecan Plan. AP
Greenville	914	2 Th, 7:30 p., Caddo M. AP
Houston	12	2 W, 7:30 p., Hobby AP FAA
Kerrville	747	1 Sa, 12 N Flying RV
Killeen	542	2 Tu, 7:30 p., (817) 699-0505
Kingsland	889	1 Su, 2 p., S. Williams AP
La Porte	712	2 Tu, 7:30 p., La Porte AP
Longview	972	2 Th, 6:30 p., Gladewater AP
Lubbock	19	2 Tu, 7:30 p., (806) 794-5961
Midland	123	3 M, 7:30 p., Skywest AP
Mt. Vernon	834	4 Th, 7:30 p., Franklin AP
New Braunfels	958	2 Sa, 10 a., N.B. AP
Plainview	1100	2 Th, 7 p., Hale Co. AP
Ranger	956	1 Sa, 6 p., Ranger EAA AP
Rio Grande V.	595	1 Su, 1:30 p., Tex. Air Mus.
Rockport	1080	1 Th, 6 p., Pac. SW Bank
San Angelo	493	3 Tu, 7 p., Mathis Field
San Antonio	35	2 Sa, 7 p., San Geronimo AP
Sherman	323	3 Tu, 7 p., Sherman AP
Sulphur Sp.	1094	3 Th, 7:30 p., TU Elec. Bldg.
Tyler	727	3 Th, 7 p., Tarrant Fld.
Uvalde	903	3 Sa, 1 p., Uvalde AP
Victoria	340	Last Sa, 9 a., Ball AP
Wharton	896	2 Th, 7 p., Wharton AP

UTAH

State, city	Chap.	Meeting
Park City	964	2 Th, 7 p., (801) 645-9543
Parowan	936	2 Sa, 2 p., Parowan AP
Salt Lake City	23	2 F, 7 p., 640 N. 2360 West

VERMONT

State, city	Chap.	Meeting
Shelburne	613	3 Su, 11 a., (802) 878-6337

VIRGINIA

State, city	Chap.	Meeting
Alexandria	186	Last Th, 7:30 p., Wash. Gas Light, Springfield
Augusta Co.	511	3 Tu, 7:30 p., Shen. Val. AP
Dublin	906	2 F, 7 p., New R. Valley AP
Fredericksb'g	1099	2 Th, 7 p., Shannon AP
New Market	1131	3 Sa, 11 a., N.M. AP
Newport News	156	2 Tu, 7:30 p., W'burg AP
Norfolk	339	1 Tu, 7:30 p., Ham. Rd. AP
Richmond	231	2 F, 7:30 p., Av. Museum
Roanoke	646	2 Tu, 7:30 p., Roanoke AP
Winchester	1031	1 Th, 7:30 p., Win. AP

WASHINGTON

State, city	Chap.	Meeting
Bellingham	404	3 Th, 7:30 p., Bell. AP
Burlington	818	2 Sa, 10 a., Bayview AP

State, city	Chap.	Meeting
Washington cont.		
Eastsound	937	2 M, 7 p., Orcas Longhouse
Ellensburg	492	2 Th, 7:30 p., (509) 962-4156
Grays Harbor	367	2 W, 7 p., (360) 289-3304
Kelso	1111	1 Th, 7 p., Aero. Unltd.
Kennewick	391	3 Th, 7 p. (509) 786-1341
Kent	441	4 M, 7:30 p., Crest AP
Moses Lake	355	3 Tu, 7 p., 4914 Road M
Olympia	684	Last Tu, 7:30 p., Olymp. AP
Port Angeles	430	Last W,7 p.,(360) 683-2138
Pt. Townsend	1026	3 Sa, 10 a., Jefferson Co. AP
Puyallup	326	2 Tu, 7 p., Pierce Co. AP
Seattle	26	2 Th, 7:30 p., Mus. of Flt.
Silverdale	406	4 W, 7 p. (360) 692-1097
Snohomish	84	2 Tu, 7:30 p., Harvey Field
Spokane	79	2 F, 7:30 p., N. Sp. Library
Vancouver	762	2 W, 7 p., Pearson Air Mus.
Walla Walla	604	2 M, 7 p., (509) 522-1767
Yakima	206	3 Th, 7:30 p., (509) 248-1413

WEST VIRGINIA

State, city	Chap.	Meeting
Fairmont	842	2 Th, 7 p., 3 Ways Inn
Glendale	738	2 Tu, 7:30 p., Fokker Fld.
Huntington	644	4 Tu, 7 p., (304) 429-1432
Parkersburg	915	1 Th, 7 p., Wood Co. AP
S'doah Jct.	1071	1 W, 7:30 p., Ch. Hangar

WISCONSIN

State, city	Chap.	Meeting
Adams	931	3 M, 7 p., 2074 11th Ave.
Appleton	444	3 M, 7 p., (414) 779-6083
Beloit	60	2 Wm 7:30 p., Rock Co. AP
Brodhead	431	1 Tu, 8 p., (608) 897-1457
Eau Claire	509	2 Th, 7 p., Chippewa AP
Fond du Lac	572	1 Sa, 8 a., F.D.L. AP
Gleason	640	2 Sa,1:30 p.,(715)359-8616
Gransburg	875	Call (715) 349-5152
Green Bay	651	2 M, 7 p., Nicolet AP
Juneau	897	2 W, 7 p., DodgeCo. AP
Kenosha	217	3 Tu, 7 p., Gateway Av.
La Crosse	307	4 Th, 7:30 p., (608) 526-9546
Madison	93	3 Th, 7:30 p., Blackhawk
Manitowoc	383	2 W, 7:30 p., M'woc AP
Marinette	535	4 W, 7 p., City Hall
Marshfield	992	3 M,7:30 p.,(715) 223-6179
Milwaukee	18	4 Tu, 7:30 p., Tim'man Fld.
Monroe	1010	2 Tu, 7:30 p., Monroe AP
Oshkosh	252	1 M, 7:30 p., EAA Av. Ctr.
Portage	371	2M, 7:30 p., Portage AP
Racine	838	2 W, 7 p., 3333 N. G.B. Rd.
Rice Lake	631	2 Tu 8 p. alt. mo. (715) 234-2464
Sheboygan F.	766	3 M, 7 p., Old Bemis Bldg.
Sturgeon Bay	630	1 M, 7 p., Cherryland AP
Thorp	1015	1 M, 8 p., (715) 669-7270
Tomah	935	1 Tu, 7 p., Bloyer Field

State, city	Chap.	Meeting
Washington Is.	982	3 Tu, 7 p., W.I. AP
Watertown	320	1 Th, 7:30 p.,Watertown AP
Wausau	243	3 Sa, 7 p., 2001 Falcon Dr.
West Bend	1158	3 W, 7 p., W.B. AP

WYOMING

State, city	Chap.	Meeting
Casper	420	2 Sa, 9 a., Good Warbirds
Cheyenne	342	1 F, 7 p., Skyview AP
Fremont Co.	1020	3 Su, 2:30 p., (307) 856-9654
Gillette	767	Call (307) 682-9648
Jackson	1049	1 Th, 7 p., Jackson Hole AP
Laramie	1008	2 Su, 7 p., Laramie
Rock Springs	1173	1 Sa, 10 a., Wh. Mt. Library

CANADIAN CHAPTERS

State, city	Chap.	Meeting
Aylesford, N.S.	1051	Last Su, 2 p., 989 Brow Mountain Rd.
Brantford, Ont.	115	1 Tu, 8 p., Gilbert Aircraft
Dartmouth, N.S.	305	1 Tu, 7;#0 p., Le Brun Ctr.
Edmond, Alta.	30	3W, 7:30 p., Av. Her. Soc.
Hamilton Ont.	65	2 F, 8 p., Hamilton AP
Montreal, Que.	266	Last Th, 7:30 p., J. Abbott Coll., Penfield Room 203
North Bay, Ont.	366	Call (705) 497-1932
Ottawa, Ont.	245	3 Th, 8 p., Nat. Av. Mus.
Peterb'ogh, Ont.	911	2 M 7:30 p., P. AP
Regina, Sask.	154	Last Sa, 1 p., Flying Club
Timmins, Ont.	1126	2 Tu, 7:30 p., Air Quebec Training Ctr.
Windsor, Ont.	185	2 Th, 7:30 p., Wind. AP

INTERNATIONAL CHAPTERS

BRAZIL

State, city	Chap.	Meeting
Porto Alegre	1068	1 Sa, 2 p., Belem Novo AP
San Leopoldo	854	Sa, Aeroclube de Sao Leo.

CHILE

State, city	Chap.	Meeting
Santiago	749	2 W, 8:30 p., Laura Tromben 5699

DENMARK

State, city	Chap.	Meeting
Spjald	655	Call (45) 569-53890

ENGLAND

State, city	Chap.	Meeting
Leeds	1168	2 Su, 1:30 p., Yorks. Aero Club, Leeds Bradford AP
Thruxton	1166	1 Sa, 5 p., Thruxton AP

GERMANY

State, city	Chap.	Meeting
Burstadt	1084	1 W, 7 p.,Worms Flugplatz
Rendsburg	1037	2 Su 11 a., alt. mo., Motel Neuwerk D-2370

GRAND CAYMAN

State, city	Chap.	Meeting
Cayman	1096	Sa, 9:30 a., Breakers @ Geo. Hawley's

State, city	Chap.	Meeting
INDIA		
Madras	962	1 W, 4 p., 560 080
JAPAN		
Kawasaki	1147	1 Su, 11a., Moriya Fl.Club
MALAYSIA		
Kuala Lumpur	1090	Call 603-776-3069
MEXICO		
Guadalajara	1039	1 W, 8 p., Sanborn's
NETHERLANDS		
Beverwijk	664	Last sa, 11 a., Pioneer Hangar Lelystad
NORWAY		
Oslo	573	Call 4751677766
POLAND		
Bielsko Biala	991	Call 46 227 22080
SOUTH AFRICA		
Halfway House	322	1 W, 7 p., Comm. Hall Midrand
Krugersdorp	973	1 Sa, 2 p., J. Taylor AP
Pietermaritzbg	357	1 Su, 1 p., Oribi AP
SWEDEN		
Stockholm	222	Th 7p., Barkarby AP

ANTIQUE CLASSIC CHAPTERS

State, city	Chap.	Meeting
Hayward, CA	29	2 Th, 7 p., Hayward AP
Sacramento, CA		2 Sa, 9 a., (707) 964-7733
Grand Jct., CO	19	1 Th, 6:30 p., (970) 257-7832
Lansing, IL	26	Call (708) 534-6240
Overland Pk., KS	16	3 F, 7:30 p., Stanley Bank
Lexington, KY	28	2 Su, 1:30 p., G'town AP
New Iberia, LA	30	1 Su, 9 a., Acadiana AP
Albert Lea, MN	13	4 Th, 7:30 p., (507) 765-2567
Minneapolis, MN	4	Call (612) 784-1172
Hampton, NH	15	2 Sa, 11 a., Hampton AP
Andover, NJ	7	1 Su, 10 a., Aeroflex AP
Mayodan, NC	3	Call (301) 872-4055
Delaware, OH	27	1 Sa, 9 a., Delaware AP
Zanesville, OH	22	3 Su (even mo.), 2:30 p, John's Landing AP
Tulsa, OK	10	4 Th.,7:30 p.,Hardesty Lib.
Houston, TX	2	3 Su., 2 p.,Dry Creek AP
Brookfield, WI	11	1 M, 7:30 p., Capitol AP

EAA IAC CHAPTERS

State, city	Chap.	Meeting
Phoenix, AZ	69	1 Su, 9:30 a., (802) 957-8376
Tucson, AZ	62	3 W., 7:30 p., 5835 N. Genematas
Delano, CA	26	2 Su, 12 N, Delano AP
Los Angeles, CA	49	Last Sa, 12 N, (310) 430-8306
San Diego, CA	36	2 Sa, 11 a., C. Hall Av.
Orlando, FL	90	1 Sa, 11 a., Mid-Fla. AP
Pompano B., FL	23	3 Th, 7 p., (954) 752-9283
Hampton, GA	3	3 F, 7 p., Manhattan Rest.
S.Simons I., GA	105	1 Sa, 10 a., 93 AP Rd.
Boise, ID	92	1 Sa, 10 a., Caldwell AP
McHenry, IL	1	Call (708) 639-4132
Louisville, KY	95	2 Sa, 7 p., Bowman AP
Houma, LA	72	3 Sa, Legros AP, Crowley
Plainville, MA	35	3 Sa, 10 a., Mansfield AP
Livonia, MI	88	Call (810) 377-0616
Kan. City, MO	15	Last W, 7:30p., D'twn. AP
Omaha, NE	80	3 M, 7:30 p., Cl. Bluffs AP
Las Vegas, NV	777	3 Tu, 7 p., N.L.V. AP Ter.
Reno, NV	53	Call (702) 883-9342
Medford, NJ	94	2 Sa, 11 a., Flying W AP
Albuquerque, NM	47	1 Su, 9 a., Coronado AP
Newburgh, NY	52	2 Sa each Quarter, 10a., Orange Co. AP, Montgom.
Apex, NC	19	2 Sda,1 12 N, call (919) 929-9322
Columbus, OH	34	3 Su, 2 p., Bolton AP
Tulsa, OK	10	Call (918) 494-8908
Portland, OR	77	Call (503) 626-8152
Harrisburg, PA	58	3 Sa,10 a.,(717) 766-3705
Houston, TX	25	1 M, 7 p., (281) 471-3703
McKinney, TX	24	2 Su, 2 p., AeroCountry
San Antonio, TX	31	2 Sa, 11 a., Zuehl AP
Richmond, VA	7	Call (804) 722-6568
Seattle, WA	67	3 Sa,10 a.,(206) 321-5629
Spokane, WA	103	1 Su, 9 a., Skyway Cafe
Fond du Lac, WI	46	Call (414) 922-6106
Land O' Lakes,WI	106	2 Sa, 12 N, L.O.L. AP
Milwaukee, WI	8	2 Tu, 7 p., Timmerman AP

State, city	Chap.	Meeting

EAA ULTRALIGHT CHAPTERS

State, city	Chap.	Meeting
Phoenix, AZ	55	1 Th, 7:30 p., Los Cerulus
Sahuarita, AZ	65	2 Sa Jan., April, July, Oct. (520) 885-5770
Tucson, AZ	61	1 W, 7:30 p., Tucson AP
Fayetteville, AR	77	1 Sa, 2 p., Wash. Co.
Ft. Lauderdale, FL	71	1 Sa, 11 a., Mac Ivor AP
Miami, FL	103	1 W, 7:30 p., JR's Grill
Cartersville, GA	68	2 Sa, 1 p. Cartersville AP
Champaign, IL	30	3 Sa, 5 p., (217) 379-4594
Winnebago, IL	18	2 Tu, 7:30 p., Cottonwood AP, Rockford
Wichita, KS	24	3 Sa,2 p., Morning Star AP
Louisville, KY	74	2 Tu, 7 p.,6300 Bardstown
Middleboro, KY	62	3 M, 7:30 p., (508) 866-5559
Fairbault, MN	66	1 Su, 2 p., 16936 Cannon City Blvd.
Minneapolis, MN	12	3 Th, 7 p., Anoka Co. AP
Dover, NH	67	4 M, 7 p., Rochester Hill Rd., Rochester
Bridgewater, NJ	53	2 W, 7:30 p.,Somerset Co. Lib.
Andover, OH	69	15th, (216) 563-5763
Osterburg, PA	64	3 F, 7:30 p., Ickes AP

State, city	Chap.	Meeting
Dallas/Ft.W, TX	78	1 Sa, 2 p., (817) 232-3379
La Crosse, WI	76	2 Tu, 7 p., La Crosse AP
Milwaukee, WI	1	2 Tu, 7:30 p. Aero Park
Oshkosh, WI	41	3 Tu, (414) 734-8682
Wausau, WI	75	3 Sa, 1:30 p., (715) 536-8828

EAA WARBIRDS OF AMERICA SQUADRONS

State, city	Chap.	Meeting
Vacaville, CA	11	2 Th, 7:30 p., Nut TreeAP
Niagara On The Lake, Ont.	9	2 W, 7 p.,Niagara Dist. AP
Wilmington, DE	8	3 Tu, 6 p., Pier 13 Rest.
Douglas, GA	10	2 Tu, 6:30 p., Douglas AP
Palos Park, IL	4	Call (630) 448-8825
Indianapolis, IN	3	2 Sa,10 a.,Mt. Comfort AP
Saugus, MA	7	Call (617) 233-9797
Creswell, OR	12	Call (503) 895-3523
Nashville, TN	1	3 Sa,12 N,(423) 453-2897
Yelm, WA	2	2 Tu, 7 p., Museum of Flt.
Milwaukee, WI	6	Bi-monthly, (414) 677-0430

YOUNG EAGLES

A Program of the EAA Aviation Foundation

The mission of the Young Eagles Program is to provide meaningful flight experience for one million young people (primarily between the ages of 8 and 17) by the year 2003, the 100th anniversary of the Wright brothers' first powered flight and the 50th anniversary of EAA.

The Young Eagles Program hopes to achieve these objectives:

- Encourage young people to become interested in flying
- Provide young people with an opportunity to gain new perspectives on their communities, their lives and the world in which they live
- Raise awareness of aviation career possibilities
- Help young people understand the knowledge necessary to become a pilot
- Respond to concerns that the nation's pool of pilots is growing smaller because of fewer new pilot starts and the number of pilots who are growing older--a situation that could cause a shortage of military and commercial pilots by the year 2000

NATIONAL
BUSINESS AIRCRAFT
ASSOCIATION, INC.

1200 Eighteenth Street, NW, Suilte 400, Washington, DC 20036
(202) 783-9000; fax (202) 331-8364; htrtp://www.nbaa.org

NBAA represents the aviation interests of more than 4,500 companies which own or operate general aviation aircraft as an aid to the conduct of their business, or are involved with business aviation. NBAA Member Companies earn annual revenues in excess of $3 trillion, and employ more than 16 million people worldwide.

Formed in 1947 by 19 companies, the organization was originally named Corporate Aircraft Owners Association. In 1953, with 214 members, it was changed to the National Business Aircraft Association, Inc. Membership reached 1,000 in 1974 and passed 2,000 six years later.

NBAA represents a range of flight departments. More than half of its member companies operate a single aircraft and employ an average of three people in their flight departments. One-quarter of the membership is composed of smaller firms with comparatively regional air travel requirements. The NBAA member fleet consists of approximately 6,000 aircraft.

50th NBAA Convention
Dallas, TX., Sept. 23-25, 1997

51st NBAA Convention
Las Vegas, NV, October 19-21, 1998

52nd NBAA Convention
Atlanta, GA, October 12-14, 1999

53rd NBAA Convention
New Orleans, LA October 10-12, 2000

NATIONAL BUSINESS AIRCRAFT ASSOCIATION
AWARD FOR MERITORIOUS SERVICE TO AVIATION

1950--**Arthur Godfrey**, for outstanding contribution to the use of aircraft as a business aid.

1951--**Col. J. Francis Taylor, Jr.**, in recognition of his development, testing and evaluation of aerial navigation aids.

1952--(no award given)

1953--**Charles Lindbergh**, for his many contributions to aviation.

1954--**Donald Stuard** (Civil Aeronautics Administration), for his development of aerial navigation systems.

1955--(no award given)

1956--**Eddie Rickenbacker**, for his contributions to military and commercial aviation.

1957--**Igor Sikorsky**, for his development of fixed- and rotary-wing aircraft.

1958--(no record of award)

1959--**James Doolittle**, for being a pioneer in instrument flight.

1960--**Donald Wills Douglas**, for his contributions to the advancement of the aeronautical sciences.

1961--**William Piper, Sr.**, in recognition of his vision and determination bringing the realm of flight to thousands of pilots.

1962--**Sen. A.S. "Mike" Monroney**, in recognition of his contributions to safety in the air.

1963--**Edwin Alber Link**, for his contribution to flying safety through his Link Trainer.

1964--**William Schulte** (FAA Asst. Administrator for General Aviation), for his untiring efforts to present business aviation's requirements to the highest government levels.

1965--**E.B. Jeppesen**, for the development of aerial cartographic and information services.

1966--**James McDonnell**, for his contributions to the progress of civil and military aviation.

1967--**Henry Dupong**, for his encouragement of the business aviation industry.

1968--**Henry Schiebel, Jr.** (Grumman Aircraft), for his contributions to the development and progress of business aviation.

1969--**Olive Ann Beech**, for her dedication and contributions to business aviation.

1970--**Juan Trippe** (Pan Am Airways), for his pioneering spirit in developing international air travel.

1971--**Michael Murphy** (Marathon Oil Co.), for creating a corporate/executive flight operation that is recognized for safety and excellence.

1972--NBAA Silver Anniversary member companies: The American Rolling Mill Co., Burlington Mills Corp., Corning Glass Works, General Electric Co., Republic Steel Co., Reynolds Metals Co., Wolfe Industries.

1973--**Dwane Wallace**, for leading general aviation to its position in the world's transportation market.

1974--**Edward Swearingen**, for perception and leadership in expanding the nation's business aircraft fleet.

1975--**William Lear, Sr.**, for contributions of incalculable value in aviation safety, navigation and operation.

1976--**George Haddaway**, for his dedication and unflagging spirit in supporting business aviation.

1977--**William Remmert, D. Robert Werner**, for their contribution to making business aviation a safe, efficient and practical transportation tool.

1978--**Dr. Richard Whitcomb**, for designing aircraft to reduce drag and increase speed without the need to add power.

Awards for meritorious service cont.

1979--Dr. Charles Stark Draper, for developing inertial navigation instruments and guidance systems.

1980--Sir Frank Whittle, Hans von Ohain, for their separate and independent development of the jet engine.

1981--Robert Holz, for his ability to motivate and promote the wisdom of advancement in air transportation, space, and defense.

1982--Charles "Kelly" Johnson, for attaining goals never before imagined.

1983--Sen. Barry Goldwater, for being an outspoken aviation advocate and bold supporter of American leadership in aeronautics.

1984--A. Scott Crossfield, for being in the right place at the right time with the right stuff in developing aviation technology.

1985--NASA space shuttle crews, they personify all that is the best of current advancements in aviation and aerospace technology.

1986--Air Traffic Control System, for dedication through 50 years of service and meeting the challenges of an ever-expanding airspace system.

1987--Dick Rutan, Jeana Yeager, Burt Rutan, for their service to aviation.

1988--Ed King, in recognition of his 40 years of service to general aviation.

1989--John Winant, in recognition of many years of service to NBAA.

1990--(no award given)

1991--Al Ueltschi, founder of FlightSafety International.

1992--James B. Taylor, for 50 years of leadership in the development, manufacture and marketing of business aircraft.

1993--Dee Howard

1996--Dr. Leonard Greene, founder of Safe Flight Instrument Corp.

1997--R. Dixon Speas, founder of three aviation consulting firms over a 50-year career.

The National Aviation Hall of Fame nominates and enshrines individuals in recognition of achievement in the development of aviation. Nomination for enshrinement into the NAHF is conducted by the 200-member Board of Nominations and its Screening Committee. Nominations can be made by any member of the NAHF. The nominee's accomplishments are reviewed by the Screening Committee and, if elected by majority vote, the nominee's name is placed on the official ballot.

The National Aviation Hall of Fame Enshrinees

Courtesy of the National Aviation Hall of Fame©
(Year of enshrinement in parenthesis)

Allen, William M.--President of Boeing, father of 747 (1971)

Andrews, Frank M.--U.S. Army General, commander of WWI air forces, assisted in establishing independent Air Force (1986)

Armstrong, Neil A.--Astronaut, first man on the moon (1979)

Arnold, Henry H. "Hap"--Commander of Army Air Forces during World War II, first 5-star General (1967)

Atwood, John Leland--Executive and engineer involved in DC-1, AT-6, P-51 and other designs (1984)

Balchen, Bernt--first pilot to fly over South Pole, founded Norwegian Airlines (1973)

Baldwin, Thomas S.--builder of first U.S. dirigible (1964)

Beachey, Lincoln--early exhibition flier (1966)

Beech, Olive Ann--co-founder and longtime head of Beech Aircraft (1981)

Beech, Walter H.--co-founder Travel Air, founder of Beech Aircraft (1977)

Bell, Alexander Graham--scientist, conducted early flight experiments (1965)

Bell, Lawrence D.--founder of Bell Aircraft (1977)

Bellanca, Guiseppe M.--Early aviator and designer (1993)

Bendix, Vincent H.--inventor, founder of Bendix Corp., provided Bendix Trophy (1991)

Boeing, William E.--founder of Boeing Aircraft, inaugurated mail and passenger service (1966)

Bong, Richard L.--World War II ace (40), Medal of Honor recipient and test pilot (1986)

Borman, Frank--Astronaut, first flight to moon, CEO of Eastern Airlines (1982)

Boyd, Albert--AAF chief of flight test, oversaw X-1 (1984)

Bradley, Mark E.-- 4-star General, Commander, Air Force Logistics Command (1992)

Brown, George "Scratchley"--4-star General, led WWII bombing raids, Korean command, Chairman, Joint Chiefs of Staff (1985)

Brukner, Clayton J.--designed and developed Waco airplanes (1997)

Byrd, Richard E.--explorer, flew over North, South Poles (1986)

Cessna, Clyde V.--co-founder Travel Air, founder of Cessna Aircraft (1978)

Chamberlain, Clarence D.--Bellanca test pilot, endurance flier (1976)

Chanute, Octave--pioneer in engineering gliders in 19th Century (1963)

Chennault, Claire L.--leader of American Volunteer Group (Flying Tigers), developed science of pursuit aviation (1972)

Cochran (Odlum), Jacqueline--record holder, first woman to fly supersonic speed (1971)

Collins, Michael--Astronaut, member of moon landing crew (1985)

Combs, Harry B.--formed Combs Aircraft, helped develop air traffic control system (1996)

Conrad, Charles Jr.--Astronaut, second moon landing (1980)

Crawford, Frederick C.--President of Thompson Products, developed sodium-cooled valve (1993)

Crossfield, A. Scott--X-15 test pilot, holder of speed records (1983)

Cunningham, Alfred A.--founder of USMC Aviation Branch (1965)

Curtiss, Glenn H.--pioneer flier, aircraft and engine builder (1964)

Dargue, Herbert A.--WWI aviator, first air/ground communication, helped develop Norden Bombsight (1997)

Davis Jr., Benjamin O.--first black West Point grad, first black pilot to earn wings, Tuskeegee Airmen, first black Air Force General (1994)

deSeversky, Alexander P.--designer of first modern fighter aircraft; founded Seversky Aircraft, which later became Republic (1970)

Doolittle, James H.--Doctor of Science degree from M.I.T., pioneer in instrument flight, high-octane gasoline, air racer, led first air raid on Tokyo in World War II, Medal of Honor winner (1967)

Douglas, Donald W.--founder of Douglas Aircraft (1969)

Draper, Charles S.--developer of inertial navigation systems (1981)

Eaker, Ira C.--Commander in Chief Allied air forces in Mediterranean, co-initiator of Allied bombing plan in WW II (1970)

Earhart (Putnam), Amelia--record holder, first woman to fly solo across the Atlantic, Pacific, first woman to fly around the world (1968)

Eielson, C. Benjamin--frst pilot to cross Arctic east-west, (1985)

Ellyson, Theodore G.--first U.S. Navy pilot, involved in floatplane testing (1964)

Ely, Eugene B.--made the first shipboard takeoff and landing (1965)

Everest, Frank K. --General, commanded troops in WWII, USAF jet speed record holder in X-2 (1989)

Fairchild, Sherman M.--pioneer designer of aircraft and cameras for aerial photography (1979)

Fleet, Reuben H.--founder of Consolidated Aircraft (1975)

Fokker, Anthony H.G.--founder of Fokker Aircraft (1980)

Ford, Henry--automobile and aircraft manufacturer (1984)

Foss, Joseph J.--World War II ace (26), Governor of South Dakota, directed Air Force Academy (1984)

Foulois, Benjamin D.--first U.S. Army aviator (1963)

Frye, William J.--funded Aerocorp, later became TWA (1992)

Gabreski, Francis S.--World War II and Korean ace (1978)

Gilruth, Robert R.--head of NACA/NASA, directed Manned Spacecraft Center and Apollo 11 (1994)

Gentile, Dominic S.--WWII "Ace of Aces" with 30 enemy planes downed (1995)

Glenn, John H. Jr.--Mercury astronaut, first American to orbit earth (1976)

Goddard, George W. --early pioneer of aerial mapping and photo reconnaissance (1976)

Goddard, Dr. Robert H.--American rocket pioneer (1966)

Godfrey, Arthur--entertainer, promoter of aviation (1987)

Goldwater, Sen. Barry M.--aviator, U.S. Senator, helped create FAA, supported NASA and NASM (1982)

Grissom, Virgil I.--Mercury astronaut, second American in space (1987)

Gross, Robert E.--Lockheed backer, financed Model 10 (1970)

Grumman, LeRoy R.--founder of Grumman Aircraft (1972)

Guggenheim, Harry F.--Naval aviator and financial supporter of aeronautics, funded Goddard's early rocket experiments (1971)

Haughton, Daniel J.--Chairman of Lockheed (1987)

Hegenberger, Albert F.--first pilot non-stop US to Hawaii, 1927; first instrument-only solo flight without check pilot (1976)

Hieinemann, Edward H.--Douglas designer (D-558-2, A-4) (1981)

Hoover, Robert A.--test pilot and exhibition pilot (1988)

Hughes, Howard R.--aviator, industrialist (Hughes Aircraft, TWA) (1973)

Ingalls, David S.--only Navy ace of WWI (1983)

James, Daniel Jr.--General, member of Tuskegee Airmen, flew 101 missions in Korea (1993)

Jeppesen, Elrey B.--mapped navigation routes (1990)

Johnson, Clarence L. "Kelly"--Lockheed designer (Electra through SR-71), originator of "Skunk Works" (1974)

Johnston, Alvin M. "Tex"--test pilot at Bell Aircraft and Boeing (1993)

Jones, Thomas V.--Lockheed executive, established Aeronautical Institute of Technology in Brazil (1992)

Kenney, George C.--Commander of 49th Fighter Group in WWII, organized SAC (1971)

Kettering, Charles F.--inventor, vice-president of GM (1979)

Kindelberger, James H. "Dutch"--co-designer of DC-3, founder of North American Aviation (1972)

Kittenger, Joseph W., Jr.--balloonist, first man to exceed speed of sound without an aircraft (1997)

Knabenshue, A. Roy--early dirigible and exhibition pilot (1965)

Knight, William J.--USAF X-15 pilot to Mach 6.72 (4,594 mph) absolute speed record (1988)

Lahm, Frank P.--Army's first pilot, winner of first Gordon Bennett Trophy (1963)

Langley, Samuel P.--Smithsonian Secretary, conducted early flight experiments (1963)

Lear, Willam P. Sr.--aviator, inventor, founder of Lear Jet (1978)

LeMay, Curtis E.--aviator, Commander of SAC, AF Chief of Staff (1972)

LeVier, Anthony W. "Tony"--Lockheed Aircraft test pilot (1978)

Lindbergh, Anne M.--aviator with husband Charles (1979)

Lindbergh, Charles A.--first person to solo Atlantic (1967)

Link, Edwin A.--inventor of flight simulator (1976)

Lockheed, Allan H.--founder of Lockheed Aircraft (1986)

Loeing, Grover--first American aeronautical engineer, manufacturer of amphibious aircraft (1969)

Luke, Frank Jr.--"balloon buster" and WWI ace (21), Medal of Honor recipient (1975)

Macready, John A.--made first transcontinental flight (1968)

Macready, Paul B.-- World Soaring Champion, designer of first man-powered airplane (1991)

Martin, Glenn L.--founder of Martin Aircraft (1966)

McCampbell, David--Navy "Ace of Aces," Medal of Honor recipient (1996)

McDonnell, James S.--founder of McDonnell Aircraft (1977)

Meyer, John C.--leading American ace in Europe, Commander in Chief of SAC (1988)

Mitchell, William "Billy"--U.S. Army pilot, pioneer of strategic bombing (1966)

Mitscher, Marc A.--U.S. Navy officer, commander of Hornet for Doolittle raid (1989)

Montgomery, John J.--early glider designer (1964)

Morrer, Thomas H.--Admiral and aviator, Chairman of Joint Chiefs of Staff, first officer to serve as Chief of both Atlantic and Pacific fleets (1987)

Moss, Sanford A.--GE engineer, turbocharging pioneer (1976)

Neumann, Gerhard--Developed Variable Stator Compressor system, led GE development of J79 jet (1986)

Nichols, Ruth R.--early woman pilot, first to fly non-stop NY to Miami, co-founder Ninety-Nines (1992)

Norden, Carl L.--developer of Norden bombsight (1994)

Northrop, John K.--founder of Northrop Aircraft (1974)

Pangborn, Clyde E.--early distance and exhibition flyer, Ferry Command--RAF in WWII (1995)

Patterson, William A.--head of United Airlines beginning in 1933 (1976)

Piper, William T. Sr.--founder of Piper Aircraft (1980)

Pitcarin, Harold F.--designer and builder of autogyros (1995)

Post, Wiley H.--altitude and distance record-holder; first to fly solo around the world (1969)

Read, Albert C.--pilot of NC-4, first transatlantic aircraft (1965)

Reeve, Robert C.--"Glacier Pilot," pioneered air mail route between Chile and Peru (1975)

Rentschler, Frederick B.--helped form Wright Aeronautical, founder of Pratt & Whitney (1982)

Richardson, Holden C.--helped develop catapult to launch seaplanes (1978)

Rickenbacker, Edward V.--World War I ace, Medal of Honor recipient, president of Eastern Airlines (1965)

Rodgers, Calbraith P.--first to make transcontinental flight (1964)

Rogers, Will--entertainer and aviation booster (1977)

Rushworth, Robert A.--Air Force test pilot, second to attain astronaut rating (1990)

Rutan, Elbert "Bert" L.-- "Voyager" designer (1995)

Ryan, T. Claude--founder of Ryan Aircraft (1974)

Schirra, Walter M.--Mercury, Gemini and Apollo astronaut (1986)

Schriever, Bernard A.--development director, Titan and Minuteman missles (1980)

Selfridge, Thomas E.--first Army officer to pilot a plane, first aviation casualty (1965)

Shepard, Alan B. Jr.--Mercury 7 astronaut, first American in space (1977)

Sikorsky, Igor I.--builder of first successful multi-engine aircraft and of first successful American helicopter (1968)

Six, Robert F.--founder and president of Continental Airlines (1980)

Slayton, Donald K.--Mercury 7 astronaut, directed program at NASA, directed first shuttle (1996)

Smith, C.R.--president of American Airlines, encouraged Douglas to build DC-3 (1974)

Spaatz, Carl A.--first Chief of Staff of USAF (1967)

Sperry, Elmer A.--inventor of gyroscope for ship navigation (1973)

Sperry, Lawrence B.--developer of gyro-stabilizer, autopilot (1981)

Stafford, Thomas P.--member of second group of NASA astronauts--Gemini and Apollo (1997)

Stanley, Robert M.--developed downward ejection seat and automatic release lap belts (1990)

Stapp, John P.--established preventative measures for high altitude bends, proved ejection seat safe at supersonic speeds (1985)

Stearman, Lloyd C.--co-founder Travel Air, founder of Stearman Aircraft, designer of Lockheed Electra (1989)

Taylor, Charles E.--assistant to Wright Brothers who helped build first Wright engine (1965)

Thomas, Lowell--explorer, journalist (1992)

Tibbets, Paul W., Jr.--pilot of "Enola Gay" which dropped first atomic bomb (1996)

Towers, John H.--one of the first U.S. Navy pilots, flew early floatplanes in test (1966)

Trippe, Juan T.--founder of Pan American Airways (1970)
Turner, Roscoe--air race pilot, WWII instructor (1975)
Twining, Nathan F.--first USAF officer to be appointed chairman of Joint Chiefs of Staff (1976)
Vandenburg, Hoyt S.--Air Force Chief of Staff, deputy commander of AEF in WWII (1991)
von Braun, Wernher--scientist, head of early U.S. space effort (1982)
von Karman, Theodore--aeronautics scientist, helped found the Guggenheim and Jet Propulsion Laboratories (1983)
von Ohain, Hans P.--inventor of first successful jet engine (1990)
Vought, Chance M.--founder of Chance Vought Aircraft (1989)
Wade, Leigh--test pilot, participant in first round-the-world flight (1974)
Walden, Henry W.--early aircraft designer and builder (1964)
Wells, Edward Curtis-- Chief Engineer, VP & GM at Boeing, directed development of B-17, B-29 (1991)
Wilson, Thornton A.--helped develop B-52, directed production of 707, 727 and 737 (1983)
Woolman, Collett E.--Pioneer in aerial application, founded first crop dusting service--which later became Delta Airlines (1994)
Wright, Orville--co-inventor and first person to fly a heavier- than-air machine (1962)
Wright, Wilbur--co-inventor of the first successful airplane (1962)
Yeager, Charles--first pilot to exceed the speed of sound, twice the speed of sound (1973)
Young, John W.--astronaut, commander of Apollo 16 to the moon (1988)

U.S. Aerobatics
U.S. Aerobatic Champions, Unlimited Class
International Aerobatic Club

Since 1974, the U.S. National Aerobatic Champion in the Unlimited category has been presented with the Mike Murphy Trophy by the International Aerobatic Club.

Year	Men	Women	Year	Men	Women*
1960	Harold Krier	Joyce Case	1977	Leo Loudenslager	
1961	Harold Krier	Joyce Case	1978	Leo Loudenslager	
1962	Duane Cole		1979	Henry Haigh	B. Stewart
1963	Frank Price		1980	Leo Loudenslager	B. Stewart
1964	Duane Cole	Mary Aikens	1981	Leo Loudenslager	
1965	Harold Krier	Joyce Case	1982	Leo Loudenslager	Patti John-son Nelson
1966	Bob Herendeen	Margaret Ritchie	1983	Kermit Weeks	
1967	Charlie Hillard	Mary Gaffaney	1984	Kermit Weeks	Julie Pfile
			1985	Kermit Weeks	Julie Pfile
1968	Harold Krier	M. Gaffaney	1986	Clint McHenry	Julie Pfile
1969	Bob Herendeen	M. Gaffaney	1987	Clint McHenry	Julie Pfile
1970	Gene Soucy	M. Gaffaney	1988	Tom Jones	
1971	Gene Soucy	M. Gaffaney	1989	Clint McHenry	
1972	Gene Soucy	M. Gaffaney	1990	Peter Anderson	
1973	Tom Poberenzy	M. Gaffaney	1991	Patty Wagstaff	
1974	Art Scholl	M. Gaffaney	1992	Patty Wagstaff	
1975	Leo Loudenslager	Betty Stewart	1993	Patty Wagstaff	
			1994	Phillip Knight	
1976	Leo Loudenslager	Betty Everest	1995	Michael Goulian	
			1996	*Not flown*	

* No separate women's championship after 1987

U.S. World Aerobatic Teams

Year--host country; team members (finishing position)
1960--Czechoslovakia; Frank Price (24)
1962--Hungary; Duane Cole, Rodney Jocelyn, Lindsey Parsons (4)
1964--Spain; Harold Krier, Robert Nance, Frank Price (8)
1966--USSR; Bob Herendeen, Charlie Hillard, Harold Krier, Art Scholl (6)
1968--E. Germany; Marion Cole, Mary Gaffaney, Bob Herendeen, C. Hillard, Harold Krier, Art Scholl (3)
1970--Great Britain; Mary Gaffaney, Bob Herendeen, Charlie Hillard, Art Scholl, Gene Soucy (1)
1972--France; Mary Gaffaney, Charlie Hillard, Tom Poberenzy, Carolyn Salisbury, Art Scholl, Gene Soucy, Bill Thomas (1)
1974--no contest held; Henry Haigh, Clint McHenry, Tom Poberenzy, Art Scholl, Gene Soucy, Bill Thomas
1976--USSR--Bob Davis, Betty Everest, Henry Haigh, Leo Louden slager, Clint McHenry, Bill Thomas (4)
1978--Czechoslovakia; Bob Carmichael, Henry Haigh, L. Loudenslager, Randall Melton, Kermit Weeks (2)
1980--United States; Tom Collier, Henry Haigh, Patti Johnson, Leo Loudenslager, Randall Melton, Paula Moore, Betty Stewart, Kermit Weeks (1) 1982--Austria; Henry Haigh, Leo Loudenslager, Linda Meyers, Patti Johnson Nelson, Jim Roberts, Brigitte de St. Phalle, Betty Stewart, Kermit Weeks, Bill Witt (2)
1984--Hungary; Alan Bush, Gene Beggs, Harold Chappell, Bob Davis, Henry Haigh, Linda Meyers, Julie Pfile, Debbie Rihn, Brigitte de St. Phalle, Kermit Weeks (1)
1986--Great Britain; Gene Beggs, Harold Chappell, Ellen Dean, Henry Haigh, Clint McHenry, Linda Meyers, Julie Pfile, Debbie Rihn, Patty Wagstaff, Kermit Weeks (2)

Category winners, U.S. National Aerobatic Championships
Source: International Aerobatic Club

	Primary	*Advanced*	
1968	Allen Cross	Gene Soucy	
1969	Gayle Bishop	Frank Christensen	
1970	J.F. Carter	Tom Poberezny	
1971	Bob Carmichael	Casey Kay	

	Sportsman	*Intermediate*	*Advanced*
1972	Gene Olsen	Bud Judy	Bob Carmichael
1973	Craig Shaw`	John Nyquist	Verne Jobst
1974	John Keplinger	Greer Parramore	Chuck Carothers
1975	Bill McCollough	Tom Adams	Chipper Melton
1976	John Morrissey	Dana Vihlen	Jerry Thomas
1977	John Morrissey	Ken Larson	Ron Cadby
1978	Verlyn Hagen	Gene Beggs	Bill Sandusky
1979	Nellie Sanchez	Ken Larson	Gene Beggs
1980	Mike Sharp	Sam Maxwell	Jack Gladish
1981	Lewis Woolery	Charlie Larkey	Skip Prest
1982	Perry Rhoades	Tom Bishop	Clint McHenry
1983	Arlyn Cook	Mike Anderson	Lonnie English
1984	Phil Sisson	*Not flown*	Don Rhynalds
1985	Chip Corley	Phil Sisson	Jimmy Goggin
1986	Wayne Fuller	Steve Van Eck	Tom Adams
1987	Dick Lewis	Mike Mays	Tom Adams
1988	Jim Eiland	Randy Henderson	Sean Tucker
1989	Rich White	Jill Butterworth	Robert Armstrong
1990	Michele Thonney	Bill Larson	Mike Goulian
1991	Matt Morrissey	Bill Williamson	Don Johnson
1992	Jan Jones	Stratton Gillis	Tom Adams
1993	Joe Dobranskyh	Gerry Molidor	Marta Meyer
1994	Karen McCullough	Bradley Vidrine	John Morrissey
1995	Patrick Clyne	Suzanne Owen	Gerry Molidor
1996	Tom Womack	*Not flown*	Gerry Molidor

Aerocryptographics

The Aresti Aerocryptographic System, or Aresti Key, was devised by Count Jose Aresti, a long-time Spanish aerobatic pilot. In addition to providing the pilot a shorthand version of the performance routine, the system provides a universal method of judging in competition.

Each figure is assigned a number for identification, and is assigned a coefficient of difficulty called the K-factor. Judges award a grade of one to ten for each maneuver depending upon the skill of its execution, and this grade is multiplied by the maneuver's K-value to determine a contestant's score. In addition, the Aresti System also assigns K-values to the harmony, rhythm and diversity of an aerobatic sequence, its "framing" within a prescribed airspace zone, and provides penalties for violation of time limits.

In national and international competition, a contestant will normally fly three programs: The Known Obligatory, about 15 maneuvers which are known several months in advance of the contest; The Unknown Obligatory, agreed upon by team members no more than 24 hours before the competition and not rehearsed; and The Freestyle, which each contestant plans for individually, and the diagram of which is provided to judges before it is flown. The two obligatory programs are limited to six and eight minutes, respectively; the Freestyle has a 10-minute time limit.

Competitions are flown in a zone, or "box" which depends on the speed of the aircraft. Slower aircraft can operated within an 800-ft. by 1,200-ft. zone, while faster aircraft require a 2,000 by 4,000-ft. space. Maximum agl upper altitude limits are 2,225 ft. for Known Obligatory, 2,600 ft. for Unknown Obligatory, and 3,250 ft. for Freestyle, and minimum altitude for all is 325 ft. All aircraft are equipped with barographs to ensure that limits are observed.

Some representative symbols of the Aresti Aerocryptographic System

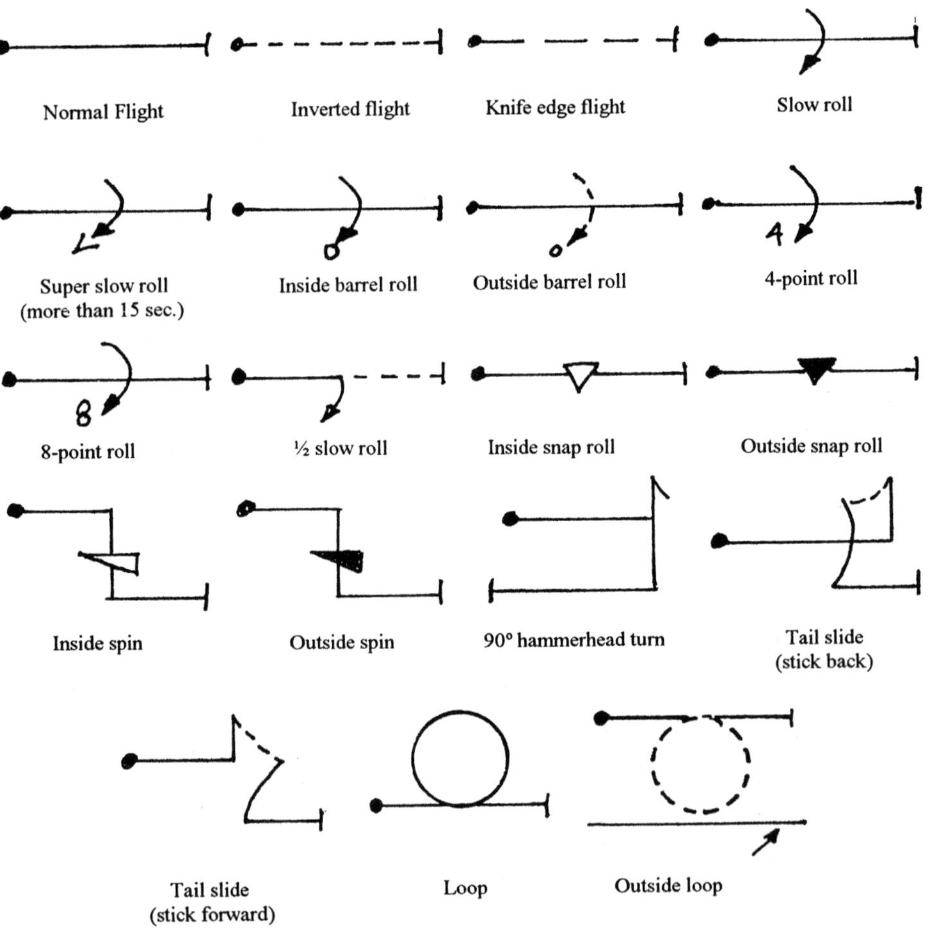

Charles A. Lindbergh popularized aviation with his epic May 1927 transatlantic flight.

AIRCRAFT MANUFACTURERS
Past and Present
Founding date, location, mergers and acquisitions

Aerocar--founded in 1948 by Molt Taylor to develop flying automobile

Aero Commander (Aero Design & Engineering Co.)--founded in 1944 in Culver City, Calif. by Ted Smith. Built first aircraft at Bethany, Okla. in 1951, acquired by Rockwell Standard Corp. in 1967, merged with North American to become North American Rockwell and later Rockwell International. Bethany facility acquired by Gulfstream Aerospace.

Aeronca (Aeronautical Corp. of America)--founded in November 1928 at Cincinnati, Ohio, changed to Aeronca Aircraft Corp in 1941 and Aeronca Manufacturing Corp. in 1952, when it ceased building aircraft.

Aerospatiale--formed by French government as a result of the January 1970 merger of Sud-Aviation , Nord-Aviation and SEREB, which had over the years absorbed such famous companies as Bleriot (1921), Breguet (1911), CAMS (1921), Caudron (1912), Dewoitine (1920), Farman (1909), Hanriot (1908), Morane-Saulnier (1911), Nieuport (1910),Potez (1922),SPAD (1913), and Voisin (1907). The U.S. marketing arms of Aerospatiale, are Eurocopter and SOCATA *(see genealogy section)*.

Agusta--formed in Italy in 1907 by Giovanni Agusta. Currently produces helicopters of its own design.

Airspeed--founded in 1931 to build Ferry aircraft, reformed at Plymouth, England in 1934 to develop its own designs. A division of de Havilland.

Air Tractor--founded 1972 in Olney, Tex. by Leland Snow, who had earlier founded Snow Aeronautical, which he sold to Rockwell Aircraft Commander Division, who built it as the Thrush Commander.

Albatros--established 1909 at Berlin, merged with Focke-Wulf in 1930.

Allison--founded as Indianapolis Speedway Team Co. in 1915, became Allison Experimental in 1919 to convert surplus Liberty engines for civil use. It was purchased by Fisher Brothers Investment in 1929 and sold to General Motors as the Allison Division. Began turbine production in 1945 with the GE J33. Now a division of Rolls-Royce Plc.

American Eagle--established by E.E. Porterfield in 1925 at Kansas City, Mo., suspended operations in 1930, merged with Lincoln Aircraft in 1931.

Antonov--Russian design bureau of Oleg Konstantinovich Antonov established in 1924.

Arado--founded in 1917 at Friedrichshafen, Germany, built the first jet bomber at end of WWII.

Armstrong Whitworth--established in 1914 in England, merged to form Hawker Siddeley Aviation in 1963 *(see genealogy section)*.

Auster--before 1946 known as Taylorcraft Aeroplanes (England).

Avro--Founded by A.V. Roe in 1911, merged with Hawker Siddeley in 1963 to form Avro-Whitworth Division..

Beech--founded by former Travel Air co-owner Walter Beech in 1932 in Wichita, Kan.; acquired by Raytheon 1980, renamed Raytheon Aircraft 1995.

Bell--Founded by former Consolidated VP and GM Lawrence D. Bell in 1935 in Buffalo, N.Y.; moved to San Diego during World War II; moved to Fort Worth, Tex. 1947; became Bell Aerospace 1960; Bell Helicopter Textron 1976.

Bellanca--1912; Roos-Bellanca 1922; Wright Bellanca 1923; Columbia Aircraft 1926; Bellanca Aircraft Corp. until 1951; reorganized under Bellanca, Inc. 1964; acquired along with Champion Aircraft by Bellanca Sales Co., a subsidiary of Miller Flying Service, Plainview, Tex. 1970.

Blériot--formed July 25, 1909 by Louis Blériot, acquired SPAD in 1920.

Blohm und Voss--German shipbuilding firm that began building maritime aircraft in early 30s.

Boeing--Founded by William E. Boeing as Pacific Aero Products, Seattle, Wash. July 15, 1916; acquired Stearman Aircraft 1934, Vertol March 31, 1960, DeHavilland of Canada, 1986, McDonnell Douglas, 1996.

British Aerospace--The English company which was formed with the merger of Hawker Siddeley Aviation and British Aircraft Corp. Hawker Siddeley included AVRO (1911). Armstrong-Whitworth (!920), Gloster (1917), Sopwith (1912), Hawker (1920), Blackburn (1910), Folland (1939), Airspeed (1931) and De Havilland (1920). BAC was made up of Bristol (1910), English Electric (1918), Vickers (1911), Supermarine (1916), Percival (1931), Scottish Aviation (1936), and Auster (1946)*(see genealogy section)*.

Canadair--formed in 1944 from Aircraft Division of Canadian Vickers Ltd. Purchased in 1946 by Electric Boat of New York and became subsidiary of General Dynamics. Company purchased by Canadian government in 1975.

Cessna--Founded December 31, 1927 by former Travel Air co-owner Clyde Cessna in Wichita, Kan.; acquired by General Dynamics in 1985; sold to Textron in 1992.

Champion--founded 1955 in Osceola, Wis. to produce Aeronca design. Acquired by Bellanca in 1964.

Convair--founded by Reuben Fleet in 1923 as the Consolidated Aircraft Corp. from remains of Gallaudet and Dayton-Wright. Located in the former Buffalo, N.Y. Curtiss factory, the company absorbed Thomas-Morse. In 1934, Consolidated moved to San Diego. In 1940, Fleet bought Hall-Aluminum Aircraft Corp., and the following year merged with Vultee, resulting in the Consolidated Vultee Aircraft Corp., later named Convair. Floyd Odlum's

Convair cont.

Atlas Corp. bought Convair in 1947, and early in 1953 it was sold to General Dynamics *(see genealogy section).*

Continental Motors Corp.--founded in 1900 in Detroit, began building aircraft engines in 1929, reorganized in 1969 as Teledyne Continental Motors.

Culver--formed in 1939 at Columbus, Ohio and took over manufacturing and sales of Dart. First product was Al Mooney-designed Cadet. Moved to Wichita, Kan. in 1941 to build pilotless aircraft. Assets acquired by Superior Aircraft in 1946.

Curtiss--established 1910, became Curtiss Airplane Division of Curtiss-Wright Corp. in 1929, terminated aircraft production in 1966 *(see genealogy section).*

Dassault--The name assumed by French designer-builder Marcel Bloch during WWII. The Dassault company began operation in 1945. Notable products are Mirage fighters and Falcon business jets.

de Havilland--Geoffrey de Havilland designed the DH series aircraft for at the Royal Aircraft Factory before founding de Havilland Aircraft in 1920 to build lightweight civil aircraft. Later designs include the WWII Mosquito and peacetime Comet jet airliner. Merged with the Hawker Siddeley Group of British Aerospace in the 70s *(see genealogy section).*

Douglas--Established by Donald W. Douglas in 1920, reorganized in 1928 at Clover Field, Santa Monica, Calif.; merged with Northrop Aircraft in 1937, and with McDonnell Aircraft April 28, 1967, bought by Boeing in 1996.

Embraer--Founded in 1970 in Brazil, Empresa Brasiliera de Aeronáutica builds several Piper types in addition to original designs including the Bandeirante turboprop commuter.

Ercoupe--An acronym for Engineering & Research Corp., formed in 1930, it built the 1937 Fred Weick-designed "easy-fly" two-control airplane. Fornaire acquired production rights in 1955, which were then bought by Alon in 1963, and Mooney Aircraft in 1967.

Fairchild--Formed by Sherman M. Fairchild, who merged Fairchild Airplane Manufacturing Corp. and Kreider-Reisner Aircraft Co. in 1925; acquired by Aviation Corp. 1929; re-acquired by Fairchild in 1931; became Fairchild Stratos 1961, Fairchild Hiller 1964; Fairchild Industries 1971; acquired Republic Aviation 1965, Swearingen Aircraft, San Antonio, Tex. 1971; that unit was purchased and reorganized as Fairchild Aircraft 1990.

Fairey--founded during World War I by Richard Fairey at Hayes, UK, moved to Harmondsworth in 1929 and merged with Westland in 1960.

Focke-Wulf--Formed in 1924 at Bremen, Germany. Reorganized after WWII, became part of VFW after 1973 merger with Wesser and Heinkel.

Garrett Corp.--Founded in 1936 as Aircraft Tool & Supply at Glendale, Calif., incorporated as Garrett Corp. in 1939. The AiResearch Division was formed in 1941. The company presently is owned by the Signal Corp.

General Dynamics--Founded May 1923 in Buffalo, N.Y. by Maj. Reuben H. Fleet as the Consolidated Aircraft Corp., the company acquired Thomas-Morse in 1929 and Hall Aluminum in 1935; it merged with Vultee in March 1943 to become the Consolidated Vultee Aircraft Corp.; its name was shortened to Convair after World War II, and became a division of the General Dynamics conglomerate March 1, 1954 *(see genealogy section)*.

General Electric--Founded in 1892, GE was an early turbocharging pioneer and the first American company to build a turbojet, in 1941.

Grumman--Founded by Leroy R. Grumman December 6, 1929 at Farmingdale, N.Y.; acquired American Aviation 1968.

Gulfstream--A division of Grumman in Savannah, Ga. which was purchased by American Jet Industries headed by Allen Paulson September 1, 1978. After selling the division to Chrysler Corp. in the early 80s, it was re-acquired by Paulson in 1989.

Ilyushin--Soviet design bureau formed in the 30s.

Israel Aircraft Industries--Established in 1953 as repair and maintenance organization, began building aircraft under license, and in 1967 acquired Rockwell's Jet Commander and manufactures and markets it as the Westwind series.

Lake Aircraft Corp.--Founded in 1959 in Sanford, Me. to produce the Colonial Skimmer amphibian. Purchased in 1980 by Armand Rivard.

Learjet--Founded by William P. Lear in 1963, at Wichita, Kan.; acquired Brantley Helicopters and was purchased by a division of the Gates Corp. and became Gates Learjet 1967; sold to Integrated Resources, Inc., 1987, acquired by Bombardier 1990.

Lockheed--Founded by Allen and Malcolm Loughead at Santa Barbara, Calif. in 1913; the company name was changed to Lockheed and moved to Hollywood in 1926, and again to Burbank in 1928; became part of Detroit Aircraft 1929; Lockheed Bros. 1931; Lockheed Aircraft Corp. 1941. Merged with Martin 1996 *(see genealogy section)*.

LTV--Created by the 1960 merger of TEMCO (Texas Manufacturing Co.--manufacturer of the Globe Swift-Temco) with electronics marketer James Ling and Chance Vought Manu-facturing *(see genealogy section)*.

Luscombe--Don Luscombe's first airplane was built by the Central States Aero Co., which became the Mono Aircraft Co. located in Moline, Ill. in 1928. Luscombe set up his own company in 1934 at Kansas City and located in Dallas, Tex. after WWII. The company was bought by Texas Engineering & Manufacturing (Temco) in the Fifties.

Lycoming Motors--founded in 1908 to supply auto engines. Its first aircraft engines were produced in 1929, and the company was acquired by AVCO in 1932.

Martin--Founded in 1909 by Glenn L. Martin in Santa Ana, Calif.; moved to Cleveland 1916, Baltimore, Md. 1929; merged with American-Marietta 1961 to form Martin Marietta, which was acquired in 1982 by the Bendix Corp. in a hostile takeover.

Maule--Founded by Belford D. Maule in 1957 at Jackson, Mich.; moved to Moultrie, Ga. 1968.

MBB--Founded in 1971 by the merger of German manufacturers Bolkow (1948), Junkers (1916), Messerschmitt (1923), HFB (1958) and Siebel AATG (1954) *(see genealogy section)*.

McDonnell--Founded July 6, 1939 at St. Louis, Mo. by James S. McDonnell; merged with Douglas Aircraft April 28, 1967, with Boeing 1996.

Meyers--Formed in 1936 at Tecumseh, Mich., acquired by Rockwell in 1965.

Mikoyan & Gurevich (MiG)--Soviet design bureau established in 1938.

Mooney--Wichita, KS, 1929-1930; Denver, Colo. 1931; Wichita, Kan. 1948-53; Kerrville, Tex. beginning in 1953; acquired Alon, Inc. 1967; acquired by American Electronics Laboratories in 1969, acquired by Butler International in bankruptcy in 1970, closed 1971-73, when Republic Steel bought assets; acquired by Armand Rivard (owner of Lake Aircraft) and Alex Couvelaire in 1984.

North American--established as a holding company in 1928, merged with Rockwell in 1967 to form North American Rockwell *(see genealogy section)*.

Northrop--Founded by John K. Northrop in California in 1929; sold to United Aircraft Corp.; Northrop Aircraft founded 1939; became Northrop Corp. 1959.

Piper--Founded by William T. Piper in 1937 at Lock Haven, Penn. from the assets of the Taylor Aircraft Corp.; acquired Stinson Div. Consolidated Vultee 1948; acquired by Bangor Punta, Lear Siegler; purchased by M.S. Millar 1987; acquired by Cyrus Eaton Group International, 1992., reorganized as the New Piper Aircraft Co.

Republic--established as Seversky in 1931, name changed to Republic in 1939, merged with Fairchild in 1965.

Robin (Avions Pierre Robin)--Formed in 1957 as Centre Est Aéronautique, named changed to Avions Pierre Robin in 1969.

Rockwell International--formed by the 1967 merger of North American Aviation and Rockwell Standard Corp. as North American Rockwell, renamed Rockwell International in 1973. Acquired Aero Commander, who in 1965 had taken over Snow Aeronautical, Meyers, and Volaircraft production, and in 1974 began selling Fuji-manufactured model as Rockwell Commander 700. In 1967, the Jet Commander design was sold to Israel Aircraft Industries, Meyers was sold in 1977, and the Snow agricultural line was acquired by Ayres.

Ryan--established 1922, became Teledyne-Ryan in 1969, ceased production in 1975.

SAAB--Svenska Aeroplane AB founded in 1937 at Trhollhätan, moved to Linköping in 1939, name changed to Saab Aktiebolag in 1965. In 1968 merged with Cania-Vabis and renamed Saab-Scania.

Seversky--established in 1931 by designer Alexander de Seversky, reorganized as Republic Aviation in 1939.

Short Bros.--The Short brothers had built balloons since 1898, and the company began building Wright biplanes under license in 1909. Although Short's is the world's oldest continually-operating aircraft manufacturer, it is now owned by the British government.

SOCATA--*Société de Construction d'Avions de Tourisme et D'Affaires* founded 1969 as a subsidiary of Sud-Aviation to produce light business and touring aircraft. Currently marketing arm of Aerospatiale general aviation aircraft *(see genealogy section)*.

Sikorsky--Founded by Igor I. Sikorsky March 5, 1923; merged with United Aircraft Corp. 1929 *(see genealogy section)*.

Stearman--founded by Lloyd Stearman in Santa Monica, Calif. in 1927. Moved to Wichita, Kan. and was acquired by Boeing in 1938.

Stinson--founded 1926 in West Detroit, Mich., acquired by E.L. Cord (Avco) in 1928, became a Vultee division in 1940.

Sukhoi--Russian design bureau established during World War II.

Taylorcraft--Formed as Taylor Brothers in 1929, acquired by W.T. Piper in 1935 and became Piper Aircraft Co. Taylor reformed as Taylorcraft in 1936 at Alliance, Ohio. The factory moved to Pittsburgh in 1954, was out of business in 1958.

Travel Air--Founded by Walter Beech, Lloyd Stearman and Clyde Cessna in 1924, purchased by Curtiss-Wright in 1930.

Vega--founded 1937, became Lockheed-Vega in 1941.

VFW-Fokker--Vereingte Flugtechnische Werke-Fokker--was founded in 1970 with the merger of German manufacturers Henschel (1934), Focke-Wulf (1924), Weser (1936), Heinkel (1921) and N.V. Fokker (1913) *(see genealogy section)*.

Vultee--founded 1939, acquired Stinson Aircraft in 1940, became Consolidated-Vultee in 1943 *(see genealogy section)*.

Wright--founded 1903, became Wright-Martin in 1916, Dayton-Wright in 1917, which was absorbed by Consolidated Aircraft in 1923, and absorbed Moth, Keystone and Travel Air and became Curtiss-Wright in 1929. The company has done no significant aviation business since 1978 *(see genealogy section)*.

GENEALOGY OF AIRCRAFT MANUFACTURERS

The charts on the following pages attempt to trace the roots of most of the world's aircraft manufacturers, by country. The founding companies are boxed, the succeeding companies are in solid black, and an attempt has been made to include the approximate date for each.

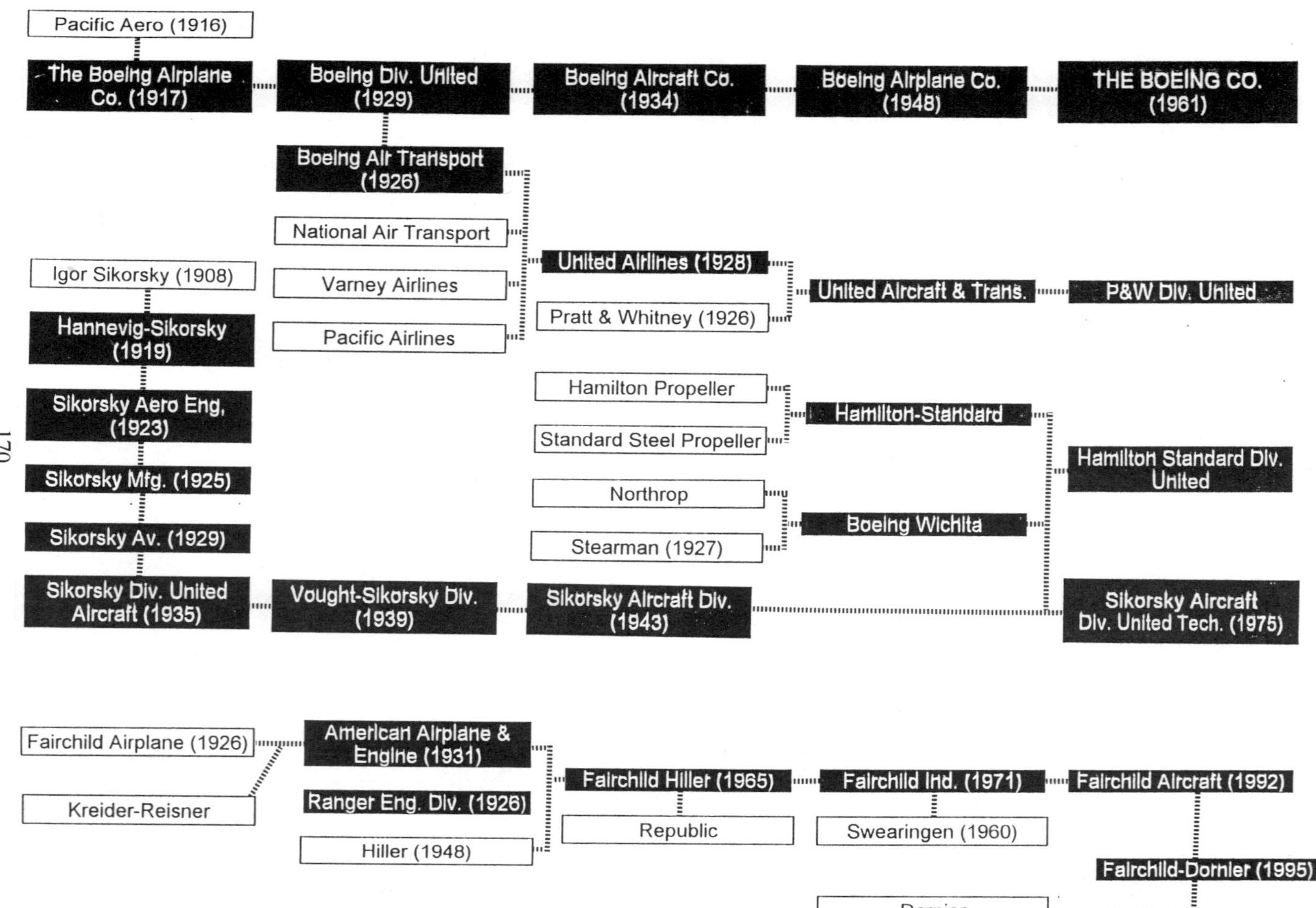

Pacific Aero (1916)
The Boeing Airplane Co. (1917)
Boeing Div. United (1929)
Boeing Aircraft Co. (1934)
Boeing Airplane Co. (1948)
THE BOEING CO. (1961)
Boeing Air Transport (1926)
National Air Transport
Varney Airlines
Pacific Airlines
United Airlines (1928)
Pratt & Whitney (1926)
United Aircraft & Trans.
P&W Div. United
Igor Sikorsky (1908)
Hannevig-Sikorsky (1919)
Sikorsky Aero Eng, (1923)
Sikorsky Mfg. (1925)
Sikorsky Av. (1929)
Sikorsky Div. United Aircraft (1935)
Vought-Sikorsky Div. (1939)
Sikorsky Aircraft Div. (1943)
Sikorsky Aircraft Div. United Tech. (1975)
Hamilton Propeller
Standard Steel Propeller
Hamilton-Standard
Hamilton Standard Div. United
Northrop
Stearman (1927)
Boeing Wichita
Fairchild Airplane (1926)
Kreider-Reisner
American Airplane & Engine (1931)
Ranger Eng. Div. (1926)
Hiller (1948)
Fairchild Hiller (1965)
Republic
Fairchild Ind. (1971)
Swearingen (1960)
Fairchild Aircraft (1992)
Fairchild-Dornier (1995)
Dornier

UNITED STATES
Lycoming Mfg. (1908)
Lycoming Aircraft Eng. Div. (1925)
Lycoming Div. AVCO (1934)
Lycoming-Spencer Div. (1941)
Lycoming Div. AVCO (1953)
AVCO Lycoming (1963)
Gallaudet (1917)
Dayton-Wright (1918)
Consolidated (1923)
Consolidated Vultee (1941)
Convair Div. Gen. Dynamics (1953)
Thomas-Morse (1918)
Vultee Aircraft (1932)
Fleet Aircraft
Hall-Aluminum
Allan & Malcom Loughead (1913)
Loughead Bros. (1916)
Loughead Aircraft (1918)
Lockheed Aircraft (1926)
Lockheed Div. Detroit Aircraft (1929)
Lockheed Aircraft (1932)
Lockheed Martin
Glenn L. Martin (1909)
Glenn L. Martin Co. (1912)
Wright-Martin (1916)
Glenn L. Martin Co. (1918)
The Martin Co. (1957)
Martin Marietta (1961)

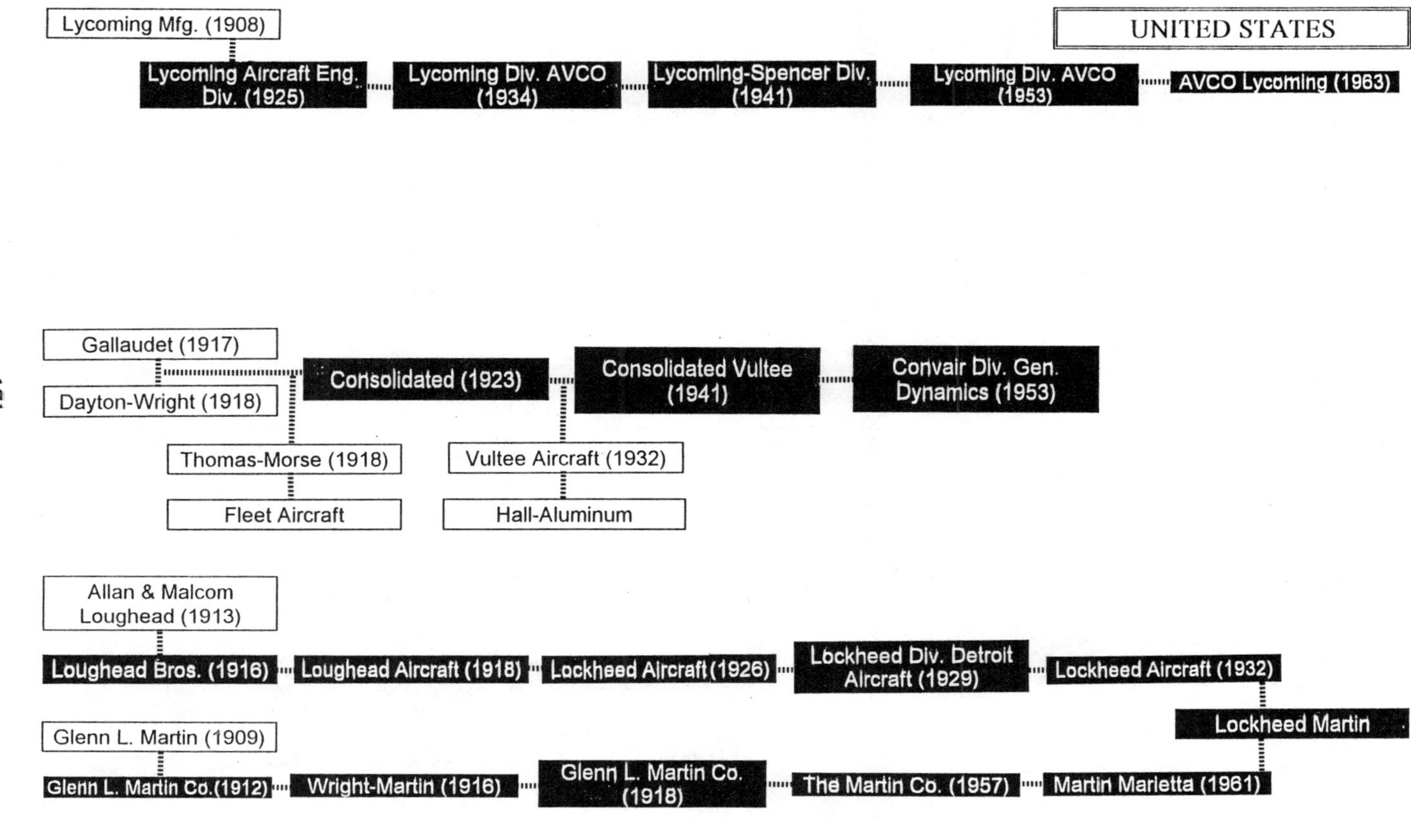

171

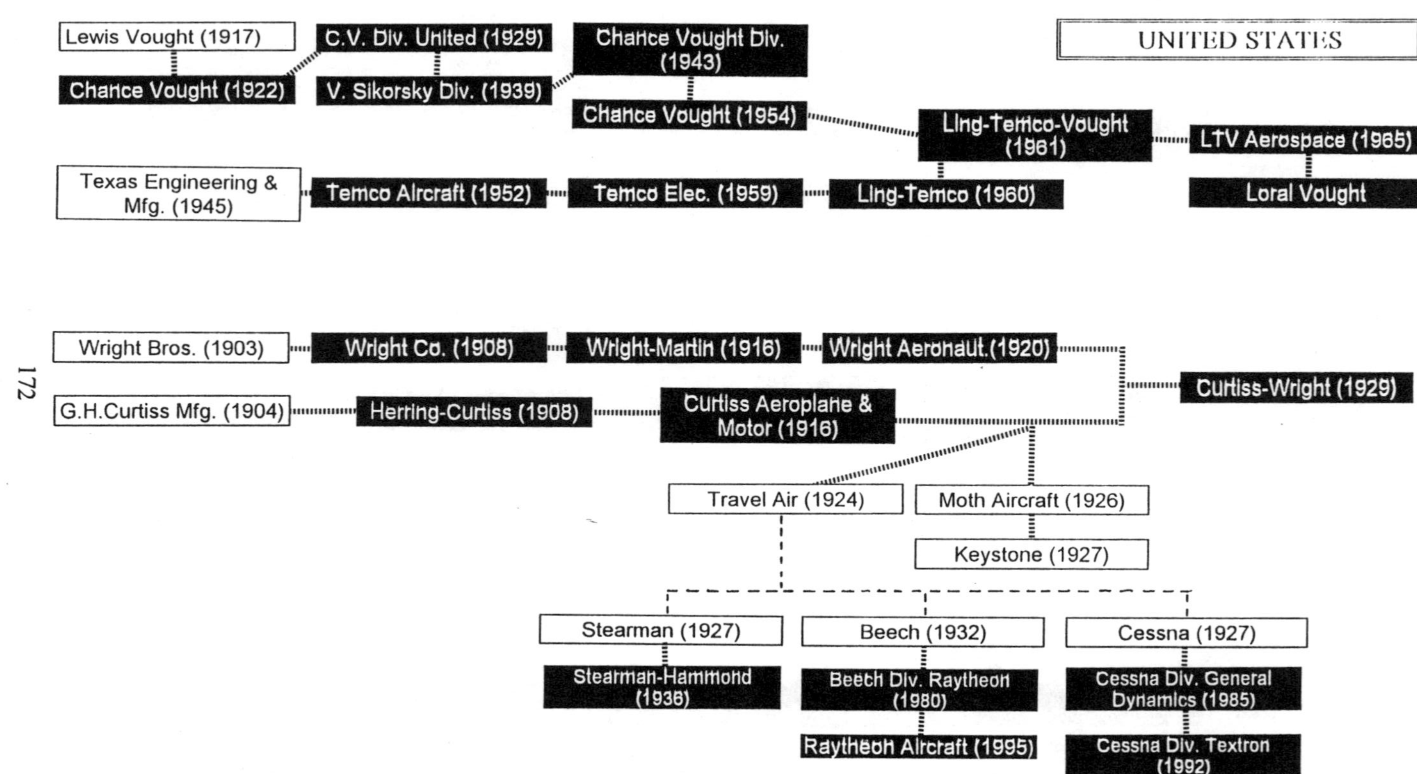
UNITED STATES
Lewis Vought (1917)
Chance Vought (1922)
C.V. Div. United (1929)
V. Sikorsky Div. (1939)
Chance Vought Div. (1943)
Chance Vought (1954)
Ling-Temco-Vought (1961)
LTV Aerospace (1965)
Loral Vought
Texas Engineering & Mfg. (1945)
Temco Aircraft (1952)
Temco Elec. (1959)
Ling-Temco (1960)
Wright Bros. (1903)
Wright Co. (1908)
Wright-Martin (1916)
Wright Aeronaut. (1920)
Curtiss-Wright (1929)
G.H.Curtiss Mfg. (1904)
Herring-Curtiss (1908)
Curtiss Aeroplane & Motor (1916)
Travel Air (1924)
Moth Aircraft (1926)
Keystone (1927)
Stearman (1927)
Beech (1932)
Cessna (1927)
Stearman-Hammond (1936)
Beech Div. Raytheon (1980)
Cessna Div. General Dynamics (1985)
Raytheon Aircraft (1995)
Cessna Div. Textron (1992)

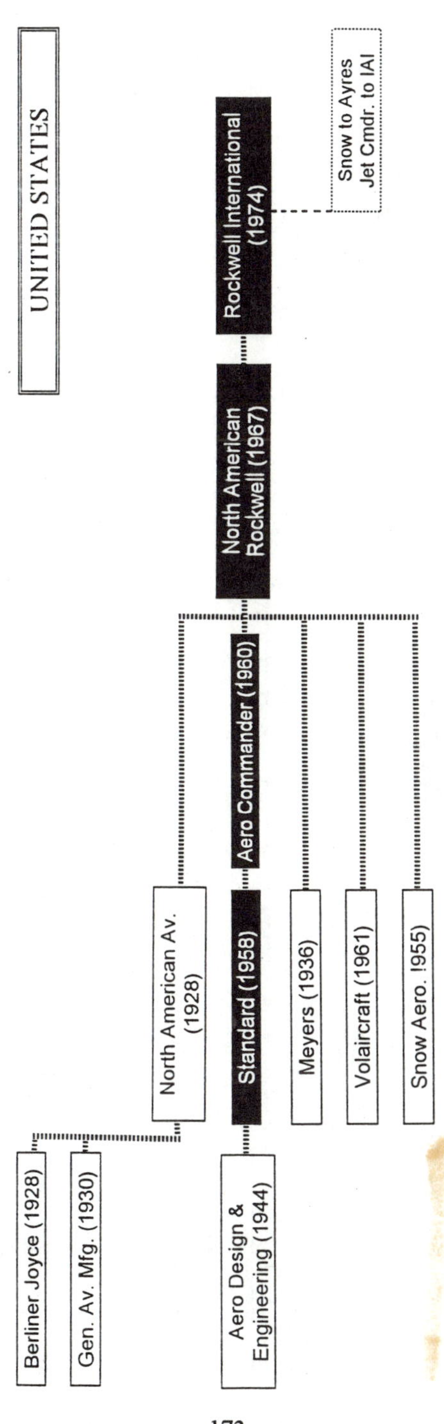

UNITED STATES
Rockwell International (1974)
Snow to Ayres
Jet Cmdr. to IAI
North American Rockwell (1967)
Aero Commander (1960)
North American Av. (1928)
Standard (1958)
Meyers (1936)
Volaircraft (1961)
Snow Aero. 1955)
Berliner Joyce (1928)
Gen. Av. Mfg. (1930)
Aero Design & Engineering (1944)

GREAT BRITAIN
A.V. Roe (Avro) (1911)
Armstrong-Whitworth (1920)
Gloustershire (1917)
Gloster (1927)
Whitworth-Gloster (1961)
Avro-Whitworth (1963)
Sopwith (1912)
H.G. Hawker (1920)
Hawker (1933)
Hawker Blackburn (1963)
Hawker Siddeley (1965)
125 bizjet to Beech
Blackburn (1910)
General (1930)
Blackburn & Gen. (1949)
Folland (1938)
Airspeed (1931)
De Havilland (1920)
DeHavilland (1953)
BRITISH AERO-SPACE PLC (1977)
Bristol Siddeley aero engines to Rolls-Royce (1964)
British & Colonial (1910)
Bristol (1920)
English Electric (1918)
British Aircraft Corp. (1960)
Pemberton-Billing (1914)
Supermarine (1916)
Vickers-Supermarine (1929)
Vickers (1940)
Vickers-Armstrong (1955)
Percival (1931)
Hunting Percival (1952)
Hunting (1957)

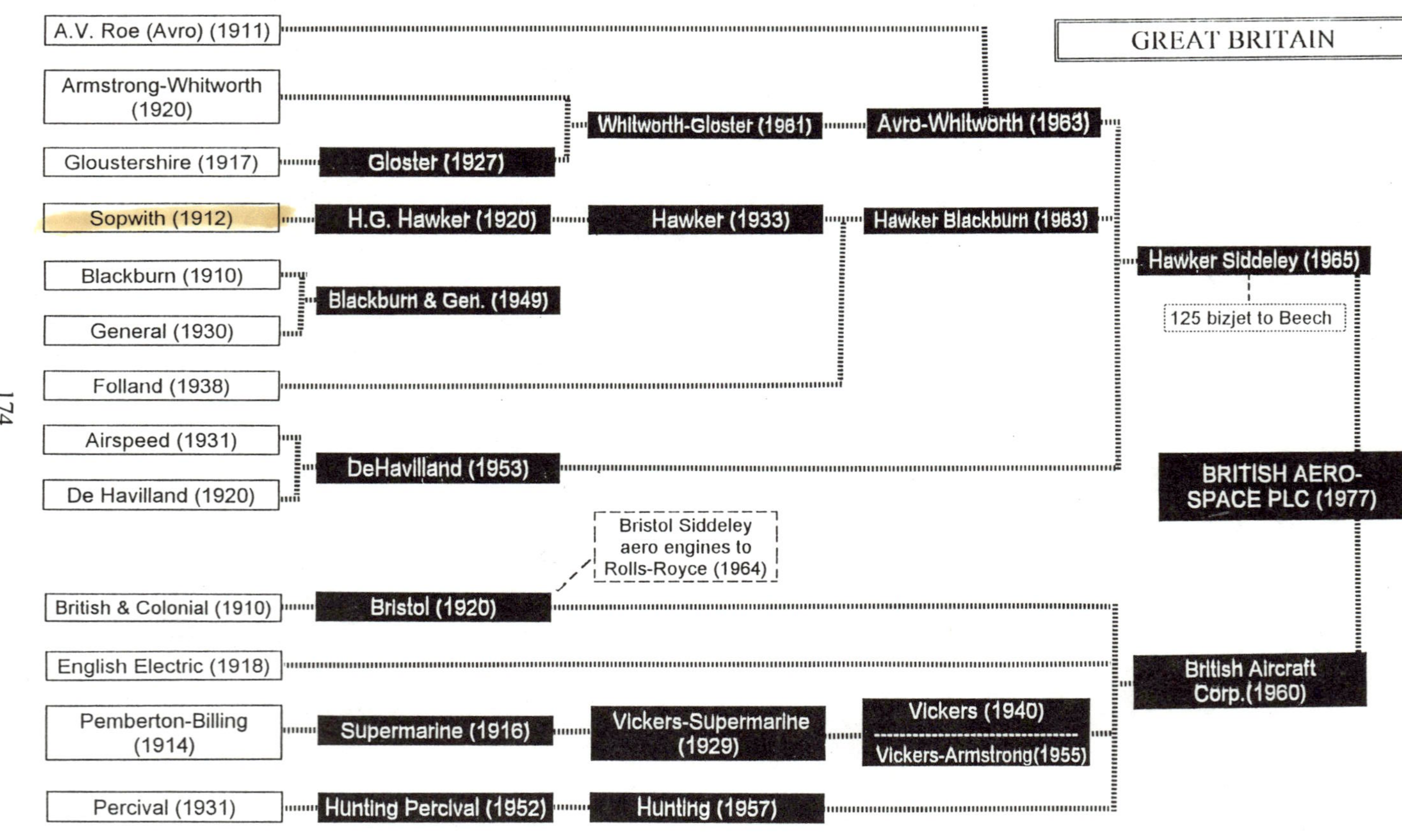

174

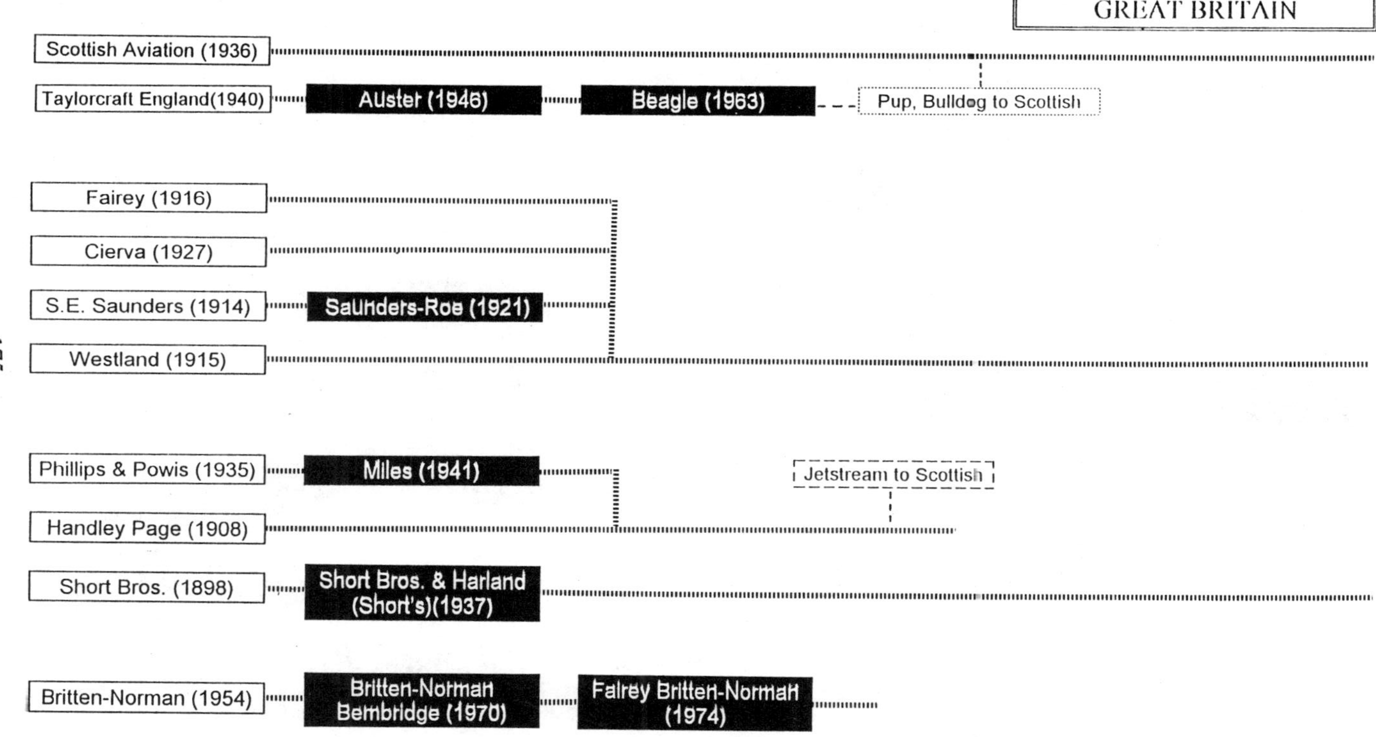
GREAT BRITAIN
Scottish Aviation (1936)
Taylorcraft England(1940)
Auster (1946)
Beagle (1963)
Pup, Bulldog to Scottish
Fairey (1916)
Cierva (1927)
S.E. Saunders (1914)
Saunders-Roe (1921)
Westland (1915)
Phillips & Powis (1935)
Miles (1941)
Jetstream to Scottish
Handley Page (1908)
Short Bros. (1898)
Short Bros. & Harland (Short's)(1937)
Britten-Norman (1954)
Britten-Norman Bembridge (1970)
Fairey Britten-Norman (1974)

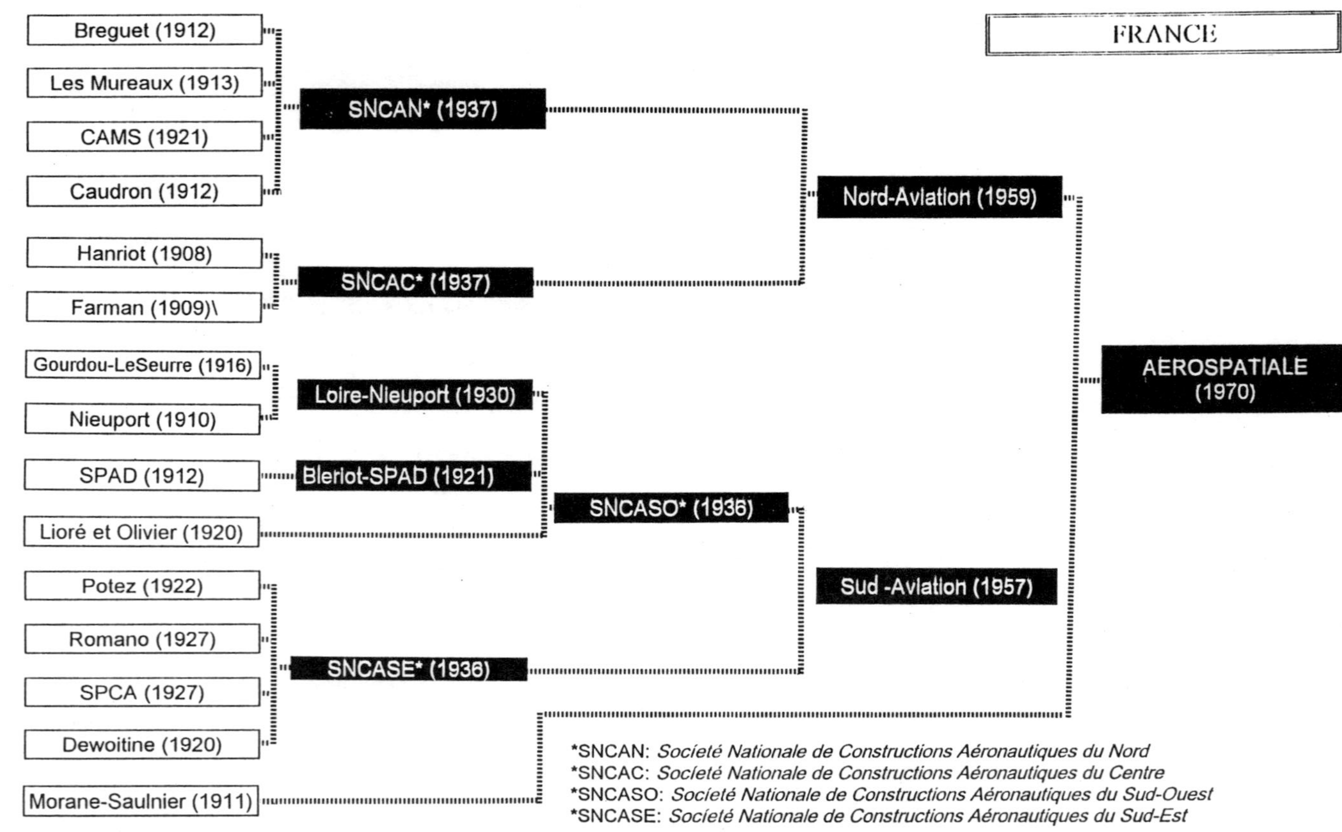
FRANCE
AÉROSPATIALE (1970)
Nord-Aviation (1959)
Sud -Aviation (1957)
SNCAN* (1937)
SNCAC* (1937)
Loire-Nieuport (1930)
Blériot-SPAD (1921)
SNCASO* (1936)
SNCASE* (1936)
Breguet (1912)
Les Mureaux (1913)
CAMS (1921)
Caudron (1912)
Hanriot (1908)
Farman (1909)\
Gourdou-LeSeurre (1916)
Nieuport (1910)
SPAD (1912)
Lioré et Olivier (1920)
Potez (1922)
Romano (1927)
SPCA (1927)
Dewoitine (1920)
Morane-Saulnier (1911)
*SNCAN: Société Nationale de Constructions Aéronautiques du Nord
*SNCAC: Société Nationale de Constructions Aéronautiques du Centre
*SNCASO: Société Nationale de Constructions Aéronautiques du Sud-Ouest
*SNCASE: Société Nationale de Constructions Aéronautiques du Sud-Est

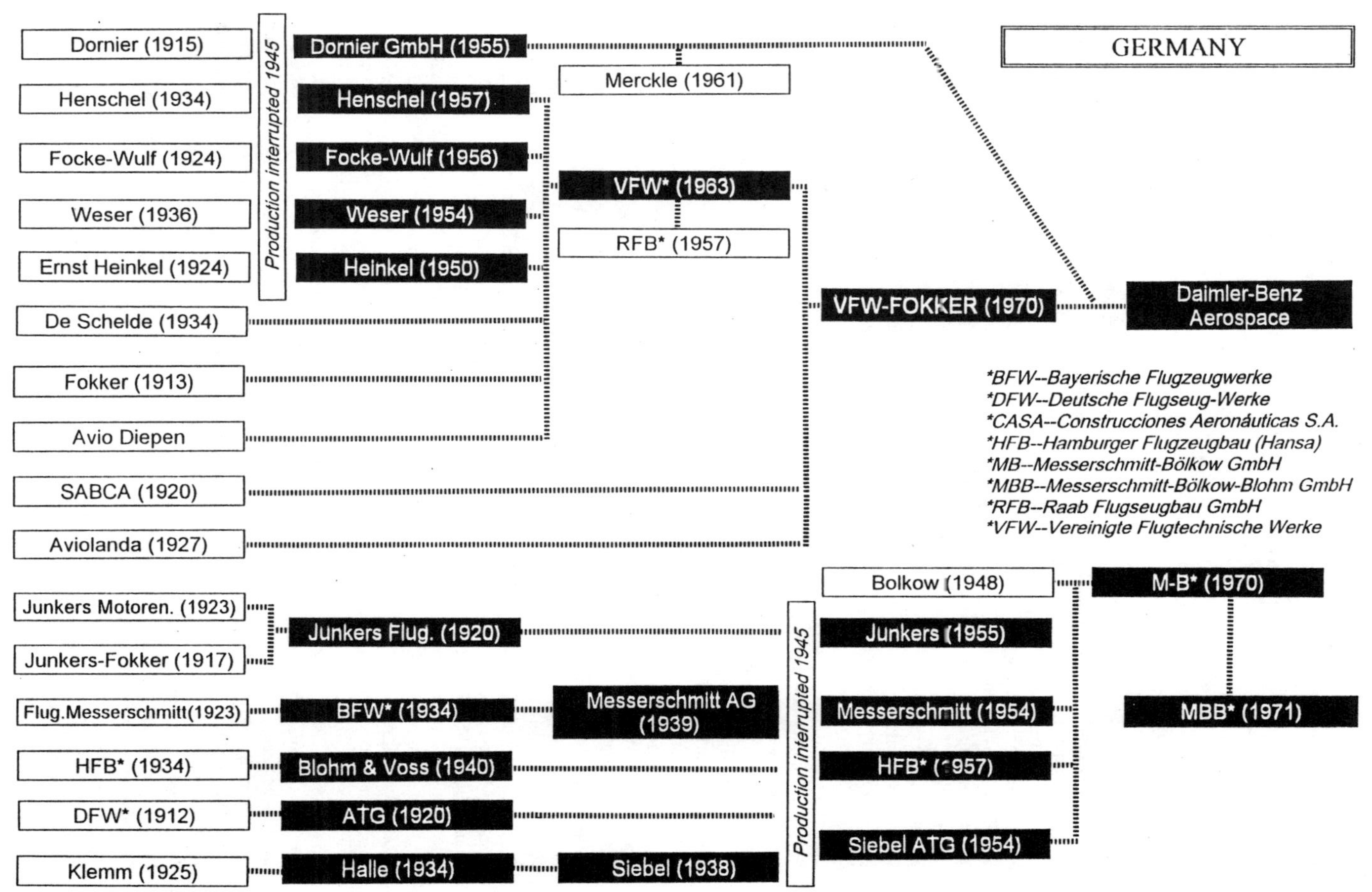

GERMANY
Dornier (1915)
Dornier GmbH (1955)
Merckle (1961)
Henschel (1934)
Henschel (1957)
Focke-Wulf (1924)
Focke-Wulf (1956)
VFW* (1963)
Weser (1936)
Weser (1954)
RFB* (1957)
Ernst Heinkel (1924)
Heinkel (1950)
De Schelde (1934)
VFW-FOKKER (1970)
Daimler-Benz Aerospace
Fokker (1913)
Avio Diepen
SABCA (1920)
Aviolanda (1927)
Production interrupted 1945
*BFW--Bayerische Flugzeugwerke
*DFW--Deutsche Flugseug-Werke
*CASA--Construcciones Aeronáuticas S.A.
*HFB--Hamburger Flugzeugbau (Hansa)
*MB--Messerschmitt-Bölkow GmbH
*MBB--Messerschmitt-Bölkow-Blohm GmbH
*RFB--Raab Flugseugbau GmbH
*VFW--Vereinigte Flugtechnische Werke
Bolkow (1948)
M-B* (1970)
Junkers Motoren. (1923)
Junkers Flug. (1920)
Junkers (1955)
Junkers-Fokker (1917)
Flug.Messerschmitt(1923)
BFW* (1934)
Messerschmitt AG (1939)
Messerschmitt (1954)
MBB* (1971)
HFB* (1934)
Blohm & Voss (1940)
HFB* (1957)
DFW* (1912)
ATG (1920)
Klemm (1925)
Halle (1934)
Siebel (1938)
Siebel ATG (1954)
Production interrupted 1945

MODERN AVIATION HISTORICAL TIMELINE

First	by, at	Date
Powered, sustained flight	Orville Wright, Kitty Hawk, NC	12/17/03
Turn	Orville Wright, Huffman Prairie, OH	9/15/04
Circle flight	Wilbur Wright, Sims Station, OH	9/20/04
Flight over 30 minutes	Orville Wright, Dayton, OH	10/4/05
Flight in Europe	Alberto Santos-Dumont, Paris, France	9/13/06
Aircraft fatality	Lt. Thomas Selfridge, Ft. Myer, VA	9/17/08
Flight in England	S.F. Cody, Farnborough	10/16/08
Crossing of English Channel	Louis Bleriot, Dover, Kent	7/25/09
Licensed woman pilot	Baroness de la Roche, Paris, France	3/8/10
Shipboard takeoff	Lt. Eugene Ely, USS Birmingham	11/14/10
Shipboard landing	Lt. Eugene Ely, San Francisco, CA	1/18/11
Licensed U.S. woman pilot	Harriet Quimby	1911
Transcontinental crossing	Calbraith P. Rodgers, Pasadena, CA	12/10/11
Speed record over 100 mph	Jules Vedrines (Deperdussin)	9/9/12
Multi-engine aircraft.	Igor Sikorsky, St. Petersburg, Russia	1913
Aerial combat.	Allied-German pilots	8/14
Regular airmail service	Washington, DC-New York	5/15/18
Atlantic crossing	Navy NC-4, NY-Lisbon, Spain	5/20/19
Non-stop Atlantic crossing	Alcock & Brown, Newfoundland-Ireland	6/15-16/19
London-Paris air service	Aircraft Travel & Transport	8/25/19
Ship sunk by aircraft	Gen. Billy Mitchell, Langley, VA	7/21/21
Speed record over 200 mph	Nieuport-Delage	9/26/21
Successful emergency bailout	Lt. Harold Harris	10/20/22
Transcontinental non-stop	Macready & Kelly, NY-San Diego	5/2-3/23
Rotary wing flight	de la Cierva, Madrid, Spain	6/9/23
Round-the-world flight	Douglas World Cruisers, Seattle, WA	9/28/24
Flight over North Pole	Floyd Bennet, Sptizbergen, Norway	5/8-9/26
Transatlantic non-stop	Charles Lindbergh, NY-Paris	5/20-21/27
Speed record over 300 mph	Macchi M.52, Venice, Italy	3/30/28
150-hour flight	Army Fokker	1/1-7/29
Instrument takeoff/landing	James Doolittle, Mitchel Field, NY	9/24/29
Rocket engine flight	Fritz von Opel, Germany	9/30/29
Paris-NY nonstop	Coste & Bellonte	9/2-3/30
Speed record over 400 mph	Supermarine S.6B seaplane, Hampshire, England	9/20/31
Non-stop Pacific crossing	Herndon & Pangborn	10/4-5/31
Round-the-world solo flight	Wiley Post, Floyd Bennet Field, NY	7/15-22/33
Successful helicopter flight	Bremen, Germany	7/4/37
Speed over 450 mph (468.94)	Messerschmitt Bf.109, Augsburg, Germany	4/26/39
Jet-powered flight	Marienehe, Germany	8/27/39
American helicopter flight	Igor Sikorsky, Stratford, CT	9/14/39
British jet	Gerry Sayer, Cranwell, England	5/15/41
American jet	Robert Stanley, Bell P-59, Muroc Dry Lake, CA	10/1/42
Speed record over 500 mph	Messerschmitt Me.262, Germany	7/6/44
Supersonic flight	Capt. Charles Yeager, Muroc Dry Lake, CA	10/14/47
Non-stop round-the-world	Boeing B-50, Ft. Worth, TX	2/27-3/2/49
Non-stop transatlantic jet	Col. David Schilling, F-84E	10/22/50
Jet airline service	BOAC De Havilland Comet	5/2/52
Transatlantic helicopter	McGovern & Moore , Sikorsky H-19	7/15-31/52
One-day transatlantic flight.	RAF Canberra	8/26/52
One-day transcontinental flight	F-86 Sabre Jet	5/21/55
Non-stop round-the-world jet	3 B-52s, March AFB, CA	1/17-18/57
Over 4,000-mph flight	Maj. William Knight, X-15A-2 , Edwards AFB, CA	10/3/67
Man-powered flight	Gossamer Condor, Shafter, CA	8/23/77
Round-the-world unrefueled	Voyager, Mojave, CA	12/14-23/87

CHRONOLOGY OF AERONAUTICAL RECORDS

Speed

Date	Pilot	Airplane	Location	MPH	Note
11/21/08	Wilbur Wright	Model A	Auvours, France	27.2	
7/28/09	Louis Bleriot	Bleriot	Reims, France	47.8	
10/29/10	Alfred Leblanc	Bleriot	Belmont Park, NY	68.2	
11/9/12	Jules Vedrines	Deperdussin	Chicago, IL	108.000	
11/29/13	Marcel Prevost	Deperdussin	Reims, France	124.500	Floatplane
6/26/19	F. Brak-Papa	Fiat BR	Italy	162.603	
10/20/20	Sadi Lecointe	Nieuport 29V	France	188.020	
9/26/21	Sadi Lecointe	Nieuport	France	206.267	
10/2/22	Russel Maughan	Curtiss R-6	Mineola, NY	219.453	
10/13/23	Billy Mitchell	Curtiss R-6	Selfridge Field	223.018	
9/16/23	Lawson Sanderson	Wright F2W-1	New York City	247.756	
11/4/23	Alford Williams	Curtiss R2C-1	Mineola, NY	266.640	
12/11/24	Florentin Bonnet	Bernard V-2	France	278.548	
9/29/31	George Stainforth	Supermarine S.6B	England	401.500	Floatplane
9/3/32	Jimmy Doolittle	Gee Bee	Cleveland, OH	294.432	
9/4/33	James Wedell	Wedell Williams	Chicago, IL	304.800	
4/10/34	Francesco Agello	Macchi M.C.72	Italy	423.820	Floatplane
12/25/34	Raymond Delmotte	Caudron	France	314.300	
9/13/35	Howard Hughes	Hughes H-1	Santa Ana, CA	352.390	
11/11/37	Herman Wuster	Bf.113R	Germany	379.630	
4/26/39	Fritz Wendel	Me.109R	Germany	469.220	Jet
11/7/45	Grp. Capt. Wilson	Gloster Meteor	England	606.250	Jet
8/25/47	Maj. Marion Carl	Douglas D-558	Muroc, CA	650.800	Jet
9/15/48	Maj. R. Johnson	F-86	Muroc, CA	670.980	Jet/rocket
10/29/53	Lt. Col. Everest	NA YF	Salton Sea, CA	755.140	Jet
8/20/55	Col. H. Hanes	F-100C	Palmdale, CA	822.270	Jet
3/10/56	L. Peter Twiss	Fairey Delta 2	England	1,132.14	Jet
12/12/57	Maj. Adrian Drew	F-101A	Edwards AFB, CA	1,207.60	Jet
5/16/58	Capt. Walter Irwin	F-104	Edwards AFB, CA	1,404.09	Jet
10/31/59	G. Mossolov	Sukhoi S-66	USSR	1,483.85	Jet
12/15/59	Maj. Joe Rogers	F-106	Edwards AFB, CA	1,525.96	Jet
4/9/60	Ivan Sukhomlin	Tu-114 Roassiya	USSR	545.07	Prop
7/7/62	Col. Georgi Mossolov	Mikoyan Ye-166	Sidorovo, USSR	1,665.89	Jet
6/27/63	Joe Walker	X-15A	Edwards AFB, CA	4,159.00	Rocket
5/1/65	Col. R. Stephens	YF-12A	Edwards AFB, CA	2,070.10	Jet
10/3/67	Man. William Knight	X-15A2	Edwards AFB, CA	4,534.00	Rocket
8/16/69	Darryl Greenamyer	F8F-2 Bearcat	Edwards AFB, CA	477.980	Piston
9/1/74	Maj. J.V. Sullivan	SR-71	Transatlantic	1,810.17	Jet
7/28/76	Capt. Eldon Jorez	SR-71	Beale AFB, CA	2,193.16	Jet
8/14/79	Steve Hinton	P-51D Red Baron	Tonopah, NV	499.047	Piston
7/30/83	Frank Taylor	P-51D Dago Red	Mojave, CA	517.100	Piston
8/21/89	Lyle Shelton	F8F Bearcat	Las Vegas, NM	529.273	Piston

Nonstop distance

Date	Pilot(s)	Aircraft	Route	Distance, st. mi.
9/14/06	Santos-Dumont	14bis	France	8.6 (yd.)
10/26/07	Henry Farman	Voisin	France	855.5 (yd.)
11/21/08	Wilbur Wright	Voisin	France	60.9
8/27/09	Henry Farman	H. Farman III	France	112
7/30/10	M. Tabuteau	M. Farman	France	363.361
9/11/11	Fourney	M. Farman	France	628.173
6/28/14	Landmann	Albatros	Germany	1,178
7/2-6/19	Maj. G.H. Scott	Airship R 34	E-W Transatlantic	3,480
6/14-15/19	Alcock-Brown	Vickers Vimy	W-E Transatlantic	1,884
4/16- 17/23	Kelly-Macready	Fokker T-2	NY-LA	2,516.67
5/20-21/27	Charles Lindbergh	Ryan NYP	NY-Paris	3,267
5/31- 6/2/28	Ferrarin-Del Prete	S.M. 64	Italy	4,764.062
8/5-7	Rossi-Codos	Fairey	NY-Syria	5,655.654
5/13-15/38	Fujita-Takahashi-Sekine	Koken	Japan	7,239.938
7/30-8/1/39	Tondi-Dagasso-Vignoli-Stagliano	S.M.79	Italy	8,038.287
9/29-10/1/46	Daives-Rankin-Reid-Tabeling	Lockheed P2V	Perth-Columbus,OH	11,235.6
6/4/59	Max Conrad	Piper Comanche	Los Angeles	7,668
1/11/62	Maj. Clyde Everly	B-52H	Okinawa-Madrid	12,532.28
11/10/69	Jim Bede	BD-2		8,973.4
12/14- 23/86	Dick Rutan-Jeana Yeager	Voyager	World	24,986.727

Round-the World

Date	Pilot(s)	Aircraft	St. Mi.	Time	Note
3/17-9/28/24	Smith-Arnold	(2) Douglas DWC	26,345	175d	1st circumnav.
7/1/31	Wiley Post/Harold Gatty	Lockheed Vega	15,474	8d 15h 51m	
7/15-22/33	Wiley Post	Lockheed	15,596	7d 18h 49m	1st solo
7/10-13/38	Howard Hughes & crew	Lockheed Electra	14,824	3d 19h 08m 10s	
10/4/45	US Air Trans. Command	Globemaster	23,279	6d 5h 44m	
4/12-16/47	Wm O. Odom	A-26	20,000	3d 6h 55m 12s	
6/17-20/47	Pan American Airlines	Lockheed Const.	22,219	4d 5h 32m	1st commercial
3/2/49	Capt. James Gallagher	Boeing B-50	23,452	3d 22h 1m	1st nonstop
1/15-18/57	USAF	(3) Boeing B-52s	24,325	1d 21h 19m	Nonstop
2/28-3-8/61	Max Conrad		25,946	8d 18h 25m 57s	
3/19-4/17/64	Jerrie Mock	Cessna 180	23,103	29 d	1st woman
5/23/66	Hank Baird & crew	Learjet 24		2d 7h 40m	1st bizjet*
6/4-7/66	Arthur Godfrey, Robt. Merrill and crew	Jet Commander	23,333	3d 14h 9m 1s	1st bizjet
6/20/66	Sheila Scott	Piper Com. 260	29,005		1st Brit
6/11-8/4/71	Sheila Scott	Piper Aztec D	34,000		Over poles
5/17-19/76	Arnold Palmer & crew	Lear 36	22,985	2d 9h 7m 12s	
10/28-31/77	Pan American Airlines	Boeing 747SP	26,382	2d 9h 25m 42s	Over poles
12/14-23/86	Dick Rutan/Jeana Yeager	Voyager	24,986	9d 3m 44s	Unrefueled
8/15-16/95		Concorde		1d 7h 27m 49s	1,114 mph ave.

*non-sanctioned

Altitude____ ______________

Date	Pilot	Aircraft	Location	Feet	Note
Nov. 1783	P. de Rozier	Montgolfier balloon	Paris	3,000	1st manned flight
Sept. 1862	H. Coxwell	Balloon	England	29,000	
11/13/08	Henry Farman	Voisin	France	82	
12/18/08	Wilbur Wright	Wright A	France	360	
12/9/09	Hubert Latham	Antoinette VII	France	1,436	
1/7/10	Hubert Latham	Antoinette VI	France	3,281	
6/14/10	W. Brookins	Wright R	Indianapolis, IN	4,379	
10/31/10	R. Johnstone	Wright Baby Grand	New York	9,600	
9/4/11	Roland Garros	Bleriot	France	12,828	
9/6/12	Roland Garros	Bleriot	France	16,269	
12/11/12	Roland Garros	M.S. G	Tunisia	18,400	
12/28/13	G. Legagneux	Nieuport	France	20,060	
7/14/14	H. Oelerich	DFW	Germany	25,725	
9/18/18	Rudolph Schroeder	Bristol	Dayton, OH	28,897	
9/18/19	R. Rohlfs	Curtiss K-12	New York	31,420	
9/18/21	John Macready	Le Pere	Dayton, OH	34,508	
7/25/27	C.C. Champion	Wright	Anacostia (US)	38.421	
5/25/29	W. Neuenhofen	Junkers	Germany	41,796	
4/11/34	G. Lemoine	Vickers	France	44,822	
9/28/36	S.R.D. Swain	Bristol	England	48,946	
5/8/37	Mario Pezzi	Captoni 113 A.O.	Italy	51,364	
10/22/38	M. Pezzi	Caproni 161 bis	Italy	56,049	
3/23/48	J. Cunningham	Vampire F	England	59,445	Jet
8/15/51	Bill Bridgeman	Douglas D-558-II	California	74,494	Rocket
9/7/56	Ivan Kincheloe	Bell X-2	California	125,907	Rocket
6/2/57	Joseph Kittinger	Man High I	Minnesota	96,000	Balloon
8/28/57	M. Randrup	Canberra	England	70,308	Jet
4/18/58	George Watkins	Grumman F11F-1	California	76,932	Jet
5/2/58	Roger Carpentier	Sud-Ouest SO 9050	France	79,452	Jet
5/7/58	Maj. H.C. Johnson	Lockheed F-104A	Palmdale, CA	91,243	Jet
6/13/58	Jean Boulet	Alouette	Buc, France	35,150	Helicopter
7/14/59	V. Ilyushin	Sukhoi T-431	Russia	94,659	Jet
11/14/59	Capt. J. Jordan	Lockheed F-104C	Edwards AFB	103,389	Jet
7/17/62	Maj. R. White	X-15	Edwards AFB	314,750	Rocket
8/22/63	Joe Walker	X-15A	Edwards AFB	354,200	Rocket
7/25/73	A. Fedotov	Mikoyan Ye-266	USSR	118,898	Jet
8/31/77	A. Fedotov MiG 24	Mikoyan Ye-266M	USSR	123,523	Jet

INVENTIONS AND DISCOVERIES RELATED TO FLIGHT

Date   Inventor (country)
Airplane, experimental--1896, Samuel Langley (U.S.)
Airplane, hydro--1911, Glenn Curtiss (U.S.)
Airplane, multi-engine--1913, Igor Sikorsky (Russia)
Airplane, powered--1903, Wright Brothers (U.S.)
Airship--1852, Giffard (France)
Airship, rigid dirigible--1900, Zeppelin (Germany)
Aluminum, refining--1886, Charles M. Hall (U.S.)
Autogyro--1920, Juan de la Cierva (Spain)
Automatic pilot--1912, Sperry (U.S.)
Balloon, hot air--1783, Montgolfier (France)
Barometer--1643, Evangelista Torricelli (Italy)
Carburetor, gasoline--1893, Maybach (Germany)
Cathode Ray Tube--1878--Crookes (England)
Circuit Breaker--1925, Hilliard (U.S.)
Circuit, integrated--1958, Kilby, Noyce:
   Texas Instruments (U.S.)
Computer, digital--1944, Harvard U./IBM
Computer, mini--1960, Digital Corp. (U.S.)
Computer, electronic--1947, von Neumann (U.S.)
Condensor Microphone--1916, Wente (U.S.)
Electric Battery--1800, Volta, (Italy)
Engine, gasoline--1872, George Brayton (U.S.),
   1889 G. Daimler (Germany)
Engine, internal combustion--1876, Nicholas Otto
   (Germany)
Engine, jet (theory)--1930, Frank Whittle
   (England)
Engine, jet (operational)--1937, Hans von
   Ohain   (Germany)
Fiberglass--1938, Owens-Corning (U.S.)
Gasoline, high octane--1930, Ipatieff (Russia)
Gasoline, lead ethyl--1922, Midgley (U.S.)
Glider--1853, George Cayley (England)
Gyrocompass--1905, Elmer Sperry (U.S.)
Gyroscope--1852, Leon Foucault (France)
Helicopter, operational--1937, Focke (Germany),
   1939, Sikorsky (U.S.)
Microphone--1877, Berliner (U.S.)
Parachute--1783, Louis Lenormand (France)
Propeller, screw--1836, Sir Francis Smith
   (England)
Radar, operational--1940, Robert Watson-Watt
   (Scotland)
Radio beacon--1928, Donovan (U.S.)
Radio, signals--1895, Marconi (Italy)
Rocket engine--1926, Robt. Goddard (U.S.)
Rotary wing--1923, Juan de la Cierva (Spain)
Rubber, vulcanized--1839, Goodyear (U.S.)
Self-starter--1911, Charles Kettering (U.S.)
Telephone, wireless--1899, Collins (U.S.)
Tire, pneumatic--1888, Dunlop (Scotland)
Transistor--1948, Bardeen-Shockley-Brittain (U.S.)
Turbine, gas--1849, Bourdin (France)
Wind tunnel--1912, Eiffel (France)

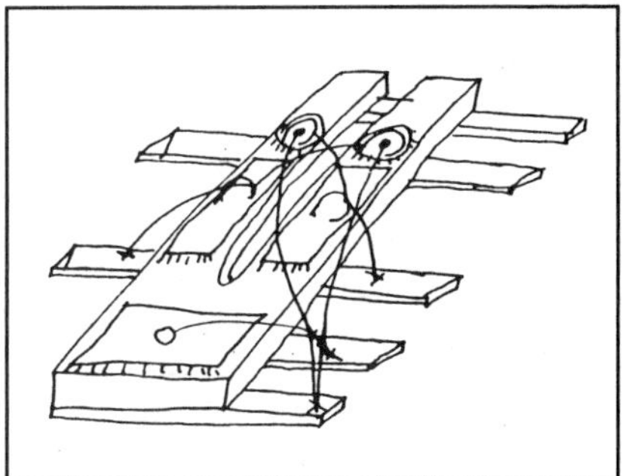

Jack S. Kilby's original sketch of an integrated circuit. The successful testing of the concept in 1958 was the beginning of the "second industrial revolution."

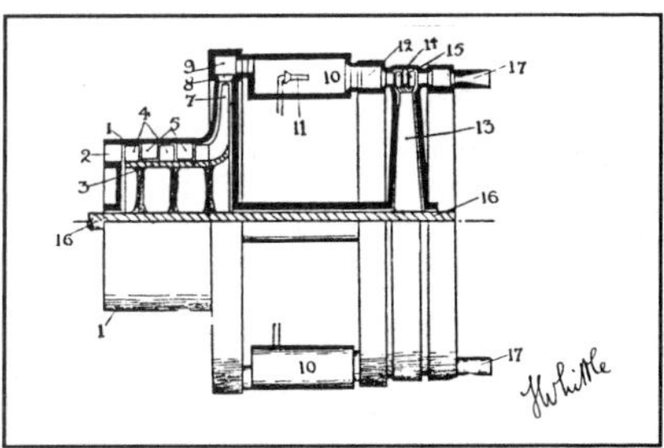

Frank Whittle's Patent drawing filed January 16, 1930 was the first in the world to describe a practical turbojet.

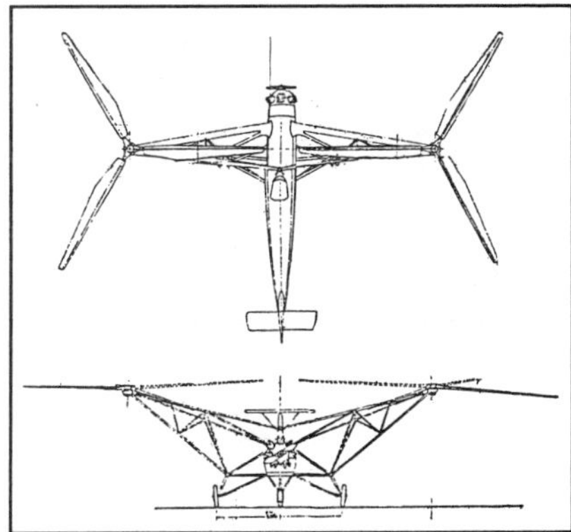

Prof. Heinrich Focke's Fw 61 of 1937 was the first helicopter which could take off and land vertically.

AVIATION MUSEUMS

Adresses, telephone, hours of operation, size/type of exhibits

*denotes some or most aircraft in collection are flown regularly

United States

ALABAMA
United States Army Aviation Museum, Fort Rucker, AL, (205) 255- 4507. Open 9:00 a.m.-5:00 p.m. Monday-Friday, 1:00 p.m.-5:00 p.m. Saturday and Sunday. Fixed wing VTOL aircraft, plus the largest helicopter collection in the world.

ARIZONA
Champlin Fighter Museum*, 4636 Fighter Aces Dr., Mesa, AZ 82505, (602) 830-4540. Open 10:00 a.m.-5:00 p.m. daily. 33 airworthy WWI, WWII and Korean fighters, including Fokker Dr.I triplane, Fw 190, and MiG-21.

Pima Air & Space Museum, 6400 E. Valencia Rd., Tucson, AZ 85706, (520) 574-9658. Open 9:00 a.m.-5:00 p.m. daily. Titan missle silo and 200 civil and military aircraft including SR-71, B-29, and the only examples of Martin Mariner, Northrop Raider and Budd Conestoga.

CALIFORNIA
The Air Museum 'Planes of Fame'*, 7000 Merrill Ave., Chino Airport, CA 91710, (909) 597-3722. Open Noon-5:00 p.m. Monday- Friday, 1-:00 a.m.-5:00 p.m. Saturday and Sunday. Approximately 140 aircraft, including Boeing P-26, Seversky P-35, P-51A, several original German and Japanese models, and the only surviving Northrop flying wing.

Castle Air Museum, Atwater, CA 95301, (209) 723-2178. Open 10:00 a.m.-4:00 p.m. daily in winter, 9:00 a.m.-5:00 p.m. summer. More than 40 military aircraft include SR-71, RB-36, F-4 Thunderbird. Indoor displays, restaurant, gift shop.

March Field Museum, March Air Reserve Base, CA 92518, (909) 697-6000. Open 10:00 a.m.-4:00 p.m. daily except New Years, Easter, Thanksgiving and Christmas. America's first jet, the Bell P-59A, prototype Northrop A-9 and SR-71 among more than 50 military aircraft.

San Diego Aerospace Museum, 2001 Pan American Plaza, Balboa Park, San Diego, CA 92101, (619) 234-8291. Open 10:00 a.m.-4:30 p.m. daily. 68 civil and military aircraft from the history of flight.

CONNECTICUT
New England Air Museum, Bradley International Airport, Windsor Locks, CT 06096, (860) 623-3305. Open 10:00 a.m.-5:00 p.m. daily. More than 70 civil, racing, and military aircraft.

DISTRICT OF COLUMBIA
National Air and Space Museum, Smithsonian Institution, Washington, DC 20560, (202) 357-3133. Open 10:00 a.m.-5:00 p.m. daily. One of the major collections in the world, with about one-fourth of the collection on display, the balance at the Paul E. Garber Facility in Silver Hill, MD, or on loan to other museums. Highlights: original Wright Flyer, Spirit of St. Louis, Winnie Mae, Bell X-1, Apollo capsule, and Voyager.

FLORIDA
National Museum of Naval Aviation, NAS, Pensacola, FL 32508-5402, (904) 452-3604.
Open 9:00 a.m.-5:00 p.m. daily except Thanksgiving, Christmas and New Years. 290,000 sq.
ft., more than 100 Naval, Marine Corps and Coast Guard aircraft dating from 1911, working
simulators and IMAX film, "The Magic of Flight.".

Fantasy of Flight*, 1400 Broadway Blvd., Polk City, FL 33868. Open 9:00 a.m.-5:00 p.m.
seven days a week. Kermit Weeks' new aviation-themed attraction featuring 30 aircraft, 8
flight simulators from WWI and WWII.

GEORGIA
Robins AFB Museum of Aviation, P.O. Box 2469, Warner Robins, GA 31099, (912) 926-
6870. Open 10:00 a.m.-5:00 p.m. every day, 85 aircraft and misslesa, GA Aviation Hall of
Fame, theater, gift store, canteen.

IOWA
Airpower Museum*, Antique Airfield, Route 2, Box 172, Ottumwa, IA 52501, (515) 938-
2773. Open 9:00 a.m.-5:00 p.m. Monday-Friday, 10:00 a.m.-5:00 p.m. Saturday, 1:00 p.m.-
5:00 p.m. Sunday. Home of the Antique Airplane Association. Inter-war civil aircraft.

KANSAS
Combat Air Museum*, P.O. Box 19142, Topeka, KS 66619, (913) 862-3303. Open 9:00
a.m.-4:30 p.m. Monday-Saturday, 10:00 p.m.-4:30 p.m. Sunday. 25 aircraft from every
conflict from World War I through Desert Storm.

Kansas Aviation Museum, 3350 George Washington Blvd., Wichita, KS 67210, (316) 683-
9242. Kansas-built civil and military aircraft, including rare 1940 Stearman "Arriel," Mooney Mite, American Eagle, examples of Swallow, Beech, Cessna, Boeing B-52, oldest Learjet.

Mid-America Air Museum,* P.O. Box 2585, Liberal, KS 67905, (316) 624-5263. Open
8:00 a.m.-5:00 p.m. Monday-Friday, 10:00a.m.-5:00 p.m. Saturday, 1:00-5:00 p.m. Sunday.
90 mostly Kansas-made civil and military aircraft from Thirties through Viet Nam.

MICHIGAN
Henry Ford Museum and Greenfield Village, The Edison Institute, 20900 Oakwood Blvd.,
Dearborn, MI 48121, (313) 271-1620. Open 9:00 a.m.-5:00 p.m. daily. Original Wright
Brothers' shop, first aircraft over the North and South poles, first Sikorsky helicopter.

MINNESOTA
The Air Museum `Planes of Fame' East*, 14771 Pioneer Trail, Eden Prairie, MN 55344,
(612) 941-2633. Open 11:00 a.m.-5:00 p.m. Saturday and Sunday during May-September by
appointment. 14 WWII military aircraft.

NEBRASKA
Strategic Air Command Museum, 2510 SAC Place, Belleville, NE 68005, (402) 292-2001
or (800) 358-5029. *Moving November 1997 to new $26 million facility on I-80 between
Omaha and Lincoln next to Mahoney State Park.* Open 8:00 a.m.-8:00 p.m. daily
Memorial Day through Labor Day, to 5:00 p.m. the rest of the year. More than 30 military
aircraft and missles, including B-29B, B-36J, SR-71 and McDonnell XF-85 Goblin

NEW MEXICO
War Eagles Air Museum*, Santa Teresa Airport, Santa Teresa, NM 88008 (near El Paso, TX), (505) 589-2000. Open 10 a.m.-4 p.m. Tuesday-Sunday. 45 civil and military WWII-Korean aircraft including (3) P-51, F4U, F-82, MiG-15, TU-2, B-26, P-38, P-40, Hawker Fury, Fieseler Storch.

NEW YORK
Cradle of Aviation Museum, Mitchel Field, Garden City, NY 11530, (516) 222-1191. Open Noon-5:00 p.m. Friday-Sunday April through October. 42 civil and military aircraft 1916-1980, including Charles Lindbergh's Jenny.

Intrepid Sea-Air-Space Museum, West 46th St. and 12th Ave., New York, NY 10036, (212) 245-2533. Open 10:00 a.m.-5:00 p.m. Wednesday-Sunday. 24 mostly Naval aircraft exhibited aboard carrier USS Intrepid.

Old Rhinebeck Aerodrome,* P.O. Box 89, Rhinebeck, NY 12572, (914) 758-8610. Open 10:00 a.m.-5:00 p.m. daily mid-May to October; air shows Saturdays and Sundays 2:30 p.m. July-October. 54 WWI and inter-war military and civil aircraft.

OHIO
United States Air Force Museum, Wright Patterson Air Force Base, OH 45433, (513) 255-3284. Open 9:00 a.m.-5:00 p.m. Monday-Friday, 10:00 a.m.-6:00 p.m. Saturday and Sunday. With over 1,000 military aircraft in inventory, one of the largest collections in the world.

Walter Soplata Collection, P.O. Box 65, Newbury, OH 44065, (216) 564-5326. Private collection, open by appointment. 50 military aircraft under restoration.

OKLAHOMA

Oklahoma Aviation and Space Hall of Fame and Museum, 2100 NE 52nd St., Oklahoma City, OK 73111, (405) 424-1443. Open 10:00 a.m.- 5:00 p.m. Monday-Saturday, 1:00 p.m.-5:00 p.m. Sunday. 17 civil aircraft, simulators and space artifacts.

SOUTH CAROLINA
Florence Air and Missle Museum, P.O. Box 1326, Florence, SC 29503, (803) 665-5118. Open 9:00 a.m.-dusk daily. 27 1940-1970 military aircraft, space artifacts.

TEXAS
Admiral Nimitz Center, P.O. Box 777, Fredericksburg, TX 78624. Open daily 8:00 a.m.-5:00 p.m. Small collection of WWII artifacts, including three original Japanese aircraft.

Cavanaugh Air Museum*, 4572 Claire Chennault, Addison, TX 75248. (214) 280-8800 or (800) 552-5264. Open 9:00 a.m. -5:00 p.m. Monday-Saturday, 11:00 a.m.-5:00 p.m. Sunday. 33 WWI through Viet Nam American, German, Russian military aircraft.

Confederate Air Force Flying Museum*, Midland International Airport, Midland, TX 79711, (915) 563-1000. Open 9:00 a.m.-5:00 p.m. Monday-Saturday, 1:00 p.m.-6:00 p.m. Sunday. Of collection of over 150 airworthy WWII Allied aircraft, 50 are on rotating display, along with American Airpower Heritage Museum.

Texas cont

C.R. Smith Museum, 4601 Hwy. 360, Fort Worth, TX, (817) 967-1560. American Airlines' collection of airline artifacts and a vintage DC-3 (Smith was AA President who gave Douglas the first order for the model).

Frontiers of Flight Museum, Love Field Terminal, Dallas, TX, (214) 350-1651. Open Monday-Saturday 10:00a.m.-5:00p.m., Sunday 1:00-5:00 p.m. Aviation history archives said to rival the Smithsonian, artifacts on display, large lighter-than-air, WWI collection.

Lyndon B. Johnson Space Center, Houston, TX 77058, 3 miles east of I-45 on NASA Road 1, (713) 483-4321. Extensive collection of space vehicles and artifacts.

Pate Museum of Transportation, P.O. Box 711, Fort Worth, TX 76101. Open TRuesday-Sunday 9:00 a.m.-5:00 p.m. Ten Naval aircraft, artifacts, vehicles from private collection.

Silent Wings Museum, Terrell Municipal Airport, Terrell, TX. Open Tuesday-Saturday 10:00 a.m.-5:00 p.m., Sunday 12:00-6:00 p.m. Military aircraft and gliders.

Southwest Aerospace Museum, Fort Worth, TX. Outdoor display at gates to General Dynamics, Carswell Air Force Base. 17 WWII and later aircraft (including Convair B-36J and B-58A) and missles.

Texas Museum of Military History, 96 BMW, Dyess Air Force Base, TX 79607, (915) 696-2121. Open 8:00 a.m.-4:00 p.m. Monday-Friday. 22 military aircraft displayed on 70-acre site.

United States Air Force History and Traditions Museum, Lackland Air Force Base, TX 78236, (512) 671-3444. Open 9:00 a.m.-4:00 p.m. Monday-Friday, 9:00 a.m.-6:00 p.m. Saturday and Sunday. 70 military aircraft in outside display.

VIRGINIA
Marine Corps Air-Ground Museum, Quantico, VA 22134, (703) 640- 2606. Open 10:00 a.m.-5:00 p.m. Tuesday-Sunday, April-November. 65 WWI to Viet Nam Marine aircraft and a Japanese Zero which participated in Pearl Harbor attack.

United States Army Transportation Museum, Fort Eustis, VA 23604, (804) 878-3603. Open 8:00 a.m.-5:00 p.m. Monday-Friday, 10:00 a.m.-5:00 p.m. Saturday, Noon-5:00 p.m. Sunday. 23 Army airplanes and helicopters from 1942 to 1964.

WASHINGTON
Museum of Flight, 9404 East Marginal Way South, Seattle, WA 98108, (206) 767-7373. Open 10:00 a.m.-5:00 p.m. daily June- September, and Tuesday-Saturday October-May; Noon-5:00 p.m. Sunday. 34 military and civil aircraft, including the first Boeing airplane.

WISCONSIN
Experimental Aircraft Association Museum*, Wittman Field, Oshkosh, WI 54903, (414) 426-4818. Open 8:30 a.m.-5:00 p.m. Monday- Saturday, 11:00 a.m.-5:00 p.m. Sunday. 150 civil, homebuilt, racing and military aircraft, including Combat Jets Museum.

Australia

Drage's Airworld*, Wangaratta Airfield, Wangaratta, Victoria, 57-218-788. Open 10:00 a.m.-5 p.m. daily. 20 Australian, British, and U.S. aircraft, including DH 89 Dragon Rapide.

Royal Australian Air Force Museum*, Point Cook, Victoria, 03-368-1373 or 1443. Open 10:00 a.m.-4:00 p.m. Wednesday and Sunday. About 40 of types flown by RAAF, including helicopters and an early Gloster Meteor.

Belgium

Royal Army Museum, Parc du Cinquantenaire/Jubelpark 3, B1040 Brussels, 734-21-57. Open 9:00-11:45 a.m., 1:00-4:30 p.m. Tuesday through Sunday. 140 pre-World War I through post-WWII military and commercial aircraft

Brazil

.**Museu Aeroespacial**, Avenue Mal Fontenelle, Campo dos Afonsos, Rio de Janerio 21740, 21-359-8880. Sixty Brazilian, US, German, French and English aircraft and memorabilia.

Canada

Canadian Warplane Heritage*, P.O. Box 35, Hamilton Civic Airport, Mount Hope, Ont. LOR 1WO, (416) 679-4141. Open 10:00 a.m.-4:00 p.m. daily. 31 civil and military aircraft, including Avro Anson and Lancaster, Hawker Hurricane, and Supermarine Seafire.

National Aviation Museum, Rockliffe Airport, Ottawa, Ont. K1A OM8, (613) 998-4566. Open 10:00 a.m.-8:00 p.m. daily May-mid- September, 10:00 a.m. 6:00 p.m. Tusday-Sunday mid-September- April. Established in 1917, this extensive collection includes 105 Canadian, British, U.S., Japanese, and German civil and military aircraft from pre-WWI years to jets.

Western Canada Aviation Museum, 958 Ferry Road, Winnipeg, Man. R3H OY8, (204) 775-8447. 54 civil and military aircraft 1928- 1957.

Czechoslovakia

Vojenske Muzeum-Exposice Letchtva a Kosmonautiky, Kbely, Praha 9, 19706, 422-27-29-65. Open 9:00 a.m.-5:00 p.m. Saturday through Thursday, 2:30-5:00 p.m. Friday. One of the largest European collections, with 200 aircraft, of which about 60 Czech, Russian, German, US, British and Italian aircraft are on display.

Narodni Technicke Muzeum, Kosteini 42, Praha 7, 373-651-9. Open 9:00 a.m.-5:00 p.m. Tuesday through Sunday. The National Technical Museum is one of Europe's oldest (1799), and its aeronautic section features more than 40 aircraft rangine from a 1905 balloon and glider through WWII Czech and European designs.

Denmark

Dansk Veteranflysamlung*, Staunning Lufthavn, Skjern 6900, 07-369044. Open 3:00-5:30 p.m. Tuesday through Friday, 1:00-5:00 p.m. Saturday and Sunday on request. Kramme and Zeuthen, Druine, Gumpert, Hollschmidt, Jurca, Mignet, Raab and Scheibe models, along with the sole surviving Miles Mercury among 30 civil and military aircraft on display.

Finland

Keski-Suomen Ilmailumuseo, PL 1, 41161 Tikkakoski, 941-752-125. Open 10:00 a.m. to 8:00 p.m. daily June 1-Aug. 20, 5:00 p.m.-7:00 p.m. Monday through Friday August 21-May 31. In addition to vintage Russian aircraft, the Finish National Collection contains eight British types, including the only surviving Gloster Gauntlet and the only remaining Martinsyde . A rare Gourdou-Leseurre GL-21 and Morane-Saulnier 60 of the Twenties is on display, and a State Aircraft Factory Pyorremyrsky fighter from 1945.

Suomen Ilmailumuseo, PL 42, 01531 Helsinki-Vantaa-Lento, 90-821-870. Open midday to 6:00 p.m. daily. 35 mainly civil aircraft from US, Britain, Germany, Russia and Scandinavia.

France

Musée de l'Air et de l'Espace, Aéroport du Bourget, 93, 837-01-73. Open 10:00 a.m.-6:00 p.m. Monday through Friday (closes at 5:00 p.m. October-April), 10:00-12:00 and 2:00-4:00 Saturday and Sunday. Founded in 1919, it is one of the three most important collections in the world. Nearly 200 aircraft on display span the history of flight: the 19th Century hang gliders of Biot and Lillienthal, Ader's steam-powered *Eole*, Prévost's 100-mph 1913 Deperdussin, a number of WWI and WWII models, and famous aircraft, including a Concorde.

Musée Aéronautique de Champagne, Aérodrome de Brienne-le-Château, 10, 3325-9284-11. Fifty homebuilt, civil and military aircraft from the early Thirties to the present.

L'Amicale Jean-Baptiste Salis*, Aérodrome de la Ferté-Alais, 91, 331-6457-5289. Collection was begun before World War II, and survivors of the occupation include an airworthy Blériot XI. In the last few years, many reeplica aircraft have been added to the collection, which consists of nearly 100 aircraft.

Germany

Deutsches Museum von Meisterwerken der Naturwissenschaft und Technik*, Museuminsel 1, 8000 Müchen 22, (089) 2179. Open 9:00 a.m.-5:00p.m. daily. Destroyed in World War II, the Museum was rebuilt in 1948, and outdoor exhibits were added at Oberschlessheim airfield. More than 100 aircraft include German civil and military models, the only Wright Model A, a 1910 Rumpler Taube, WWII Focke-Achgelis Fa 330A-1 helicopter, and the Do 31, the only VTOL transport ever produced.

Hubschrauber Museum, Postfach 1310, Sableplatz, Bückeburg 3062, 057-22-5533. Open 9:00 a.m.-5:00 p.m. daily. Housed in the 16th-century home of Baron von Münchausen, the Hubschrauber is a private rotor-wing collection which shows the history of vertical flight, with 30 examples on display.

Hungary

Kozlekedesi Muzeum, Varosligeti Korut 11, Budapest XIV, 420-565. Open 10:00 a.m. to 6:00 p.m. Tuesday-Sunday. One of the oldest transporation museums (1896) the aviation collection features about 35 US, Russian, German and Hungarian designs.

Italy

Museo Aeronautico Caproni di Taliedo, Via Durini 24 Milano 20122, 78-19-15. Founded by Count Gianni Caproni di Taliedo, the private collection includes his first airplane, the Caproni Ca.1, as well as a dozen later models, and other predominently Italian aircraft.

Italy cont.

Museo Storico Dell'Aeronautica Militare, Aeroporto vi Vigna di Valle, Vigna di Valle 00062. 6-9024-034. Open Tuesday-Sunday June-September 9:00 a.m.-6:00 p.m., October-May 9:00 a.m.-4:00 p.m. More than 70 Italian designs, including Schneider Trophy winning Macchi M.39.

Netherlands

Aviodome, Schiphol Centrum, Amsterdam 1118AA, 020-173640. Open daily 10:00a.m.-5:00 p.m. April-October, Tuesday-Sunday in winter. Nearly 50 French, British, US, designs from 1911 plus eight vintage Fokkers 1928-78.

Militaire Luchtvaart Museum, Kamp van Zeist, Soesterberg 3769ZK, 03404-34222. Open April-October from 10:00 a.m. to 4:00 p.m. daily. 20 examples of Dutch, French, US and English aircraft operated by the Dutch armed forces.

New Zealand

Museum of Transport and Technology, Great North Road, Western Springs, Auckland 2, 9-860-198. Open 9:00 a.m.-5:00 p.m. daily. A major tourist attraction in New Zealand, its more than 30 aircraft include Short Solent and Sunderland flying boats, an Avro Lancaster, the only known Vickers Vildebeest, and three crop dusting airplanes.

Norway

Kongelige Norsk Ljuftforssvaret Musett, Gardermoen Lufthavn, Gardermoen 2062, 06-94-10-00. Open 10:00 a.m.-2:00 p.m. Saturday and Sunday May through October. A display of 30 of the 70 aircraft preserved by the air force include an Heinkel 111, Fw 190, a 1917 Avro 540K, 1920 Farman F.46, Henschel Hs 293, and a Gloster Gladiator.

Poland

Muzeum Lotnictwa I Astronautyki, 30-969 Krakow 28, 44-71-81. Open 10:00 a.m.-2:00 p.m. daily May through October. One of the major collections in Europe, it was founded on the battered remnants of German aircraft found before the end of the war and hidden from authorities by a group of Polish citizens. One survivor is the historic Me 109 VI which set a world speed record at 466 mph in 1939. Numbering more than 100 aircraft, the collection is a little-known treasure trove of WWI, and inter-war and WWII Polish aircraft.

Portugal

Museu Do Ar, Alverca do Ribatejo 2615, 258-27-22. Open 10:00 a.m.-6:00 p.m. Tuesday through Sunday. The history of Portuguese aviation from 1912 is traced through 40 British, French, and US models.

Russia

Air Force Museum, 141170 Monino, Moscow, 2445624, ext. 2186 or 2225. Open 10:00 a.m.-5:00 p.m. Monday-Thursday, 10:00 a.m.-2:30 p.m. Saturday. One of the major collections of military aircraft in the world. In addition to prototypes and production examples of most Soviet aircraft from post-WWI to the Cold War years, there is a 1911 Farman, a Sikorsky S-58 helicopter and a Sopwith Triplane.

National Aviation and Space Museum, 24a Leningradsky Prospekt, Moscow, 125040, (7095) 213-88-53. Open daily 10:00 a.m. to 7:00 p.m. Located on the airport where the first Russian flew in 1910, this outdoor museum is informal, and consists of about 30 mostly jet military aircraft since WWII.

Spain

Museo Del Aire, Cuatro Vientos, Madrid. Open 10:00 a.m.-1:00 p.m. Tuesday-Sunday. One of Juan de la Cierva's first autogyros, the only He 111e-1 in existence, and a 1944 Dornier DO 24 flying boat headline a display of about 40 civil and military aircraft.

Sweden

Flygvapenmuseum Malmen, Linköping, 013-29-92-70. Open midday-3:00 p.m. Sunday through Friday. The collection is virtually in the back yard of SAAB Aircraft, and more than a dozen of its models are displayed. Also among the 80 aircraft in the Malmen Air Force Museum are the last of their kind: a 1919 Macchi M.7 flying boat and Phönix 122, an Albatros 120 (circa 1925), and Sparman S-1A trainer.

Switzerland

Verkehrshaus der Schweiz, Lidostrasse 5, Luzern CH-6006, 041-31-44-44. Open 9:00 a.m. to 6:00 p.m. daily March through November, 11 to 4 Tuesday-Saturday December-February. The 50-plane collection includes the first successful Swiss airplane from 1910, a host of Swiss, French, German and US designs, and the world's only Lockheed Orion, a 1931 Altair.

United Kingdom

Fleet Air Arm Museum*, Royal Naval Air Station, Yelovilton, Ilchester, Somerset BA22 8HT, 0935-840551, ext. 521. Open 10:00 a.m. to 5:30 p.m. (or dusk, if earlier) daily. A major collection of Britain's finest, plus a few American built models, spanning the history of aviation. Practically the only museum in the world where one can view a Gloster Sea Gladiator, Sopwith Baby, Westland Wyvern, Fairey Swordfish or Albacore alongside a Supermarine Walrus and a Concorde.

Imperial War Museum*, Lambeth Road, London, SE1 6HZ, 01-735-8922. Open 10:00 a.m.-5:30 p.m. Monday-Saturday, 2:00-5:30 p.m. Sunday. Site of the main aircraft collection: Duxford Airfield Duxford, Cambridgeshire CB2 4QR, 0223-833963. Open 10:30 a.m.-5:30 p.m. daily March 15 through November 2. In addition to an astounding array of some of the most significant aircraft in English history, the collection includes one of the world's two flying B-29s and an impressive array of British jet bombers and fighter aircraft.

Mosquito Aircraft Museum, Salisbury Hall, London Colney, near St. Albans, Hertfordshire AL2 1BU, 0727-22051. Open 10:30 a.m. to 6:00p.m. Sunday, Easter through October, 2-5:30 p.m. Thursdays July-September. Located in the 17th century house where the deHavilland SH 98 Mosquito was built, the original prototype and two production variants are on display at Salisbury Hall, along with 13 other deHavilland models, a Blériot XI and 1918 Royal Aircraft Factory BE.2e.

Museum of Army Flying, Army Air Corps Centre, Middle Wallop, Stockbridge, Hampshire SO20 8DY, 0264-62121. Open 10:00 a.m.-4:00 p.m. daily. About 30 aircraft of the type used by the Army: Austers, Cessnas, helicopters and troop carrying gliders.

Science Museum, Exhibition Road, South Kensington, London SW7 2DD, 01-589-3456. Open 10:00 a.m.-6:00 p.m. Monday-Saturday, 2:30-6:00 Sunday. Britain's equivalent of America's NASM, this display now houses a cross-section of aviation: Lilienthal's 1896 glider, the Vickers Vimy which was the first airplane to cross the Atlantic non-stop, a 1910 Antoinette VII, the first man-powered airplane, a Messerschmitt Me 163 Komet, and the Schneider Trophy-winning Supermarine S.6B.

United Kingdom cont.

Shuttleworth Collection*, Old Warden Aerodrome, Biggleswade, Bedfordshire SG18 9ER, 0767-27-288. Open 10:30 a.m.-5:30 p.m. daily. A collection begun in the Thirties which has grown to include about 40 unique, historic British civil and military aircraft.

Royal Air Force Museum Collection

Aerospace Museum, RAF Cosford, Wolverhampton, West Midlands WV73EX, 090-722-4872. Open 10:00 a.m.-4:00 p.m. daily April-October, 10-6 Monday-Friday November-March. More than 60 aircraft, including prototypes and research designs such as the prototype Meteor, the Fairey Delta 2, Bristol 188 stainless steel jet and one of two remaining TSR-2s. The British Airways Collection contains a range of post-1945 airliners.

Royal Air Force Museum, Battle of Britain Museum and **Bomber Command Museum**, Hendon, London NW9 5LL, 01-205-2266. Open 10:00 a.m.-6:00 p.m. Monday-Saturday, 2-6 p.m. Sunday. Three separate collections feature aircraft and artifacts from RAF history, the Battle of Britain, and the bombers of two wars.

RAF Battle of Britain Memorial Flight*, RAF Coningsby, Lincolnshire LN4 4SY, 0526-42581. Consists of four Spitfires, one Lancaster, and two Hurricanes, all maintained for exhibition flights.

High Flight

by John Gillespie Magee, Jr., Pilot Officer RCAF
*Written in a letter to his parents a few months before his death in a Spitfire
December 11, 1941*

Oh! I have slipped the surly bonds of earth

And danced the skies on laughter-silvered wings;

Sunward I've climbed, and joined the tumbling mirth

Of sun-split clouds--and done a hundred things

You have not dreamed of--wheeled and soared and swung

High in the sunlit silence. Hov'ring there

I've chased the shouting wind along, and flung

My eager craft through footless halls of air.

Up, up the long delirious, burning blue,

I've topped the windswept heights with easy grace

Where never lark, or even eagle flew--

And, while with silent lifting mind I've trod

The high untrespassed sanctity of space,

Put out my hand and touched the face of God.

<table><tr><td>1990</td></tr></table>

The Department of Defense announces a new high technology radio navigation system called GPS, or **Global Positioning System**, which would utilize a network of 24 orbiting satellites. The system is planned to be in place by 1993.

Beech Aircraft begins deliveries of its all-composite **Starship**. The Rutan-designed twin turboprop is four years behind schedule.

Mooney Aircraft introduces its 270 hp TLS model.

Piaggio's P-180 turboprop pusher begins FAA flight certification testing in Italy, and the first American delivery is made in October at NBAA in New Orleans.

Basler Turbine Conversions, Oshkosh, Wis., is approved for installation of PT6 turboprop engines in DC-3s.

An **SR-71** set for retirement flies from Beale AFB, Calif. to Washington, D.C. in 68 minutes, 17 seconds, averaging Mach 3.2.

The August 2 Iraqi invasion of Kuwait and the Mideast oil embargo that follows drive fuel prices up by as much as 25 cents a gallon.

An Avianca Boeing 707 enroute from Columbia to New York's JFK Airport runs out of fuel and crashes, killing 72 of 161 passengers.

Dee Howard Co. shows the first of its Rolls-Royce **Tay-powered BAC 111** at Britain's Farnborough Air Show.

The last example of a total of 226 of **Falcon 10/100s** built is delivered, while plans for the newest Falcon, the 2000, are revealed. In England, the **BAe 1000** long-range bizjet is flown.

Two Boeing 747-200Bs are delivered to the 89th Military Airlift Wing to replace two 707-320Cs in use since 1962 and 1972 as **Air Force One.**

The fifth production Concorde emerged after an 18-month 40,000-hour inspection and overhaul. The aircraft logged 11,650 hours since introduction to the Air France fleet in 1976.

Aerospatiale and Messerschmitt Boelkow Blohm GmbH agree to unite their helicopter divisions to form **Eurocopter**. In the U.S., **Learjet** is purchased for $75 million by Canada's Bombardier Group.

Fairchild Aircraft Corp. is purchased for a reported $66.5 million by an investment group headed by Carl Albert.

Piper Aircraft tries restructuring its management, laying off workers, and even selling some of its Type Certificates in an effort to stay in business.

The summer airshow season claims the lives of eight performers, including former U.S. Aerobatic Champion **Tom Jones**. At the National Air Races in Reno, Lyle Shelton piloted his Grumman F8F-02 "Rare Bear" to an Unlimited victory at 468.26 mph.

Robinson Helicopter's four-place R44 makes its first flight in Torrance, Calif.

Longtime AOPA president **John Baker** resigns, and **Clarence L. "Kelly" Johnson**, 80, the legendary designer who headed Lockheed's famous "Skunk Works," dies in Burbank, Calif.

Cessna announces plans for its Mach 0.9 Citation X, with flight testing beginning in 1993 and deliveries starting in 1995.

1991

Operation Desert Storm begins January 18 with the air attack of Iraqi command and control centers in Baghdad. In two days, more than 3,100 sorties are flown by Coalition forces.

The Williams-engined **Swearingen SJ30** light twin-jet flies at San Antonio.

Piper Aircraft Co. files for Chapter 11 bankruptcy protection, and after talks early in the year with Aerospatiale, owner M. Stuart Millar finally sells his company to Cyrus Eaton Group International.

Bell Helicopter Textron flies its new **Model 230** in Fort Worth, Tex.

The 50-passenger **Saab 2000** is unveiled at Linkoping, Sweden before the King and Queen, while in Britain, Queen Elizabeth attends rollout ceremonies for the British Aerospace **Jetstream 41**.

In separate incidents on successive days, **Sen. John Heniz** (R-PA) and former Texas **Sen. John Tower** are killed in aircraft accidents.

Capping a six-year process, the team of **Boeing Helicopter and Sikorsky Aircraft** has won the US Army's light helicopter (LH) competition, and a $2.8 billion award to develop six prototypes.

In Wichita, Cessna makes the first flight of its new **CitationJet**, and the **Learjet 60** begins its flight test program. The first delivery of the turbine-powered single-engine **TBM 700** is made in San Antonio, Tex. Pilatus' single-turboprop **PC-12** flies in Switzerland.

The FAA predicts that General Aviation will recover from a "Short and relatively mild recession" in 1991, with moderate but steady growth through 2002.

The Air Force's **F-117** "stealth" strike fighter makes a surprise appearance at the Paris Air Show.

Canadair's 50-passenger **RJ** regional jet makes its maiden flight.

The freshly certificated **Bae 1000** business jet debuts at NBAA in Houston, and Canadair reveals its **Global Express**.

McDonnell Douglas' **520N** becomes the first NOTAR (No tail rotor)-equipped aircraft to receive FAA certification.

FAA Administrator James Busey resigns.

The first Boeing 727 ever built is retired to Seattle's Museum of Flight after more than 64,000 flying hours with United Air Lines. Boeing also announces termination of production of the 707, after 37 years of continuous manufacture.

The Confederate Air Force moved its headquarters from the Rio Grande Valley city of Harlingen, to Midland, Tex.

The FAA prohibits flight of the **Piper PA-46** into instrument conditions (later modified to icing conditions and turbulence) following the inflight breakup of six Malibus and one Mirage in the U.S., Mexico and Japan.

Beech Aircraft finishes a record year, topping $1.1 billion in sales.

1992

A jury in U.S. District Court in Rochester, N.Y. finds Teledyne Continental Motors liable for a 1986 Bonanza crash which resulted in the deaths of a local family, and awards the plaintiffs' estate **$107 million** in damages, a record judgment. The decision is later overturned.

1992 cont.

Textron, Inc. pays $600 million to General Dynamics for its **Cessna Aircraft** division. In a joint venture, Bombardier and the provincial government of Ontario purchase Boeing's **deHavilland (Canada)** division for $220 million.

Piper Aircraft is allowed to complete 65 aircraft from parts in inventory by the bankruptcy court. Later in the year, the company will attempt to sell off its Lakeland, Fla. facility to help pay debts.

Butler Aviation merges with **Page Avjet Airport Services** to become the largest FBO chain in the world, with 42 operations at 38 U.S. airports.

McDonnell Douglas announces plans to build the **MD-12**, a four-engine airliner which could seat up to 600 passengers.

Trans World Airlines files Chapter 11 bankruptcy to seek protection while it tries to recover debt of almost $2 billion.

The only flying **Lockheed YF-22** prototype makes a hard landing at Edwards Air Force Base, Calif., resulting in unrepairable damage. No plans are made to build another development aircraft, and the program will later be abandoned.

The 50-58-seat **Saab 2000** commuter turboprop makes its first flight at Linköping, Sweden.

Sporty's Pilot Shop owner **Hal Shivers** announces he intends to build a single-engine aircraft, a clone of the Cessna 172 which will be called the Liberty. Deliveries are expected to begin in 1993.

Heikki Nummela flies **372 consecutive loops** in a Fournier RF-4 in Finland. While the feat is no longer recognized in official record books, Nummela beat the previous number by 80 loops.

The FAA lifts its icing/turbulence flight ban on the Piper PA-46 after testing reveals no structural problems with the design.

A Gulfstream G-IV demonstrator flies 4,776 nm from Tokyo to Las Vegas in a world record time of ten hours, four minutes. Carrying a crew of three, nine passengers and cargo, the aircraft averaged 474.48 kts.

A Rolls-Royce RB.211-524B4 jet engine is removed from the tail of a Delta L-1011 after **24,267 hours**, setting a new record for reliability. It had been in continuous service for more than five years.

Taiichi Ishida, president of Ishida Industries, who is proposing to build the **TW-68 tilt-wing** in Fort Worth, Tex., predicted that the four-engine, two-prop aircraft will fly on March 9, 1996, "at nine a.m."

Ernest K, Gann, 81, passes away at his home in Washington state. He was the author of more than 20 books, including *Fate of the Hunter*, *The High and the Mighty*, and *Island in the Sky*

Edna Gardner Whyte, 89, who got a private certificate in 1931 and spent a lifetime instructing (35,000 hours) and participating in air races (125 trophies), passes away in Roanoke, Tex. **Ed Heinemann**, 83, dies at his home in Rancho Mirage, Calif. Heinemann was responsible for the design of the A-4 Skyhawk, SBD Dauntless, A-20, A-24, A-26, AD Skyraider and A-3 Skywarrior while at Douglas Aircraft.

1993

The first of the FAA's new **Primary Aircraft Category Type Certificate** is awarded to Quicksilver Enterprises' GT500 at the EAA Convention in Oshkosh, Wis. In addition, the Katana DA20 and Zenith CH 2000 were awarded provisional certificates.

Dassault's **Falcon 2000** debuts at Merignac, France.

1993 cont.

Pilatus Aircraft Ltd. signs a letter of intent to acquire the assets of Piper Aircraft Co.

An April 19 MU-2 crash near Dubuque, Iowa, claims six, including **George Mickelson**, governor of South Dakota.

The NTSB criticizes flight procedures of the FAA after one of the latter's King Airs crashes near Front Royal, Va. while on a navaid inspection.

Cessna flies its version of the JPATS trainer, a variant of its CitationJet, on Dec. 20, and the next day its **Citation X** makes its first flight, both at Wichita.

At NBAA in Atlanta, the **Falcon 2000** makes its first American appearance; Gulfstream Aerospace announces design configuration of its **Gulfstream V**, planning on a maiden flight in 1997, and Israel Aircraft Industries officially kicks off its **Astra Jet IV** program. The National Business Aircraft Association launches its "No Plane, No Gain" advocacy program.

Veteran race pilot **Rick Brickert** is killed in the crash of the Pond Racer, an experimental auto engine-powered aircraft, during practice for the National Air Races at Reno.

The medical certificate of aerobatic legend **R.A. "Bob" Hoover** is revoked when two FAA inspectors file a complaint some months after a June 1992 airshow in Oklahoma City, alleging the 72-year-old performer is hampered by a "cognitive deficit."

Raytheon Corp. acquires British Aerospace's Corporate Jets subsidiary for approximately $372 million. The company will be called **Raytheon Corporate Jets**, and will produce the Hawker 800 and 1000 jets in tandem with its Beech Aircraft business.

Former Denver Mayor **Federico Peña** is appointed Secretary of Transportation by President Clinton, and **David Hinson**, co-founder of Midway Airlines, is named FAA Administrator.

The General Aviation Manufacturers Association reports deliveries at the lowest level since World War II, with only of 899 new aircraft worth $1.836 billion sold. Of that total, 363 units valued at $626 million were for the export market. Beech Aircraft was the production leader with a total 305, and its Bonanza A36 was the best-selling model with 61 delivered.

Floods in the midwest along the Mississippi, Missouri, Illinois, and Kansas rivers damages airports and aircraft in eight states.

Jim Bede's BD-10J two-place single-jet kitplane makes its first flight at Mojave, Calif.

Bill "Tiger Destefani flies his P-51 "**Strega**" to his second straight Reno National Championship, averaging 455.38 mph in the Gold Final.

NASCAR race driver **Alan Kulwicki** dies in the April crash of a Merlin 300, and driver -- **Davey Allison** is killed three months later when the MD 500 helicopter he is piloting crashes in the infield at the Talledega (Ala.) Speedway.

Fred Weick, designer of the Ercoupe and several Piper models including the Cherokee and Pawnee, dies at his home in Vero Beach, Fla. Weick was 93.

Olive Ann Beech, 89, co-founder of Beech Aircraft with her husband Walter, dies in Wichita, Kan., and aviation pioneer **James H. Doolittle**, 96, passes away in California

---

1994

President Clinton predicts that a **corporatized ATC system** will go in service soon.

The Polish-built **PZL Koliber II** Model 150A gains U.S. certification. The Pilatus single-engine PC-12 turboprop is jointly certified in Europe and the U.S.

The cause of an Aeroflot crash that kills 75 in Russia is found to be that the pilot was giving his 15-year-old son **impromptu flight lessons** in the Airbus at the time.

Boeing rolls out its latest seven-series transport, the 777, in ceremonies at Seattle, Wash.

1994 cont.

On August 17, President Clinton signs the **General Aviation Revitalization Act of 1994**, limiting manufacturers' product liability to 18 years. The next day, Cessna Chairman Russ Meyer announced the company's plans to begin production of single-engine models.

An American Eagle ATR-72 crashes in Indiana after encountering icing, killing 68 persons. Both U.S. and Canadian operators are restricted from operating the -72 and smaller ATR-42 in known icing conditions until testing can confirm the safety of the airplane.

AlliedSignal Inc. acquires the **Lycoming Turbine Engine Division** of Textron Inc. for a reported price of $375 million plus assumption of certain liabilities.

Marion Jayne, Roanoke, Tex., wins the Around The World Air Race in a 1970 Piper Twin Comanche.

741,750 people attend Sun 'n Fun in Lakeland, Fl. In ceremonies during EAA Oshkosh '94, air racer **Steve Wittman** donates his famous racer "Bonzo" to the EAA Air Museum. At Reno, Lyle Shelton's Bearcat, "Rare Bear," wins for the seventh time in its 26-year racing career. Two racers died during the 1994 races, 70-year-old T-6 pilot Ralph Twombly, and 48-year-old Unlimited racer William Speer, Jr.

Darryl Greenamyer announces plans to rescue a B-29 which was abandoned in 1947 some 180 miles north of Thule, Greenland. The former Lockheed test pilot and Reno racer plans to fly the airplane off the ice cap in late summer.

The fly-by-wire **Boeing 777** makes its first flight June 12.

Mooney Aircraft announces the **Ovation**, a 280 hp M-20 variation which will sell between the $161,400 MSE and the $261,000 TLS.

Ron Bower of Austin, Tex., set a world record when he flew his Bell 206B-3 JetRanger around the world in 24 days, four hours, 36 minutes.

American General Aircraft Corp., manufacturer of the Grumman Tiger design, has filed for Chapter 11 bankruptcy.

A leaky valve at a Chevron refinery in California blended Jet A with 100LL gasoline. The company will pay for inspections and repairs to any affected piston engines.

The U.S. Department of Justice **Drug Enforcement Administration** (DEA) moves into new headquarters at Alliance Airport near Fort Worth, Tex.

The FAA kills the implementation of **Microwave Landing System** (MLS) which it had been promoting for nearly a decade, opting to wait for development of a precision GPS system.

An exhibit at the National Air and Space Museum exhibit of the *Enola Gay*, the B-29 which dropped the first atomic bomb on Hiroshima, Japan in 1945, is criticized by veteran's organizations for being a sanitized, revisionist version of history.

On Sept. 15, Beech Aircraft and Raytheon Corporate Jets merged to become Raytheon Aircraft Co.

French pilot Jacques Lemaigre du Breuil becomes only the second person to fly solo non-stop from New York to Paris in a single-engine prop-driven airplane (Lindbergh was the first) when he makes the crossing June 25 in a TBM 700 in 10 hours, 54 minutes.

1995

Rolls-Royce purchases former General Motors subsidiary Allison Engines for $525 million.

A class-action lawsuit is brought against Mobil alleging its **AV-1 synthetic oil** caused damage to Teledyne Continental aircraft engines. An out-of-court settlement is made with an estimated 2,000 to 3,000 owners to provide inspection and/or repair.

1995 cont.

After numerous delays, **Denver International Airport** opens Feb. 28. The airport, the largest in the U.S., is nearly two years late and $3.5 billion over budget.

The U.S. Air Force selects Raytheon's **Beech Mk II** turboprop (aka Pilatus PC-9) as winner of its $7 billion JPATS (Joint Primary Aircraft Training System) contract which calls for the manufacture of more than 700 units.

Cessna Aircraft flies its PW530A-powered **Citation Bravo**.

The **Cirrus SR20** makes its first flight at Duluth, Minn. The four-place composite single is expected to be certificated in 1996.

Embraer rolls out its 50-seat **EMB-145** prototype in Brazil. In Savannah, Gulfstream Aerospace debuts its global bizjet (6,500 nm), the **Gulfstream V**.

Piper Aircraft announces plans to sell most of its assets, and the FAA transfers type certificates to **The New Piper Aircraft, Inc.**

Socata buys the Type Certificate and unveils plans to build the Grumman American GA7 light twin. Out of production since 1979, the airplane would be marketed as the TB-320 Tangara.

Four Mexican Air Force jets crash in the Mexico City suburb of Cuajimalpa following a mid-air collision during ceremonies marking the 185th anniversary of the country's independence from Spain. According to witnesses, a F-5 hit a T-33, and debris from the collision caused the crash of two other T-33s.

French engine manufacturer Turbomeca is ordered to pay a **$350 million judgment** to the family of a Rocky Mountain Helicopter pilot who died in a 1993 Kansas City Life Flight medical evacuation operation. An earlier trial by the family of a patient who also died in the crash netted an additional $70 million.

Charles Philip Arthur George Windsor, Prince of Wales, has decided to give up pilot privileges on RAF Royal Squadron aircraft after a 1994 incident in a Bae 146 he was piloting overshot a runway in Scotland. Damage to the aircraft was about $1.6 million, but there were no injuries.

206 Cessna Citations participate in a program to carry more than 1,700 participants to the **Special Olympics**, held in July at Yale University. Cessna said it was the largest one-day peacetime airlift in history.

The world's largest twin-engine aircraft, the **Boeing 777**, begins service June 7 for United Airlines flying between London and Washington, DC. The half-million-lb. widebody carries 292 passengers.

After a one-month ban on operations in known icing, the FAA announces that ATR 42 and 72 airliners are in full compliance with safety regulations. The agency had conducted 30 hours of airborne tests behind a tanker spraying water on an ATR 72.

Cessna Aircraft announces plans for construction of a new single-engine manufacturing facility at Independence, Kan., 160 miles southeast of Wichita, and that the first new aircraft will roll off the line in late 1996.

The annual **Aircraft Owners and Pilots Assn.** convention is held in Atlantic City, N.J., and 8,500 members register.

1996

The **Learjet 45** debuts in Wichita, and the out-of-production Grumman/Rockwell Commander GA7 Cougar is resurrected by French manufacturer SOCATA under the new name TB 320 Tangara. The **Gulfstream V** makes its first flight at Savannah.

1996 cont.

Bill "Tiger" Destefani's P-51 "Strega" nips former Reno Champion "Rare Bear" for the Unlimited Championship. Destefani averages 467.029 to Rare Bear pilot John Penney's 465.159 mph.

Ralph Harmon, 81, the engineer responsible for the concept and development of the Beech Bonanza and later vice president of engineering at Mooney Aircraft, dies in Kerrville, Tex. Aerocar designer Molt Taylor, passes away at his Washington home.

Pioneer homebuilt designer and air racer **Steve Wittman**, 91, and his wife Paula die in the crash of a Wittman Tailwind while returning to Oshkosh, Wis. from their winter home in Ocala, Fla.

Cessna formally opens their new single-engine manufacturing facility at Independence, Kan.--appropriately on Independence Day--and announces prices of their resurrected **172** and **182** models at $124,500 and $190,600 respectively, two months later at the EAA Fly-In at Oshkosh, and the company's **Citation Excel** makes its first flight.

Sir Frank Whittle, British-born inventor of the turbojet engine, dies at age 90 at his home in Columbia, Md., and 1972 World Aerobatic Champion **Charlie Hillard** perishes in a freak accident in his Hawker Sea Fury at the Sun 'n Fun Airshow in Florida.

Golfer **Arnold Palmer** takes delivery on the first production Citation X.

During the presidential campaign, Bill Clinton proposes a **$225-per-flight fee** on turbine business aircraft, saying the revenue would be deposited in the Airport and Airway Trust Fund to help pay for a reading program, although there was no clear method of transferring funds to the latter. FAA Administrator **David Hinson** resigns three days after the elections.

An **Air Force C-130** carrying Presidential vehicles and related gear crashes near Jackson Hole, Wyo., killing eight crew members and a Secret Service agent.

Cessna inks an order for 25 Citation Xs and 20 Citation VIIs with Executive Jet Aviation, and Continental Express agrees to purchase 25 EMB-145 regional jets from beleaguered Brazilian manufacturer Embraer.

Canadair's **Global Express** long-range bizjet flies at Downsview, Ontario.

Bill Destefani wins his fifth Unlimited title at the Reno Air Races, flying his P-51 *Strega* to an average 467.948 mph.

The Rutan-designed VisionAire **Vantage** single-engine jet flies Mojave, Calif., and Swearingen's SJ30-2 makes its first flight at San Antonio, Tex. Raytheon Aircraft takes the wraps off its new mid-size **Hawker Horizon** jet at the NBAA Convention in Orlando, Florida.

Teledyne Continental wins a $9.5 million grant to develop a diesel engine for aircraft.

Gulfstream Aerospace receives a provisional type certificate on its **Gulfstream V** December 13, and Embraer wins certification for its EMB-145 twinjet regional airliner three days later.

Oshkosh '96 by the numbers
Estimated attendance: 800,000 (including 2,258 from 78 nations outside the US); 11,000 aircraft--2,478 in the show; 40,000 campers; 5,000 volunteers

---

Burt Rutan comes to the EAA Oshkosh Convention in his latest radical design, an "asymmetric" twin-engine pressurized aircraft called "Boomerang." The unusual layout carries a 210 hp Lycoming and a passenger cabin in the fuselage, and a 200 hp engine in a parallel boom. Rutan claims a 264-kt cruise speed.

Hamilton Standard receives an FAA production certificate to build the 360 hp FV4000-2TC V-8 piston aero engine, a twin turbocharged derivative of the Lexus 4-liter auto engine.

1996 cont.

Fairchild Aircraft takes over Daimler-Benz Aerospace subsidiary **Dornier**, and Dutch manufacturer **Fokker** succumbs to bankruptcy. In December, **Boeing** announces it has acquired **McDonnell Douglas** in a stock swap.

Despite ongoing protests from national and local pilots and aviation associations, the City of Chicago attempts to close **Meigs Field**, its downtown lakefront business airport. In December, the State of Illinois takes over the property to foil Mayor Richard Daley's plans to turn the airport into a city park.

1997

Michael Raisler, a 26-year-old flight instructor from Clermont, Fla., wins the AOPA 1996 Sweepstakes Grand Prize, the first new Cessna 182 off the production line.

The Gulfstream V receives FAA certification and celebrates with a 6,012 nm nonstop flight from Tokyo to New York..

Mooney Aircraft delivers its 100th Ovation in Kerrville, Tex.

SabreTech, Inc., the Orlando, Fla. maintenance company implicated in the crash of **ValueJet 592** in the Florida Everglades, surrenders its repair station certificate and closes.

Meigs Field gains a five-year reprieve from closure. Meanwhile, a private airstrip in River Falls, Wis. is closed by a judge who found it violated local zoning laws.

The General Aviation Manufacturers Association reported that 1996 was the best year in history for total billings, as 1,132 units worth $3.1 billion were delivered, representing 5.1% more aircraft worth 10.6% more than those delivered the previous year. Meanwhile, FAA figures for 1995 indicate an 18% growth in corporate aviation flight hours.

A shortage of $1.2 billion is discovered in the Aviation Trust Fund when the Treasury Department admits it kept transferring money to the Fund while unaware Congress had suspended collections of aviation excise taxes at the end of 1996.

Cessna Aircraft is named winner of the **Collier Trophy** for the Citation X, and renewal of its single-engine-engine production receives *Flying* magazine's Editor's Choice award.

The congressionally mandated National Civil Aviation Review Commission draws flak when it is revealed there are no members from the general aviation business.

The EAA's fourth annual "**International Young Eagles Day**" June 14 provides more than 15,000 young people on five continents the world with their first airplane experience. Launched in 1992, the goal of the program is to provide one million free demonstration flights by the year 2003--the 100th anniversary of powered flight and 50th anniversary of the EAA.

Gone West. **Max Karant**, veteran pilot and founder and editor of *AOPA Pilot* magazine, passed away February 1 in Gaithersburg, Md. **Warren Basler**, the Oshkosh modifier who converted DC-3s to turbine power, was killed along with three of his employees in a March mid-air collision during a photo mission near Sheboygan, WI.

Executive Jet Aviation (EJA) placed the largest single order ever recorded for general aviation aircraft when it agreed to pay $210 million to Raytheon Aircraft for 20 Hawker 800 XPs which will be delivered over the next three years.

Two Texas pilots commemorated historic record flights; San Antonioan Linda Finch flew around the world retracing Amelia Earhart's 1937 route in a replica Lockheed 10E, and Dallas pilot Bill Signs relived Lindbergh's epic adventure on the same dates (and starting at the same times) 70 years later in a Cessna 210.

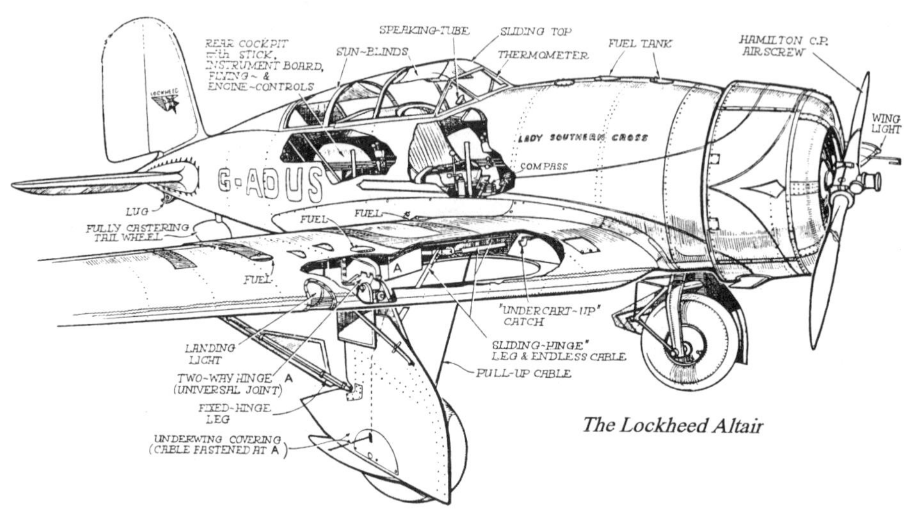

The Lockheed Altair

False reports of a successful 1927 transatlantic flight by Frenchmen Nungesser and Coli sparked a premature celebration in Paris. The pair disappeared without a trace, and two weeks later, Charles Lindbergh became the first person to cross the Atlantic nonstop.

The Gee Bees were perhaps the most famous--and potentially the most dangerous--racing airplanes of all. Even sixty years after the last Gee Bee raced, Delmar Benjamin thrills air show crowds throughout the country with his replica R-2.

GORDON BENNETT CUP

James Gordon Bennett, who was publisher of the *New York Herald*, lived in Paris, was an early supporter of both the automobile and flying. From 1900 to 1905, the Cup was given for road racing; in 1906, it began a long tradition in ballooning, and in 1909, it was first offered as a speed prize for airplanes.

1909, August 29, Reims, France--20 km (12.43 mi.)

Pilot	Aircraft (powerplant)	mph
Glenn Curtiss (US)	Curtiss Reims Racer (Curtiss V-8)	46.91
Louis Bleriot (F)	Bleriot XI (ENV V-8)	46.82
H. Hubert Latham (GB)	Antionette (Antoinette V-8)	42.49
Eugene Lefebvre (F)	Wright (Wright 4)	37.7
Cockburn (GB)	Farman (Gnome)	--

1910, October 29, Belmont, NY--100 km (62.137 mi.)

Pilot	Aircraft (powerplant)	mph
Claude Grahame-White (GB)	Bleriot XI bis (Gnome 14c)	61.25
John Moisant (F)	Bleriot XI (Gnome 7c)	31.50
Alec Ogilvie (GB)	Wright C (Wright 4c)	29.40
H. Hubert Latham (GB)	Antoinette (Antoinette V-16, 100 hp)	17.8
Leblanc (F)	Bleriot (Gnome 14)	--
Drexel (F)	Bleriot XI (Gnome 7)	--
J. Radley (US)	Bleriot XI bis Gnome 14)	--
Brookins (US)	Baby Wright (Wright V-8)	--

1911, Eastchurch, Isle of Sheppey, England--150 km (94 mi.)

Pilot	Aircraft (powerplant)	mph
Charles Weymann (US)	Nieuport monoplane (Gnome 7c)	78.08
Alfred Leblanc (F)	Bleriot (Gnome 7c)	75.84
Edouard Nieuport (F)	Nieuport (Gnome 7c)	75.08
Alec Ogilvie (GB)	Wright (N.E.C. V-8)	51.31
M. Chevalier (F)	Nieuport (Nieuport 7)	--
Hamel (GB)	Bleriot (Gnome 7)	--

1912, December 29, Chicago, IL--201 km (124.8 mi.)

Pilot	Aircraft (powerplant)	mph
Jules Vedrines (F)	Deperdussin (Gnome)	105.46
Maurice Prevost (F)	Deperdussin (Gnome 14c)	103.82
Andre Frey (F)	Hanriot	--

1913, September 29, Reims, France--200 km (124.3 mi.)

Pilot	Aircraft (powerplant)	mph
Maurice Prevost (F)	Deperdussin (Gnome)	124.82
Emile Vedrines (F)	Bleriot XI	123.00
Eugene Gilbert (F)	Deperdussin (Gnome 14c)	119.49
Henri Crombez (F)	Deperdussin	106.9

1920, September 28, Etampes, France--300 km (186.451 mi.)

Pilot	Aircraft (powerplant)	mph
Joseph Sadi-Lecointe (F)	Nieuport 29V (Hispano-Suiza V-8)	168.73
Bernard de Romanet (F)	Spad S.20 bis (Hispano Suiza V-8)	112.87
Kirch (F)	Nieuport (Hispano-Suiza V-8)	--
Raynham (GB)	Martinsyde Semiquaver (Hispano-Suiza V-8)	--
R.W. Schroeder (US)	Verville Packard (Packard)	--
Howard Rinehart (US)	Dayton Wright (Hall-Scott 6)	--
Roland Rohles (US)	Curtiss Texas Wildcat (Curtiss C -12)	--

HENRY DEUTSCH DE LA MEURTHE GRAND PRIX D'AVIATION

1912, May 1, Paris--200 km (124.3 mi.)

Pilot	Aircraft (powerplant)	mph
Emmanuel Helen (F)	Nieuport (Gnome 70)	77.9

1913, October 27, Paris--200 km (124.3 mi.)

Pilot	Aircraft (powerplant)	mph
Eugène Gilbert (F)	Deperdussin (Gnome 160)	101.5

1919, September 2-January 24, Paris--190.399 km (118.333 mi.)

Pilot	Aircraft (powerplant)	mph
Sadi Lecointe (F)	Nieuport 29V (Hispano-Suiza 42)	165.5

DEUTSCH DE LA MEURTHE CUP

1921, October 1, Paris--300 km (186.451 mi.)

Pilot	Aircraft (powerplant)	mph
Georges Kirsch (F)	Nieuport-Delage (Hispano-Suiza 42)	175.7
Fernand Lasne (F)	Nieuport-Delage 29V (Hispano Suiza 42)	160.9

1922, September 30, Paris--300 km (186.451 mi.)

Pilot	Aircraft (powerplant)	mph
Fernand Lasne (F)	Nieuport-Delage 29V (Hispano-Suiza 42)	180.1

LONDON AERIAL DERBY

1912, June 8, Hendon--81 mi. (130.329 km)

Pilot	Aircraft (powerplant)	mph
T.O.M. Sopwith (BG)	Blériot XI (Gnome 70)	58.4

1913, September 20, Hendon--95 mi. (152.855 km)

Pilot	Aircraft (powerplant)	mph
Gustav Hamel (GB)	Morane-Saulnier (Gnome 80)	75.1

1914, May 23, Hendon--94.5 mi. (152.05 km)

Pilot	Aircraft (powerplant)	mph
William L. Brock (US)	Morane-Saulnier (Gnome 80	71.8

1919, June 21, Hendon--189 mi. (304.101 km)

Pilot	Aircraft (powerplant)	mph
Gerald Gathergood (GB)	Airco DH-4R (Napier Lion II)	132.3
R.H. Nisbet (GB)	Martinsyde F.4 Buzzard (Rolls-Royce Falcon III)	124.9
Marcus D. Manton (GB)	Airco DH-4 (Rolls-Royce Eagle VIII)	118.7

1920, July 24, Hendon--200 mi. (321.8 km)

Pilot	Aircraft (powerplant)	mph
Francis T. Courtney (GB)	Martinsyde Semiquaver (Hispano-Suiza 42)	154.6
James H. James (GB)	Nieuport & General L.C.1 (ABC Dragonfly)	145.2
Cyril F. Uwins (GB)	Bristol 32 Bullet (Bristol Jupiter I)	136.1

1921, July 16, Hendon--200 mi. (321.8 km)

Pilot	Aircraft (powerplant)	mph
James H. James (GB)	Gloucestershire Mars I (Napier Lion II)	163.5
Cyril F. Uwins (GB)	Bristol 32 Bullet (Bristol Jupiter I)	142.9
Walter H. Longton (GB)	S.E.5a (Hispano-Suiza Wolseley Viper)	119.7

London Aerial Derby cont.

1922, August 7, Croydon--198 mi. (318.582 km)

Pilot	Aircraft (powerplant)	mph
James H. James (GB)	Gloucestershire Mars I (Napier Lion II)	178.7
Rollo A. de Haga Haig (GB)	Bristol 32 Bullet (Bristol Jupiter II)	150.2
Frederick P. Paynham (GB)	Martinsyde F.6 (Hispano-Suiza Wolseley Viper)	111.5

1923, August 6, Croydon--200 mi. (321.8m)

Pilot	Aircraft (powerplant)	mph
Larry L. Carter (GB)	Gloucestershire Gloster I (Napier Lion III)	195.6
Walter H. Longton (GB)	Sopwith 107 Hawker (Bristol Jupiter III)	166.5
C.D. Barnard (GB)	deHavilland DH-9A (Napier Lion Ia)	149.2

LOUIS D. BEAUMONT CUP

1924, October 14, Marseille--300 mm (186.451 mi.)

Pilot	Aircraft (powerplant)	mph
Sadi Lecointe (F)	Nieuport-Delage 42 (Hispano-Suiza 42)	197.4

1925, October 18, Marseille--300 km (186.451 mi.)

Pilot	Aircraft (powerplant)	mph
Sadi Lecointe (F)	Nieuport-Delage 42 (Hispano-Suiza 42)	194.1

SCHNEIDER TROPHY for seaplanes

Coupe d'Aviation Maritime Jacques Schneider was trials and a race for seaplanes which was open to any FAI-affiliated club. Hosted in the country of the previous winner after 1914, any country winning three times within five consecutive contests had permanent possession of trophy.

dns=did not start; dnf=did not finish; * Eliminated in sea trials

1913, April 16, Roquebrune Bay, Monaco--280 km (174 mi.)

Pilot	Aircraft (powerplant)	mph
Maurice Prevost (F)	Deperdussin (Gnome 160)	45.7
Roland Garros (F)	Morane-Saulnier (Gnome 80)	--
Charles Weyman (US)	Nieuport (Gnome 160)	dnf
Gabriel Espanet (F)	Nieuport (Gnome 100)	dnf
Georges Chemet (F)	Borel Monaco (Gnome 160)	*
Henri Breguet (F)	Breguet (Salmson/Canton-Unne 110/115)	*
Rene Moineau (F)	Breguet (Salmson/Canton-Unne 200)	*

1914, April 20, Roquebrune Bay, Monaco--280 km (174 mi.)

Pilot	Aircraft (powerplant)	mph
C. Howard Pixton (GB)	Sopwith Tabloid (Gnome 100)	83.73
Ernest Burri (Sz)	FBA (Gnome 100)	83.67
Gabriel Espanet (F)	Nieuport (Gnome 100)	dnf
Pierre Levasseur (F)	Nieuport (Gnome 100)	dnf
Charles Weyman (US)	Nieuport (Gnome 100)	dnf
Lord Carbery (GB)	Morane-Saulnier (Gnome 100)	dns
William Thaw (US)	Curtiss Triad (Gnome 100)	dns
Stoefler (G)	Aviatik (Benz 120)	dns
M. Prevost (F)	Deperdussin	*
Roland Garros (F)	Morane-Saulnier (Gnome 80)	*
Brindejonc dxes Moulinais	Morane-Saulnier (Gnome 100)	*
Janoir	Deperdussin	*
Moineau	Breguet (Salmson 120)	*
Bertin	Nieuport (Gnome 100)	*

Schneider Trophy cont.

1919, September 24, Bournemouth, England--370.4 km (230.205 mi.)

Pilot	Aircraft (powerplant)	mph
Guido Jannello (I)	Savoia S.13 bis (Isotta-Fraschini V-6 205)	Race voided
Vincent Nicholl (GB)	Fairey III N.10 (Napier Lion 450)	
Basil Hobbs (GB)	Supermarine Sea Lion I (Napier Lion)	
Buido Janello (I)	Savoia S.13 S (Isotta-Fraschini V-6 205)	
Jean Caslale (F))	Nieuport 29 (Hispano-Suiza 42 275)	
Sadi Lecointe (F)	Spad S.20 Hispano-Suiza 42 260)	
Harry Hawker (GB)	Sopwith (Cosmos Jupiter 450)	

1920, September 21, Venice, Italy--371.17 km (230.683 mi.)

Pilot	Aircraft (powerplant)	mph
Luigi Bologna (I)	Savoia S.12 bis (Ansaldo V-12)	105.97

1921, August 7, Venice, Italy--394.1 km (244.938 mi.)

Pilot	Aircraft (powerplant)	mph
Giovanni De Briganti (I)	Macchi M.7 bis (Isotta 6 200)	117.85
Arturo Zanetti (I)	Macchi M.19 (Fiat A-14)	dnf
Piero Corgnolino (I)	Macchi M.7 (Isotta-Fraschini 250)	dnf

1922, August 12, Naples, Italy--370.689 km (230.385 mi.)

Pilot	Aircraft (powerplant)	mph
Henry Biard (GB)	Supermarine Sea Lion II (Napier Lion 450)	145.72
Alessandro Passaleva (I)	Savoia S.M.51 (Hispano-Suiza 300)	142.67
Arturo Zanetti (I)	Macchi M.17 bis (Isotta-Fraschini)	132.78
Piero Corniglio (I)	Macchi M.17bis	90.6

1923, September 28, Cowes, Isle of Wight, England--344.472 km (214.090 mi.)

Pilot	Aircraft (powerplant)	mph
David Rittenhouse (US)	Curtiss CR-3 (Curtiss D-12 465)	177.27
Rutledge Irvine (US)	Curtiss CR-3 (Curtiss D-12 465)	173.38
Henry Biard (GB)	Supermarine Sea Lion II (Napier 525)	157.06
Hurel (F)	CAMS 38a (Hispano-Suiza 360)	dnf
Frank Wead (US)	Wright TR-3A (Wright-Hispano E-4A 290)	dnf
Georges Pelletier d'Oisy	CAMS 38bis (Hispano-Suiza 360)	dnf
Duhame (F)	Latham No. 1 (2 Lorraine-Dietrich 400)	dns

1925, October 26, Bay Shore Park, MD--350 km (217.48 mi.)

Pilot	Aircraft (powerplant)	mph
James Doolittle (US)	Curtiss R3C-2 (Curtiss V-1400)	232.57
Hubert Broad (GB)	Gloster IIIA (Napier Lion VII)	199.17
Giovanni De Briganti (I)	Macchi M.33 (Curtiss D-12A)	168.44
George Cuddihy (US)	Curtiss R3C-2 (Curtiss V-1400)	dnf
Ralph Ofstie (US)	Curtiss R3C-2 (Curtiss V-1400)	dnf
Henry Biard (GB)	Supermarine S.4 Napier Lion VII 700)	dns

1926, November 13, Hampton Roads, VA--250 km (217.48 mi.)

Pilot	Aircraft (powerplant)	mph
Mario De Bernardi (I)	Macchi M.39 (Fiat A.S.2 V-12)	246.49
Christian Schlit (US)	Curtiss R3C-2 (Curtiss V-1400)	231.36
Adriano Bacula (I)	Macchi M.39 (Fiat A.S.2 V-12)	218.05
William Tomlinson (US)	Curtiss F6C-3	137.02
George Cuddihy (US)	Curtiss R3C-4 (Packard V-1550)	dnf
Arturo Ferrarin (I)	Macchi M.39 (Fiat As.2)	dnf
S.N. Webster (GB)	Supermarine S.5 (Napier Lion VIIG)	281.65
O.F. Worseley (US)	Supermarine S.5 (Napier Lion VIIG)	273.07
Guazzetti (I)	Macchi M.52 (Fiat As.2)	dnf

1927 Schneider Trophy cont.

1927, September 26, Venice, Italy--350 km (217.48 mi.)

Pilot	Aircraft (powerplant)	mph
Kinkead (GB)	Gloster IVB (Napier Lion VIIB)	dnf
deBernardi (I)	Macchi M.52 (Fiat As.3 1,000)	dnf
Arturo Ferrarin (I)	Macchi M-52 (Fiat As.3 1,000)	dnf

1929, September 7, Crowes, Isle of Wight, England--350 km (217.48 mi.)

Pilot	Aircraft (powerplant)	mph
H.R.D. Waghorn (GB)	Supermarine S.6 (Rolls-Royce R 1,900)	328.62
Tomaso Dal Molin (I)	Macchi M.52R (Fiat As.3 V-12)	284.20
d'Arcy Greig (GB)	Supermarine S.5 (Napier Lion VIIG)	282.11
Richard Atcherly (GB)	Supermarine S.6 (Rolls-Royce R 1,900)	dnf
Remo Cadringher (I)	Macchi M.67 (Isotta-Fraschini Asso 1,800)	dnf
Giovanni Monti (I)	Macchi M.67 (Isotta-Fraschini Asso 1,800)	dnf

1931, September 13, Calshot, Lee-on-Solent, England--350 km (217.48 mi.)

Pilot	Aircraft (powerplant)	mph
John Boothman (GB)	Supermarine S.6B (Rolls-Royce R 2,600)	340.10

THE PULITZER TROPHY

Sponsored by Ralph, Joseph Jr. and Herbert Pulitzer, publishers of the *New York World* and St. Louis *Post-Dispatch*, open to any civil or military aircraft.

1920, Nov. 25, Mitchel Field, NY

Pilot	Aircraft (powerplant)	mph
Lt. Corliss Moseley	Verville-Packard (Packard V-12 638)	156.50
Capt. Harold Hartney	Thomas-Morse Scout MB-3 (Wright 300)	148.00
Bert Acosta	Ansaldo "Balilla" (SPA 6)	134.50
Lt. St.Clair Streett	Orenco Model D (Wright 300)	133.00
Lt. A. Lavvrents	Vought VE-7 (Wright-Hispano V-8)	125.0
Lt. John Roullot	de Havilland DH-4 (Liberty)	124.0
Willis Taylor	SVA-9 (SPA 6)	117.0
Capt. Maxwell Kirby	SE-5A (Wright-Hispano V-8)	116.7
Charles Colt	Morane-Saulnier (LeRhone)	95.0
Lt. B.G. Bradley	Leoning Special (Wright-Hispano V-8)	150.0

1921, Nov. 3, Omaha, NE--153.5 mi.

Pilot	Aircraft (powerplant)	mph
Bert Acosta	Curtiss CR-1 (Curtiss CD-12 465)	176.70
Clarence Coombs	Curtiss-Cox "Cactus Kitten" (Curt. C-12 435)	170.26
James Macready	Thomas-Morse MB-6 (Wright V-8)	160.70
Lloyd Bertaud	Ansaldo "Balialla" (Curtiss C-12)	149.70
Harold E. Hartney	Thomas-Morse MB-7 (Wright V-8)	--
James Curran	SVA-9 (PA 6)	--

1922, Oct. 14, Detroit, MI--155.34 mi.

Pilot	Aircraft (powerplant)	mph
Lt. Russell Maughan	Curtiss R-6 Army Racer (Curtiss D-12 460)	205.80
Lt. Lester Maitland	Curtiss R-6 Army Racer (Curtiss D-12 460)	198.80
Lt. Harold Brow	Curtiss R-2 Navy Racer (Curtiss D-12 405)	193.80
Lt. A.J. Williams	Curtiss R-1 Navy Racer (Curtiss D-12 405)	186.70
Lt. E.H. Barksdale	Verville-Sperry R-3 (Curtiss D-12 500)	181.20

1922 Pulitzer cont.

Pilot	Aircraft (powerplant)	mph
Capt. Corliss Moseley	Verville-Packard VCP-1 (Packard 1A-2025)	179.00
Lt. F.B. Johnson	Verville-Sperry R-3 (Wright H-3)	178.00
Lt. E.C. Whitehead	Loening R-4 (Packard 1A-2025)	170.2
Schulz	Leoning R-4 (Packard 1A-2025)	160.9

1923, Oct. 4, St. Louis, MO

Pilot	Aircraft (powerplant)	mph
Lt. Al Williams	Curtiss R2C-1 (Wright T-3 750)	243.67
Lt. Harold Brow	Curtiss R2C-1 (Wright T-3 750)	241.78
Lt. Lawson Saunderson	Navy-Wright F2W (Wright T-3)	230.06
Lt. Stephen Callaway	Navy-Wright F2W (Wright T-3)	230.00
Lt. Walter Miller	Army-Curtiss R-6 (Curtiss D-12)	218.00
Lt. John Corkill	Army-Curtiss R-6 (Curtiss D-12)	216.45
Lt. A. Pearson	Verville-Sperry R-3 (Curtiss D-12)	--

1924, Oct. 4, Dayton, OH

Pilot	Aircraft (powerplant)	mph
Lt. H.H. Mills	Verville-Sperry R-3 (Curtiss CD-12)	216.72
Lt. Wendell Brookley	Army Curtiss R-6 (Curtiss D-12)	214.45
Lt. Rex Stoner	Curtiss P-W-8A (Curtiss D-12)	168.00
Capt. Burt Skeel	Army Curtiss R-6 (Curtiss D-12)	--

1925, Oct. 13, Mitchel Field, NY--124.28 mi.

Pilot	Aircraft (powerplant)	mph
Lt. Cyrus Bettis	Curtiss RC3-1 Army Racer (C.V-1400)	248.99
Lt. Al Williams	Curtiss RC3-1 Navy Racer (C.V-1400)	241.70
Lt. Dawson	Curtiss P-1 (Curtiss V-1150)	169.90
Lt. Norton	Curtiss PW-8 (Curtiss D-12)	168.80
Capt. Cook	Curtiss PW-8 (Curtiss D-12)	167.40
Lt. C.T. Cuddihy	Curtiss PW-8 (Curtiss D-12)	--

NATIONAL AIR RACES

1926--Kansas City Rotary Club Trophy (military), Philadelphia--10 laps, 200 mi.

Pilot	Aircraft (powerplant)	mph
Lt. George Cuddihy	Boeing FB-3 (Packard 2A-1500)	180.49
Lt. L.G. Elliott	Curtiss P-2 (Curtiss V-1400)	178.61
Capt. Ross Hoyt	Curtiss P-2 (Curtiss V-1400)	170.91
Lt. C.C. Nutt	Curtiss P-2 (Curtiss V-1400)	170.76
Kt. H.T. McCormick	Curtiss P-2 (Curtiss V-1400)	169.59
Lt. H.D. Barner	Boeing FB-3 (Packard 2A-1500)	163.57
Lt. L.H. Sanderson	Boeing FB-3 (Packard 2A-1500)	163.36
Lt. A.B. Ballard	Curtiss P-1 (Curtiss D-12)	159.25
Lt. W. McKierman	Curtiss P-1 (Curtiss D-12)	Disq.

1927--Spokane Spokesman-Review Trophy, Spokane--10 laps, 120 mi.

Pilot	Aircraft (powerplant)	mph
Lt. E.C.Batten	Curtiss XP-6A (Curtiss V-1570-1)	201.24
Lt. A.J. Lyon	Curtiss XP-6 (Curtiss V-1570-1)	189.61
Lt. Thomas Jeter	Boeing FB-5 (Packard 2A-1500)	176.94
Lt. H.E. Regan	Boeing FB-5 (Packard 2A-1500)	175.93
Lt. Gerald Bogan	Boeing FB-5 (Packard 2A-1500)	172.87
Lt. C.H. Beverly	Boeing PW-9C (Curtiss D-12D)	169.73
Lt. F.C. Rogers	Curtiss F6C-4 (P&W Wasp)	161.56
Lt. W.L. Cornelius	Curtiss P-1B (Curtiss D-12)	161.50
Lt. I.A. Woodring	Curtiss P-1B (Curtiss D-12)	159.18

1927 National Air Races cont.

Pilot	Aircraft (powerplant)		
Lt. L.C. Mallory	Curtiss P-1B (Curtiss D-12)		disq.

1928--Military free-for-all, Los Angeles--60 mi., 6 laps

Pilot	Aircraft (powerplant)		*mph*
Lt. Thomas Jeter	Boeing XF4B-1 (P&W Wasp)		172.26
Lt. Edgar Cruise	Boeing F2B-1 (P&W Wasp)		159.86
Lt. Harrigan	Boeing F2B-1 (P&W Wasp)		151.60
Lt. Burroughs	Boeing F2B-1 (P&W Wasp)		150.3
Lt. Crommelin	Boeing F2B-1 (P&W Wasp)		149.8
Lt. Williamson	Boeing F2B-1 (P&W Wasp)		146.00

1928--New York to Los Angeles

Class A--to 510 cu. in.--$5,000 to winner

Pilot	Aircraft (powerplant)	*(hr.)*	*mph*
Earl Rowland	Cessna AW (Warner)	27:00:31	
Robt. Drake/Theodore Taney	American Moth (Warner)	28:18:43	
W.H. Emery, Jr.	Travel Air (Warner)	28:48:28	

Class B--510-800 cu. in.--$7,000 to winner

Pilot	Aircraft (powerplant)	*(hr.)*	*mph*
John Livingston	Waco-10 (Whirlwind)	22:56:59	
E.E. Ballough	Laird (Wright Whirlwind)	23:16:24	
John P. Wood	Waco-10 (Whirlwind)	24:31:08	

Class C--larger than 800 cu. in.--$5,000 to winner

Pilot	Aircraft (powerplant)	*(hr.)*	*mph*
Robert Cantwell	Lockheed Vega (P&W Wasp)	24:09:01	
Capt. C.B.D. Collyer	Fairchild	27:10:45	
Edward J. Brooks	Fokker	27:24:53	

1929--Cleveland, August 24-September 2

Women's Derby--Santa Monica to Cleveland--$3,000 to winner

Pilot	Aircraft (powerplant)	*hrs.*	*mph*
Louise Thaden	Travel Air Speedwing (Wright J-5)	20:02:02	135.97
Gladys O'Donnell	Waco 220 (Wright J-5)	21:21:43	127.52
Amelia Earhart	Lockheed (Wright J-5)	22:12:42	122.64

Oakland to Cleveland Derby--$3,000 to winner

Pilot	Aircraft (powerplant)	*hrs.*	*mph*
Loren Mendell	Buhl (Whirlwind 300)	17:43:16	
W.J. Barrows	Fairchild 71 (P&W Wasp)	17:46:45	
J.O. Donaldson	Travel Air (Wright 300)	18:17:18	

Portland to Cleveland Derby--$3,000 to winner

Pilot	Aircraft (powerplant)	*hrs.*	*mph*
T.H. Wells	Travel Air (Wright J-5)	14:44:10	
Tex Rankin	Waco Taperwing (Wright 225)	15:26:24	
Snydor Hall	Travel Air (Wright J-5)	17:26:00	

Miami to Cleveland Derby--$1,550 to winner

Pilot	Aircraft (powerplant)	*hrs.*	*mph*
Robert Dake	Davis (Warner)	13:06:02	
Charles Meyers	Great Lakes (Cirrus)	13:10:58	
C.A. Burrows	Fleet (Kinner K-5)	13:27:33	

1929 National Air Races cont.

Philadelphia to Cleveland Derby--$1,636 to winner

Pilot	*Aircraft (powerplant)*	*hrs.*	*mph*
Errett Williams	Bullet (Whirlwind 165)	6:31:31	
Ike Stewart	Monocoach (Whirlwind 165)	7:29:21	
Howard Young	Bellanca (Wright J-5)	7:30:09	

Non-stop Los Angeles to Cleveland--$5,000 to winner

Pilot	*Aircraft (powerplant)*	*hrs.*	*mph*
Harry Brown	Lockheed (P&W Hornet)	13:15:07	156.20
Lee Shoenhair	Lockheed Vega (P&W Wasp)	13:51:11	149.65

Free-for-all speed contest--5 laps of 10 miles--$750 to winner

Pilot	*Aircraft (powerplant)*	*min.*	*mph*
Doug Davis	Travel Air Mystery S (Whirlwind 300)	14:05.9	194.90
Lt. R.G. Breene	Curtiss P-3A (P&W Wasp)	14:42.4	186.84
Roscoe Turner	Lockheed Vega (P&W Wasp)	16:48.7	163.44
Lt. J.J. Clarke	Curtiss Hawk		

Thompson Trophy--Free-for-all closed course speed dash sponsored by Thompson Products, Inc., limited to fastest qualifying aircraft and engine displacement less than 1,860 cu. in.

Bendix Trophy--Cross-country race to site of National Air Races sponsored by Bendix Aviation Corp., open to all aircraft.

Greve Trophy--established by Louis W. Greve, president of Cleveland Pneumatic Tool Co.and president of the National Air Races for fastest qualifying aircraft with engine displacement under 549 cu. in.

1930--Chicago--August 23-September 1

Thompson Trophy Race--20 laps of five miles--$5,000 to winner

Pilot	*Aircraft (powerplant)*	*Time (min.)*	*mph*
Charles Holman	Laird Solution (P&W Wasp)	29:43.0	201.91
James Haizlip	Travel Air R (Wright Whirlwind)		199.80
Ben Howard	Howard DGA-3 "Pete" (Wright Gipsy)		162.80
Paul T. Adams	Travelair Speedwing (W. Whirlwind)		142.64
Arthur Page	Curtiss XF6C-6 (Curtiss Conqueror)		DNF
Errett Williams	Wedell 92 (Wright Whirlwind)		DNF
Frank Hawks	Travel Air R (Wright Whirlwind)		DNF

Fastest lap: 219 mph, Page

1931--Cleveland--August 29-September 7

Bendix Transcontinental Trophy Race--Los Angeles to Cleveland (2,043 mi.)--$7,500 to winner

Pilot	*Aircraft (powerplant)*	*time (hr.)*	*mph*
J.H. Doolittle	Laird Super Solution (P&W Wasp Jr.)	9:10:21*	233.03
Harold Johnson	Lockheed Orion (P&W Wasp)	10:14:22	198.81
Beeler Blevins	Lockheed Orion (P&W Wasp)	10:49:33	188.99
Ira C. Eaker	Lockheed Altair (P&W Wasp)	10:59:45	186.07
Arthur Goebel	Lockheed Altair (P&W Wasp)	11:55:48	171.50
James G. Hall	Lockheed Altair(P&W Wasp)	12:51:16	159.16

DNF: Louis Reichers, Lockheed Altair

**Doolittle continued on to Newark, NJ for a transcontinental time of 11 hours, 16 minutes and 10 seconds (216.958 mph), breaking existing marks.*

1931 National Air Races cont.

Thompson Trophy Race--10 laps of 10 miles--$7,500 to winner

Pilot	Aircraft (powerplant)	time (min.)	mph
Lowell Bayles	Gee Bee Z (P&W Wasp Jr.)	25:24.0	236.24
J.R. Wedell	Wedell-Williams (P&W Wasp Jr.)		227.99
Dale Jackson	Laird Solution (Wright J6)		211.18
Robert L. Hall	Gee Bee Y (P&W Wasp C)		201.51
Ira C. Eaker	Lockheed Altair (Wright Whirlwind)		192.82
Ben O. Howard	Howard DGA-3 "Pete" (Curtiss Gipsy)		163.51
William Ong	Laird Speedwing (Wright J6)		153.04
J.H. Doolittle	Laird "Super Solution" (P&W Wasp Jr.)		DNF

Fastest lap: 241 mph, Bayles

1932--Cleveland--August 27-September 5

Bendix Transcontinental Trophy Race--Burbank to Cleveland (2,043 mi.)--$6,750 to winner

Pilot	Aircraft (powerplant)	time (hr.)	mph
J.H. Haizlip	Wedell-Williams (P&W Wasp Jr.)	8:19:45*	245.00
J.R. Wedell	Wedell-Williams (P&W Wasp Jr.)	8:47:31	232.00
Roscoe Turner	Wedell-Williams (P&W Wasp Jr.)	9:03:25	226.00
Lee Gehlbach	Gee Bee R-2 (P&W Wasp)	9:41:39	

DNF: Clair Vance, Vance Flying Wing;. DNS: Frank Hawks, Northrop Gamma 2A; Robert Hall, Hall Bulldog; J.H. Doolittle Laird Super Solution

** Haizlip continued to New York for a transcontinental time of 10 hours, 19 minutes, breaking the existing speed record.*

Thompson Trophy Race--10 laps of 10 miles--$4,500 to winner

Pilot	Aircraft (powerplant)	time (min.)	mph
J.H. Doolittle	Gee Bee R-1 (P&W Wasp)	23:44.69	252.68
J.R. Wedell	Wedell-Williams (P&W Wasp Jr.)		242.49
Roscoe Turner	Wedell-Williams (P&W Wasp Jr.)		233.04
J.H. Haizlip	Wedell-Williams (P&W Wasp Jr.)		231.30
Lee Gehlbach	Gee Bee R-2 (P&W Wasp Jr.)		222.09
Robert Hall	Hall "Bulldog" (P&W Wasp)		215.57
William Ong	Howard "Ike" (Menasco B6)		191.07
Ray Moore	Rider R-1 (Menasco C6S)		DNF

Fastest lap: 266 mph, Doolittle

Shell Speed Dashes for World Record over 3 Kilometer Course		*mph*
J.H. Doolittle	Gee Bee R-2 (P&W Wasp)	296.28
J.R. Wedell	Wedell-Williams (P&W Wasp Jr.)	277.05
Roscoe Turner	Wedell-Williams (P&W Wasp Jr.)	266.67
J.H. Haizlip	Wedell-Williams (P&W Wasp Jr.)	266.44
Lee Gehlbach	Gee Bee R-1 (P&W Wasp Jr.)	247.33
Robert Hall	Hall "Bulldog" (P&W Wasp)	243.71
Ray Moore	Keith Rider (Menasco)	237.73
Ben Howard	Howard "Ike" (Menasco B6S)	213.85
L.L. Bowen	Israel (Menasco)	202.49

1933--Los Angeles--July 1-4

Bendix Transcontinental Trophy Race--New York to Los Angeles (2,450 mi.)--$5,050 to winner

Pilot	Aircraft (powerplant)	time (hr.)	mph
Roscoe Turner	Wedell-Williams (P&W Wasp Jr.)	11:30:00	214.78
J.R. Wedell	Wedell-Williams (P&W Wasp Jr.)	11:58:18	209.23

DNF: Russell Boardman+, Gee Bee R-1; Russell Thaw, Gee Bee R-1, DNS: Arthur Knapp, Laird Sol-ution; Lee Gehlbach, Wedell-Williams; Amelia Earhart, Lockheed Vega; Ruth Nichols, Lockheed Orion; Claire Vance, Vance Flying Wing; Alex de Seversky, Seversky SEV-3. *+ killed in crash at Indianapolis*

1933 National Air Races cont cont.

Thompson Trophy Race--10 laps of 10 miles--$3,375 to winner

Pilot	Aircraft (powerplant)	time (min.)	mph
J.R. Wedell	Wedell-Williams (P&W Wasp Jr.)	15:08.0	237.95
Lee Gehlbach	Wedell-Williams (P&W Wasp Jr.)		224.94
Roy Minor	Howard "Mike" (Menasco B6S)		199.87
George Hague	Keith-Rider (Menasco C4S)		183.20
Z.D. Granville	Gee Bee Y (P&W Wasp)		173.07
Roscoe Turner	Wedell-Williams (P&W Wasp Jr.)		241.03 (disq.)
F. Klingensmith+	Gee Bee Y (P&W Wasp Jr.)		-

+ killed in crash. Fastest lap: 265 mph, Turner

Shell Speed Dash--unlimited, 3 kilometers--$1,125 to winner

		mph
Roscoe Turner	Wedell-Williams (P&W Wasp Jr.)	280.24
J.R. Wedell	Wedell-Williams (P&W Wasp Jr.)	278.92
Lee Gehlbach	Wedell-Williams (P&W Wasp Jr.)	251.93
Roy Minor	Howard "Mike" (Menasco B6S)	241.61
Ray Moore	Keith-Rider (Menasco)	231.70

1934--Cleveland--August 31-September 3

Bendix Transcontinental Trophy Race--Los Angeles to Cleveland (2,043 mi.)--$4,500 to winner

Pilot	Aircraft (powerplant)	time (hr.)	mph
Doug Davis	Wedell-Williams (P&W Wasp)	9:26:41	216.23
J.A. Worthen	Wedell-Williams (P&W Wasp)	10:03:00	203.21

DNF: Lee Gehlbach, Gee Bee QED;. DNS: Roscoe Turner, Wedell-Williams; Harold Neumann, How-ard DGA-6; Murray Dilley, Vance Flying Wing; Roy Minor, Gee Bee R-1/2; James Granger+, Keith Rider R-3; Lee Miles, Wedell-Williams; Jacqueline Cochran, Northrop Gamma 2G.

+ killed in crash at Burbank

Thompson Trophy Race--12 laps of 12 1/2 miles--$4,500 to winner

Pilot	Aircraft (powerplant)	time (min.)	mph
Roscoe Turner	Wedell-Williams (P&W Hornet)	24:10.36	248.12
Roy Minor	Brown B-2 (Menasco C6S)	27:54.58	214.92
J.A. Worthen	Wedell-Williams (P&W Wasp)	28:47.65	208.27
Harold Neuman	Howard "Ike" (Menasco B6S)	28:58.59	207.06
Roger Don Rae	Keith-Rider R-1 (Menasco)	29:13.04	205.35
Art Chester	Chester "Jeep" (Menasco)	31:18.94	191.59
Lee Miles	Miles & Atwood (Menasco C4S)		DNF
Doug Davis+	Wedell-Williams		-

+ killed in crash while leading

Greve Trophy Race--best record in three races--$765 to winner

		mph
Lee Miles	Miles & Atwood (Menasco C4S)	206.24

Shell Unlimited

		mph
J.A. Worthen	Wedell-Williams (P&W Wasp)	302.03

1935--Cleveland--August 30-September 2

Bendix Transcontinental Trophy Race--Los Angeles to Cleveland (2,043 mi.)--$4,500 to winner

Pilot	Aircraft (powerplant)	time (hr.)	mph
Ben Howard	Howard DGA-6 "Mr. Mulligan" (P&W Wasp)	8:33:16	238.70
Roscoe Turner	MacMillan Special (P&W Hornet)	8:33:39	238.55
Russell Thaw	Northrop Gamma 2D-2 (Wright Cyclone)	10:06:45	201.92
Roy Hunt	Lockheed Orion 9D-1 (P&W Wasp)	11:41:03	174.76
Amelia Earhart	Lockheed Vega (P&W Wasp)	13:47:06	149.57

DNF: Royal Leonard, Gee Bee QED; Cecil Allen+, Gee Bee R-1/2; Earl Ortman, Rider R-3; Jacqueline

1935 Bendix cont.

Cochran, Northrop Gamma 2G. DNS: William Ong, Wedell-Williams; Howard Hughes, Hughes Racer; Seward Pulitzer, Northrop Delta 1D; Maurice Rossi, Renault Racer; Vance Breese, Breese-Dallas Transport.
+ killed in crash at Burbank

Thompson Trophy Race--15 laps of 10 miles--$6,750 to winner

Pilot	*Aircraft (powerplant)*	*time (min.)*	*mph*
Harold Neumann	Howard Mr. Mulligan (P&W Wasp)	40:52.38	220.19
Steve J. Wittman	Wittman Bonzo (Curtiss D-12)		218.68
Roger Don Rae	Keith-Rider R-4 (Menasco C6S)		213.94
Joe Jacobs	Howard "Pete" (Menasco B6S)		209.10
Lee Miles	Seversky SEV-3 Amphibian (Wright Cyclone)		193.59
Marion McKeen	Brown B-2 (Menasco C6S)		188.85
Roscoe Turner	Wedell-Turner (P&W Hornet)		DNF

Fastest lap: 240 mph, Turner

Greve Trophy Race--best record in three races--$1,125 to winner

Pilot	*Aircraft (powerplant)*	*mph*
Harold Neumann	Howard "Mike" (Menasco C6S)	212.71
Don Rae	Keith-Rider R-1 (Menasco C6S)	210.1
McKeen	Brown B-2 (Menasco C6S)	206.4
Art Chester	Chester Jeep (Menasco C4S)	199.1
Lee Miles	Miles & Atwood Spl. (Menasco C4S)	189.6
Steve Wittman	Chief Oshkosh (Cirrus Hermes)	189.4
Elmendorf	Wedell-Williams	175.1

1936--Los Angeles--September 4-7

Bendix Transcontinental Trophy Race--New York to Los Angeles (2,450 mi.) $7,000 to winner

Pilot	*Aircraft (powerplant)*	*time (hr.)*	*mph*
Louise Thaden	Beech C17-R (Wright 420)	14:55:01	165.34
Laura Ingalls	Lockheed Orion (P&W Wasp)	15:39:38	157.46
William Gulick	Vultee V-1A (P&W Wasp)	15:45:25	156.49
George Pomeroy	Douglas DC-2 (P&W Wasp)	16:16:51	151.47
Amelia Earhart	Lockheed Electra 10E (P&W Wasp)	16:34:53	

DNF: Joe Jacobson, Northrop Gamma 2A; Ben Howard Howard DGA-6 "Mr. Mulligan"; Lee Miles, Gee Bee QED. DNS: Roscoe Turner, Turner Laird LTR-14; Roscoe Turner/Frank Clark, Wedell-Williams; Steve Wittman, "Bonzo"; George Haldeman, Bellanca 28-70; Jacqueline Cochran, Breese-Dallas Transport.

Thompson Trophy Race--15 laps of 10-miles--$9,500 to winner

Pilot	*Aircraft (powerplant)*	*time (min.)*	*mph*
Michel Detroyat (F)	Caudron C-460 (Renault)	34:03.42	264.26
Earl Ortman	Rider R-3 (P&W Wasp)		248.04
Roger Don Rae	Rider R-4 (Menasco B6S)		236.55
Harold Neumann	Folkerts SK-2 (Menasco C4S)		233.07
Marion McKeen	Brown B-2 (Menasco C6S)		230.46
Harry Crosby	Crosby C6R3 (Menasco C6S)		226.07
Lee Miles	Granville QED (P&W Wasp)		DNF

Fastest lap: 301 mph, Detroyat

Greve Trophy Race--20 laps of 10 miles--$4,900 to winner

Pilot	*Aircraft (powerplant)*	*time (min.)*	*mph*
Michael Detroytat (F)	Caudron (Renault)		247.30
Harold Neumann	Folkerts SK-2 (Menasco C4S)		225.85
Art Chester	Chester Jeep (Menasco C4S)		224.68
R.A. Kling	Kling-Rider (Menasco)		215.33
Joe Jacobson	Howard DGA-4 (Menasco C6S)		214.42
Roger Don Rae	Rider R-4 (Menasco B65-489)		212.32
McKeen	Brown B-2 (Menasco C6S)		204.5
Miles	Miles & Atwood Spl. (Menasco C4S)		

National Air Races cont

1937--Cleveland--September 3-6

Bendix Transcontinental Trophy Race--Los Angeles to Cleveland(2,043 mi.)--$13,000 to winner

Pilot	Aircraft (powerplant)	time (hr.)	mph
Frank Fuller	Seversky Sev-S2 (P&W Twin Wasp)	7:54:27*	258.24
Earl Ortman	Keith Rider R-3 (P&W Twin Wasp Jr.)	9:49	224.83
Jacqueline Cochran	Beech D17W (P&W Wasp Jr.)	10:29	194.74
Frank Sinclair	Seversky Sev-S2 (P&W Twin Wasp)	11:02	184.92
Milo Burcham	Lockheed 12A (2 P&W Twin Wasp)	11:03	184.52
Eiler Sundorph	Sundorph (Jacobs L-5)	12:17	166.21

DNF: Roscoe Turner, Turner-Laird LTR-14; Robert Perlick, Beech A17F; Joseph Mackey, Wedell-Williams. DNS: Arthur Davis, Wedell-Williams; Alex de Seversky, Seversky AP-2; Alex Pampana, Bellanca 28-92; Reg Robbins, Wedell-Williams.

** Fuller continued on to Bendix, NJ to set a new transcontinental record in 9:35:00, 255 mph*

Thompson Trophy Race--20 laps of 10 miles--$9,000 to winner

Pilot	Aircraft (powerplant)	time (min.)	mph
Rudy Kling	Folkerts SK-3 (Menasco C6S)	46:42.54	256.91
Earl Ortman	Keith Rider R-3 (P&W Twin Wasp Jr.)	46:43	256.85
Roscoe Turner	Laird Turner (P&W Twin Wasp)	47:16	253.80
Frank Sinclair	Seversky Sev-S2 (P&W Twin Wasp)	47:33	252.36
S.J. Wittman	Wittman "Bonzo" (Curtiss D-12)	47:58	250.10
Ray Moore	Seversky Sev-S2 (P&W Twin Wasp)	50:20	238.41
Gus Gotch	Schoen-Rider R-4 (Menasco B65-489)	55:05	217.81
Joe Mackey	Wedell-Turner (P&W Hornet)		DNF
Marion McKeen	Brown B-2 (Menasco C6S)		DNF

Fastest lap: 279 mph, Kling

Greve Trophy Race--20 laps of 10 miles--$4,500 to winner

Pilot	Aircraft (powerplant)	time (min.)	mph
Rudy Kling	Folkerts SK-3 (Menasco C6S4)	25:49	232.27
S.J. Wittman	Wittman "Chief Oshkosh" (Menasco CS4)	25:51	231.99
C.H. Gotch	Schoenfeldt Rider R-4 (Menasco C6S4)	25:54	231.59
Roger Don Rae	Folkerts (Menasco C4S)	26:45	224.19
Marion McKeen	Brown B-2 (Menasco C6S)	26:49	223.64
Haines	Haines H-3		177.71
McArthur	Delagado Flash		

1938--Cleveland--September 3-5

Bendix Transcontinental Trophy Race--Burbank to Cleveland (2,043 mi.)--$12,500 to winner

Pilot	Aircraft (powerplant)	time (hr.)	mph
Jacqueline Cochran	Seversky Sev-S2 (P&W Twin Wasp)	8:10:31	249.77
Frank Fuller, Jr.	Seversky Sev-S2 (P&W Twin Wasp)	8:33:29	238.60
Paul Mantz	Lockheed Orion 9C (Wright Cyclone)	9:36:25	206.57
Max Constant	Beech D17W (P&W Wasp Jr.)	10:14:39	199.33
Ross Hadley	Beech D17S (P&W Wasp Jr.)	11:13:46	181.84
John Hinchey	Spartan Executive 7W (P&W Wasp Jr.)	11:30:27	177.44

DNF: George Armistead, Gee Bee QED; Lee Gehlbach, Wedell-Williams; Frank Cordova, Bellanca 28-92; Bob Perlick, Beech A-17F. DNS: Bernarr MacFadden, Northrop Gamma 2G.

Thompson Trophy Race--30 laps of 10 miles--$22,000 to winner*

Pilot	Aircraft (powerplant)	time (min.)	mph
Roscoe Turner	Turner-Laird "Meteor" (P&W Twin Wasp)	63:30.61	283.41
Earl Ortman	Marcoux-Bromberg Rider R-3 (P&W Tw. Wasp Jr.)	66:44.18	269.71
S.J. Wittman	Wittman "Bonzo" (Curtiss D-12)	69:26:87	259.18
Leigh Wade	Hawks "Time Flies" (P&W Twin Wasp)	72:02.74	249.84

1938 Thompson Trophy cont.

Joe Mackey	Wedell-Williams (P&W Hornet)	72:06.44	249.62
Joe Jacobson	Keith Rider R-6 (Menasco C6S4)		214.57
Art Chester	Chester "Goon" (Menasco C6S4)		DNF
Harry Crosby	Crosby CR-4 (Menasco C6S4)		DNF

*Fastest lap: 293 mph, Turner. *includes $4,000 for new record*

Greve Trophy Race--20 laps of 10 miles--$12,000 to winner

Pilot	*Aircraft (powerplant)*	*time (min.)*	*mph*
Tony LeVier	Schoenfeldt "Firecracker"Rider R-4 (Menasco C6S4)	47:49.89	250.88
Art Chester	Chester "Goon" (Menasco C6S4)	47:55.22	250.41
Joe Jacobson	Rider R6 (Menasco C6S4)	54:57.03	218.27
Earl Ortman	Marcoux-Bromberg Rider R-5 (Menasco B6S)	56:06.18	192.50
Crosby	Crosby CR-4 (Menasco C6S4)		
Dory	Bushey-McGrew		

1939--Cleveland--September 2-4

Bendix Transcontinental Trophy Race--Burbank to Cleveland (2,043 mi.)--$12,500 to winner

Pilot	*Aircraft (powerplant)*	*time (hr.)*	*mph*
Frank Fuller	Seversky Sev-S2 (Twin Wasp)	7:14:19	282.19
Arthur Bussy	Bellanca 28-92 (Ranger, 2 Menasco)		244.48
Paul Mantz	Lockheed Orion 9C (Wright Cyclone)		234.87
Max Constant	Beech D-17W (P&W Wasp Jr.)		231.36
Arlene Davis	Spartan 7W (P&W Wasp Jr.)		196.84
William Maycock	Beech D-17S		187.18

DNS: Jacqueline Cochran, Seversky AP-7

Thompson Trophy Race--30 laps of 10 miles--$16,000 to winner

Pilot	*Aircraft (powerplant)*	*time (min.)*	*mph*
Roscoe Turner	Turner-Laird "Meteor" (P&W Twin Wasp)	63:43.0	282.53
Tony La Vier	Schoenfeldt "Firecracker" Rider R-4 (Menasco C6S4)		272.53
Earl Ortman	Marcoux-Bromberg Rider R-4 (Twin Wasp Jr.)		254.43
Harry Crosby	Crosby CR-4 (Menasco C6S4)		244.52
S.J. Wittman	Wittman "Bonzo" (Curtiss D-12)		241.36
Joe Mackey	Wedell-Williams (P&W Hornet)		232.92
Art Chester	Chester "Goon" (Menasco C6S4)		DNF

Greve Trophy Race--

Pilot	*Aircraft (powerplant)*	*time (min.)*	*mph*
Art Chester	Chester "Goon" (Menasco C6S4)		263.4
LeVier	Schoenfield-Rider R-4 (Menasco C6S4)		
Crosby	Crosby CR-4 (Menasco C6S4)		
Williams	Brown B-2 (Menasco CS5)		

1946 Bendix Transcontinental Trophy Race--Los Angeles to Cleveland--$10,000 to winner

Pilot	*Aircraft (powerplant)*	*time (hr.)*	*mph*
Paul Mantz	North American P-51C (Packard V-1650)	4:42:14	435.5
Jacqueline Cochran*	North American P-51B (Packard V-1650)		420.9
Thomas Mayson	North American P-51C (Packard V-1650)		408.2
William Eddy	North American P-51D (Packard V-1650)		373.3
James Harp, Jr.	Lockheed P-38 (F-5G-6) (2 Allison V-1710-89/91)		370.4
Donald Husted	Douglas A-26C (2 P&W R-2800 Double Wasp)		367.9
Charles Tucker	Bell P-63C-5 (Allison V-1710-93)		367.1
Harvey Hughes	Lockheed F-5G-6 (2 Allison V-1710)		356.4
Walter Bullock	Lockheed F-5G-6 (2 Allison V-1710)		355.9
Harold Johnson	Lockheed F-5G-6 (2 Allison V-1710)		343.3
John Carroll	Lockheed F-5G-6 (2 Allison V-1710)		337.8
H. Marshall	Lockheed F-5G-6 (2 Allison V-1710)		335.9

1946 Bendix cont.

Pilot	Aircraft (powerplant)	mph
Rex Mays	Lockheed F-5G-6 (2 Allison V-1710)	327.5
William Lear, Jr.	Lockheed F-5G-6 (2 Allison V-1710)	327.1
Thomas Call	FG-1D Corsair (P&W R-2800 Double Wasp)	325.6
William Fairbrother	Lockheed P-38L (2 Allison V-1710)	325.2
Andrew Grant	Lockheed F-5G-6 (2 Allison V-1710)	261.6

DNF: Spiro Dilles, Bell P-63C; John Schields, P-38L; H. Calloway, P-38L; Herman Salmon, F5G-6; J. Yandell, F-5G-6. DNS: J. Redwine, F-5G; Paul Franklin, P-51D; Nadene Ramsey, F-5G-6; Earl Ortman, F-5G-6, James Philpott, P-47; W. Clayson, FG-1D; J. Alexander, P-38M; William Odom, YP-47M; Robert Swanson, P-51K.

Jet Division		Time (hr.)	mph
Col. Leon Gray	Lockheed FP-80A Shooting Star	4:08:25	494.779
Maj. Rudell	Lockheed P-80A	4:18:51	474.836

1946 Thompson Trophy Race--10 laps of 30 miles, 300 mph minimum qualifying speed

Pilot	Aircraft (powerplant)	mph
Tex Johnston	Bell P-39Q (2 Allison V-1710G)	373.91
Tony LeVier	Lockheed P-38L (2 Allison V-1710F)	370.19
Earl Ortman	North American P-51D (Packard V-1650)	367.63
Bruce Raymond	North American P-51D (Packard V-1650-7)	364.66
Robert Swanson	North American P-51D (Packard V-1650-7)	362.05
Cook Cleland	Goodyear FG-1D (P&W R-2800 Double Wasp)	357.47
Woody Edmundson	North American P-51D (Packard V-1650-3)	354.40
S.J. Wittman	Bell P-63C (Allison V-1710-117)	341.23
Howard Lilly	Bell P-63A (Allison V-1710-93)	328.15
H.L. Pemberton	Bell P-63F (Allison V-1710-135)	304.41
George Welch	North American P-51D (Packard V-1650-7)	DNF
Charles Tucker	Bell P-63C (Allison V-1710-117)	DNF

Fastest lap: 389 mph, Johnston

Jet Division--6 laps of 30 miles		mph
Maj. Gus Lindquist	Lockheed P-80 A1 Shooting Star	515.85
Maj. Robin Olds	Lockheed P-80 A1 Shooting Star	514.71
Capt. A.M. Fell	Lockheed P-80 A1 Shooting Star	509.38
Capt. J.E. Sullivan	Lockheed P-80 A1 Shooting Star	470.04
Lt. Col. R.L. Petit	Lockheed P-80 A1 Shooting Star	Disq.
Maj. R.O. Chilstrom	Lockheed P-80 A1 Shooting Star	DNF

1947 Bendix Transcontinental Trophy Race Los Angeles to Cleveland--$10,000 to winner

Pilot	Aircraft (powerplant)	hr.	mph
Paul Mantz	North American P-51C (Packard V-1650-3)	4:26:57	460.4
Joe DeBona	North American P-51D (Packar5d V-1650-7		458.2
Edmund Lunken	North American P-51D (Packard V-1650-7)		408.7
Bruce Gimbel	North American P-51B (Packard V-1650-7)		404.1
William Eddy	North American P-51D (Packard V01650-7)		376.5
Thomas Mayson	North American P-51C (Packard V-1650-3)		376.1
Frank Whitton	Goodyear FG-1D Corsair (P&W R-2800 Double Wasp)		320.0
Wm. Lear, Jr.	Lockheed F-5G-6 (2 Allison V-1710)		292.6
Jane Hlavcek	Lockheed F-5G-6 (2 Allison V-1710)		247.8

DNF: James Ruble, P-38F; Dianna Cyrus, A-26B; Joseph Kinkella, P-63C. DNS: William Odom, YP-47M

Jet Division		hr.	mph
Col. Leon Gray	Lockheed FP-80-A5 Shooting Star	4:02:18	507.255
Maj. Clay Allbright	Lockheed P-80A Shooting Star		486.28
Capt. W.S. Patterson	Lockheed P-80A Shooting Star		463.968

National Air Races cont

1947 Thompson Trophy Race--20 laps of 15 miles, 300 mph minimum qualifying speed

Pilot	Aircraft (powerplant)	mph
Cook Cleland	Goodyear F2G-1 (P&W R-4360-4 Wasp Major)	396.13
Dick Becker	Goodyear F2G-1 (P&W R-4360 Wasp Major)	390.13
Jay Demming	Bell P-39Q (Allison V-1710G)	389.84
Steve Beville	North American P-51D (Packard V-1650-7)	360.84
Tony LeVier	Lockheed P-38L (Allison V-1710F)	357.49
William Bour	Bell P-63A (Allison V-1710-93)	327.28
Ron Puckett	Goodyear F2G-1 (P&W R-4360)	DNF
Joe Ziegler	Curtiss P-40Q (Allison V-1710-121)	DNF
Woody Edmundson	North American P-51A (Packard V-1650-3)	DNF
Paul Penrose	North American P-51D (Packard V-1650)	DNF
Charles Walling	Lockheed P-38J (Allison V-1710)	DNF
Jack Hardwicke	North American P-51C (Packard V-1650-7)	DNF
Tony Janazzo+	Goodyear F2G-1 (P&W R-4360 Wasp Major)	

+ killed in crash 1 on lap 7. Fastest lap: 404 mph, Cleland and Penrose

Jet Division--8 laps, 180 miles

Lt. Col. R.L. Petit	Lockheed FP-80A Shooting Star	515.85
Lt. J. Howard	Lockheed FP-80A Shooting Star	497.94
Lt. Col. Dunham	Lockheed FP-80A Shooting Star	494.65
Capt. W. Gates	Lockheed FP-80A Shooting Star	484.87
Lt. Col. Preston	Lockheed FP-80A Shooting Star	443.16
Capt. L. Powers	Lockheed FP-80A Shooting Star	430.23

1947 Goodyear Trophy Race

Pilot	Aircraft (powerplants all Continental C-85)	mph
William Brennand	Wittman "Buster"	165.9
Paul Penrose	Chester "Sweet Pea"	165.4
Herman Salmon	Cosmic Wind Spl.	158.8
Tony LeVier	Cosmic Wind "Little Toni"	157.9
Siem	Loose Siem	151.3
Robinson	Brown B-1	143.9

1948 Bendix Transcontinental Trophy Race--Los Angeles to Cleveland--$10,000 to winner

Pilot	Aircraft (powerplant)	hr.	mph
Paul Mantz	North American P-51C	4:33:48	448.1
Linton Carney	North American P-51C		446.1
Jacqueline Cochran	North American P-51B		445.8
Edmund Lunken	North American P-51D		441.6
Jesse Stallings	de Havilland DH.98 Mosquito Mk. 25		341.1

DNF: Doe DeBona, P-51B. DNS: Jane Hlavacek, YP-47M; Donald McVicar, DH 98; Dianna Cyrus, DH 98; Jack Becker, F-5G-6.

Jet Division

Ens. F.E. Brown	North American FJ-1 Fury	4:10:34	489.26
Cmdr. E.P. Aurand	North American FJ-1 Fury	4:13:04	484.67
Lt. E.R. Hanks	North American FJ-1 Fury	4:15:53	479.36
Ens. R.E. Oechelin	North American FJ-1 Fury	4:20:44	470.44

1948 Thompson Trophy Race--20 laps of 15 miles, 300 mph minimum qualifying speed

Pilot	Aircraft (powerplant)	mph
Anson Johnson	North American P-51D (Packard V-1650-25)	383.77
Bruce Raymond	North American P-51D (Packard V-1650-7)	365.23
Wilson Newhall	Bell P-63C (Allison V-1710-117)	313.57
Charles Brown	Bell P-39Q (Allison V-1710G-2)	DNF
Charles Walling	North American P-51D (Packard V-1650-7)	DNF

1948 Thompson cont.

Pilot	Aircraft (powerplant)	
Woody Edmundson	North American P-51D (Packard V-1650-7)	DNF
M.W. Fairbrother	North American P-51D (Packard V-1650-7)	DNF
Robert Eucker	Bell P-63A (Allison V-1710-93)	DNF
Cook Cleland	Goodyear F2G-1 (P&W R-4360 Wasp Major)	DNF
Dick Becker	Goodyear F2G-1 (P&W R-4360 Wasp Major)	DNF

Fastest lap: 413 mph, Brown

1948 Goodyear Trophy Race

Pilot	Aircraft (powerplant)	mph
Herman Salmon	Cosmic Wind "Minnow"	169.7
Steve Wittman	Wittman Spl.	168.9
Art Chester	Chester "Sweet Pea II"	168.2
William Brennand	Wittman "Buster"	167.1
Robinson	Cosmic Wind "Little Toni"	165.1
Quigley	Pitts Spl.	164.9
Downey	Cosmic Wind "Ballerina"	161.5
LeFevers	Falcon Spl.	156.6

1949 Bendix Transcontinental Trophy Race--Rosamond Dry Lake, CA to Cleveland--$10,000 to winner

Pilot	Aircraft (powerplant)	hr.	mph
Joe De Bona	North American P-51B	4:16:17	470.1
Stanley Reaver	North American P-51C		450.2
Herman Salmon	North American P-51C	449.2	
Donald Bussart	de Havilland DH.98 Mosquito Mk. 25	343.8	

DNF: Leland Cameron, Douglas B-26C; Vincent Perron, Republic AT-12

Jet Division

Pilot	Aircraft (powerplant)	hr.	mph
Maj. Vernon Ford	Republic F-84E Thunderjet	3:45:51	529.61
Capt. J.W. Newman	Republic F-84E Thunderjet	3:47:00	524.62
Lt. Col. L.E. Moon	Republic F-84E Thunderjet	3:48:02	524.55
Capt. Harry M. Lester	Republic F-84E Thunderjet		514.74

1949 Thompson Trophy Race--15 laps of 15 miles, 300 mph minimum qualifying speed

Pilot	Aircraft (powerplant)	hr.	mph
Cook Cleland	Goodyear F2G-1 (P&W R-4360 Wasp Major)		397.07
Ron Puckett	Goodyear F2G-1 (P&W R-4360 Wasp Major)		393.53
Ben McKillan	Goodyear F2G-1 (P&W R-4360 Wasp Major)		387.59
Steve Beville	North American P-51D (Packard V-1650-7)		381.21
Charles Tucker	Bell P-63C (Allison V-1710-117)		378.34
Jim Hagerstrom	North American P-51D (Packard V-1650-7)		372.72
Wilson Newhall	North American P-51K (Packard V-1650-7)		372.32
James Hannon	North American P-51A (Packard V-1650-7)		300.40
Anson Johnson	North American P-51D (Packard V-1650-25)		DNF
Bill Odom+	North American P-51C (Packard V-1650)		

+ killed in crash on lap 2
Fastest lap: 406 mph, Cleland

Jet Division--10 laps, 150 miles

Pilot	Aircraft	mph
Capt. Bruce Cunningham	North American F-86A Sabre Jet	586.17
Capt. Martin Johansen	North American F-86A Sabre Jet	580.15
Capt. Vern Henderson	North American F-86A Sabre Jet	DNF

1949 Goodyear Trophy Race

Pilot	Aircraft (powerplant)	mph
William Brennand	Wittman "Buster"	177.3
Keith Sorenson	"Deerfly"	176.7

1949 Goodyear cont.

Pilot	Aircraft		mph
Steve Wittman	Wittman "Bonzo"		176.2
Vincent Ast	Cosmic Wind "Ballerina"		176.0
Herman Salmon	Cosmic Wind "Minnow"		175.7
Mone	Williams "Estrellita"		175.0
Downey	Mercury Air		171.4
Johnson	Long LA-1		167.3
Kistler	Kistler Spl.		153.4
Foss	"Jinny"		

1951 Bendix Transcontinental Trophy Race--Muroc, CA to Detroit--USAF jets only

Pilot	Aircraft	hr.	mph
Col. Keith Compton	North American F-86A Sabre Jet	3:26:56	553.76
Co. Emmertt Davis	Republic F-84E Thunderstreak	3:35:17	
Lt. Col. Geo. Thabault	North American B-45 Tornado	3:36:11	
Maj. Gilbert Pederson	Republic F-84E Thunderstreak	3:36:32	
Capt. B.W. Watts	North American B-45 Tornado	3:39:33	
Capt. Ed Johnson		3:44:35	
Maj. Leo Dykes, Jr.	North American B-45 Tornado	3:47:04	

1951 Thompson Trophy--Detroit, MI--one lap, 62.14 mi. (100 km)

Pilot	Aircraft		mph
Col. Fred Ascani	North American F-86E Sabre Jet		628.696

1953 Bendix Transcontinental Trophy Race--Muroc, CA to Dayton--USAF jets only

Pilot	Aircraft	hr.	mph
Maj. Wm. Whisner, Jr.	North American F-86F Sabre Jet	3:05:45	603.54
Maj. Ed Jonson	North American F-86F Sabre Jet	3:05:49	
Col. Clay Tice, Jr.	North American F-86F Sabre Jet	3:07:17	

1954 Bendix Transcontinental Trophy Race--Muroc, CA to Dayton--USAF jets only

Pilot	Aircraft	hr.	mph
Capt. Ed Kenny	Republic F-84F Thunderstreak	3:01:56	616.2
Maj. Harry Evans	Republic F-84F Thunderstreak	3:04:38	
Maj. Robt. Stephens	Republic F-84F Thunderstreak	3:04:44	
Capt. Glen Dunnway	Republic F-84F Thunderstreak	3:09:16	
Capt. H. Stockman	Republic F-84F Thunderstreak	3:09:51	
Maj. Ray Hunt	Republic F-84F Thunderstreak	3:09:53	
Maj. Geo. Loving	Republic F-84F Thunderstreak	3:10:02	
Capt. H. Cunningham	Republic F-84F Thunderstreak	3:09:56*	
Lt. Col. T. Robertson	Republic F-84F Thunderstreak	DNF	
Capt. Jesse Green	Republic F-84F Thunderstreak	DNF	

*dropped wing tank at finish line

1955 Bendix Transcontinental Trophy Race--Victorville, CA to Philadelphia--USAF jets only

Pilot	Aircraft	hr.	mph
Col. Chas. Talbott	North American F-100C Super Sabre	3:48:24	610.72
Capt. A, Moorman	North American F-100C Super Sabre		3:50:45
Lt. Col. M. Long	North American F-100C Super Sabre		3:55:16

1956 Bendix Transcontinental Trophy Race--Victorville, CA to Oklahoma City--USAF jets only

Pilot	Aircraft	hr.	mph
Capt. M. Fernandez	North American F-100C Super Sabre	1:40:38	666.66
Capt. Robt. Madden	North American F-100C Super Sabre	1:42:14	
Capt. A. Iddings	North American F-100C Super Sabre	1:42:18	
Capt. A.C. Edinburg	North American F-100C Super Sabre	1:45:34	
Lt. John Niemela	North American F-100C Super Sabre	1:43:37	
Capt. R. Farnsworth	North American F-100C Super Sabre	1:43:54	

1957 Bendix Transcontinental Trophy Race--Chicago to Washington, DC--USAF jets only

Pilot	Aircraft	hr.	mph
Capt. Kein Chandler	Convair F-102 Delta Dagger	0:54:45	679.05
Col. Robt. Gould	Convair F-102 Delta Dagger	0:55:16	
Capt. Leroy Scendsen	Convair F-102 Delta Dagger	0:55:17	
Capt. Martin Detlie	Convair F-102 Delta Dagger	0:55:44	
Lt. Col. Chas. Rigney	Convair F-102 Delta Dagger	0:55:50	
Maj. Chas. Macefield	Convair F-102 Delta Dagger	0:56:42	

1961 Bendix Transcontinental Trophy Race--Ontario, CA to Brooklyn, NY--Navy jets only

Pilots	Aircraft	hr.
Lt. Richard Gordon/ Lt.(jg) Bobbie Young	McDonnell F4H-1 Phantom II	2:47:17
Lt. Cdr. L. Lamoreaux/ Lt. Thomas Johnson	McDonnell F4H-1 Phantom II	2:57:00
Cdr. Julian Lake/ Lt.(jg) E. Cowart	McDonnell F4H-1 Phantom II	3:03:00
Lt. Cdr. P. Spence/ Lt. Jim Wagner	McDonnell F4H-1 Phantom II	3:47:00

1962 Bendix Transcontinental Trophy Race--Los Angeles to New York--USAF jets only

Pilots	Aircraft	hr.	mph
Capt. Robt. Sowers/			
Capt. Robt. MacDonald/Capt. John Walton	Convair B-58 Hustler	2:00:56	1,214.71

OTHER TRANSCONTINENTAL TROPHY DASHES

1964 Harold's Club--St. Petersberg, FL to Reno, NV, 2,254 miles

Pilot	Aircraft	hr.	mph
Wayne Adams	North American P-51D		318.8
Charles Lyford	North American P-51D		307.7
C.E. Crosby	North American P-51D		276.9
Richard Snyder	North American P-51D		264.2
Jack Shaver	North American P-51D		245.0
Stan Hoke	North American P-51D		239.1

DNF: Ellis Weiner, P-51D; Howard Olsen, P-51D. DNS: Clay Lacy, P-51D.

1965 Harold's Club--St. Petersburg, FL to Reno, NV, 2,254 miles

Pilot	Aircraft	hr.	mph
Ellis Weiner	North American P-51D		347.9
Clay Lacy	North American P-51D		341.4
Wayne Adams	North American P-51D		330.6
Richard Kestle	North American P-51D		288.9
John Gower	North American P-51D		279.1
Jack Shaver	North American P-51D		269.1
Doug Wood	North American P-51D		256.3

DNF: Thomas Green, Cessna 310 Riley Rocket; James Fugate, P-51D.

1966--St. Petersburg. FL to Palm Springs, CA. 2,038 miles

Pilots	Aircraft	hr.	mph
Ellis Weiner	North American P-51D		363.9
Richard Kestle	North American P-51D		343.5
Michael Carroll	Hawker Sea Fury F.B. Mk. 11		325.2

1967--Palm Springs, CA to Celeveland, OH, 2,020 miles

Pilot	Aircraft	hr.	mph
Ellis Weiner	North American P-51D		411.0
Michael Carroll	Hawker Sea Fury F.B. Mk. 11		386.0

Palm Springs to Cleveland cont.

Pilot	Aircraft	hr.	mph
Robert Guilford	North American P-51D		224.0

DNF: James Ventura+, P-51D. DNS: Clay Lacy, P-51D; Robert Garrison, P-51D; John Gower, P-51D
+crashed fatally near Minden, NE

1967 Harold's Club--Rockford, IL to Reno, NV, 1,620 miles

Pilot	Aircraft	hr.	mph
Michael Carroll	Hawker Sea Fury F.B. Mk. 11		417.3
Ellis Weiner	North American P-51D		399.3
Richard Kestle	North American P-51D		304.7
Thomas Kuchinsky	North American P-51D		275.3
James Fugate	North American P-51D		286.8*

DNF: H.F. Rupp, P-51D. DNS: John Church, F8F-2. *penalized for late start and arrival

1968 Harold's Club--Milwaukee, WI to Reno, NV, 1,667 miles

Pilot	Aircraft	hr.	mph
Ellis Weiner	North American P-51D		361.1
Richard Kestle	North American P-51D		278.4
Burns Byram	Noarth American P-51D		264.6
Robert Kucera	Grumman F8F-2		264.2
Thomas Kuchinsky	North American P-51D		245.7
Robert Guilford	North American P-51D		189.2

DNF: Mark Foutch, P-51D; Carl Koeling, P-64. DNS: Walter Ohlrich, F8F-2; John Sandberg, P-51D; John Silberman, P-51D; Wendel Trogdon, P-51D; Peter Brucia, F6F; James Fugate, P-51D; Michael Carroll+, Sea Fury; Gunther Balz, F8F-1.
+crashed fatally before start

1969 Florida Cross-Country Race, Frederick, MD to Ft. Lauderdale, FL, 929 miles

Pilots	Aircraft	hr.	mph
Ed Bowlin	North American P-51D		310.9
Richard Kestle	North American P-51D		294.0
William Hogan	North American P-51H		276.0
John Sliker	North American P-51D		254.7
Wallace Garrick	North American P-51D		207.9

DNF: Paul Finefrock, P-51D

1969 Harold's Club--Milwaukee, WI to Reno, NV, 1,667 miles

Pilots	Aircraft	hr.	mph
Richard Kestle	North American P-51D		313.1
John Sliker	North American P-51D		312.2
Thomas Kuchinsky	North American P-51D		286.9
Burns Byram	North American P-51D		283.2
Charles Doyle	North American P-51D		279.8
Howard Keefe	North American P-51D		275.3
Gunther Balz	Grumman F8F-1		265.4
Walter Ohlrich	Grumman F8F-2		254.9

DNF: Richard Thomas, G4U-4; Judy Wagner, Beech Bonanza. DNS: Darryl Greenamyer, P-51D; Ellis Weiner, P-51D; William Cooper, Sea Fury; Robert Merrill, MM1; Robert Guilford, P-51D.

1970 Harold's Club--Milwaukee, WI to Reno, NV, 1,667 miles

Pilot	Aircraft	hr.	mph
Richard Kestle	North American P-51D		283.6
Gunther Balz	Grumman F8F-1		273.4
John Sliker	North American P-51D		254.9
Howard Keefe	North American P-51D		254.7
Burns Byram	North American P-51D		246.0

DNF: Jack Huismann, P-51D; Ron Reynolds, F8F-2. DNS: William Cooper, Sea Fury; Paul Finefrock, P-51D.

1971 Alton P-51 Tournament--Milwaukee, WI to East Alton, IL, 309 miles

Pilot	Aircraft	hr.	mph
Clay Lacy	North American P-51D		353.8
Sherm Cooper	Sea Fury F.B. Mk. 11		353.5
Gunther Balz	Grumman F8F-1		337.1
Leroy Penhall	North American P-51D		329.5
Howard Keefe	North American P-51D		329.0
William Hogan	North American P-51H		317.3
Burns Byram	North American P-51D		312.8
Paul Finefrock	North American P-51D		311.2
Wallace Oakes	North American P-51D		301.8
Richard Foote	North American P-51D		275.4

NATIONAL CHAMPIONSHIP AIR RACES, Reno, Nevada

FI=Formula One--minimum weight of 500 lb, minimum wing area of 66 sq ft, fixed gear, fixed-pitch propeller, 200 cu in Continental engine.
B=Biplane--minimum weight of 500 lb, no less than 30% of the 75 sq ft wing area on smaller of two wings, fixed gear, fixed-pitch propeller, an engine no larger than320 cu in.
AT-6=unmodified North American advanced trainer.U=Unlimited

Year	class	Pilot	Aircraft	mph
1964	FI	Bob Porter	"Little Gem"	193.44
	B	Clyde Parsons	Parsons Twister	144.57
	U	Mira Slovak	"Miss Smirnoff" (Grumman F8F-2 Bearcat)	376.84
1965	FI	Bob Porter	"Deerfly"	202.14
	B	Bill Boland	Boland Mong	148.68
	U	Darryl Greenamyer	Grumman Bearcat (Grumman F8F-2 Bearcat)	375.10
1966	FI	Bill Falck	"Rivets"	193.10
	B	Chuck Wickliff	Clark Dollar Spl.	147.72
	U	Darryl Greenamyer	"Smirnoff" (Grumman F8F-2 Bearcat)	396.22
1967	FI	Bill Falck	"Rivets"	202.70
	B	Bill Boland	Boland Mong	151.64
	U	Darryl Greenamyer	"Smirnoff" (Grumman F8F-2 Bearcat)	392.62
1968	FI	Ray Cote	"Shoestring"	214.61
	AT-6	Hendrick Otzen	"Condor"	181.32
	B	Dallas Christian	"Mongster"	175.13
	U	Darryl Greenamyer	Grumman Bearcat (F8F-2)	388.65
1969	FI	Ray Cote	"Shoestring"	225.55
	AT-6	Ben Hall	"Miss Meridian Pavers"	190.90
	B	Dallas Christian	"Mongster"	184.02
	U	Darryl Greenamyer	"Conquest I" (Grumman F8F-2 Bearcat)	412.63
1970	FI	Ray Cote	"Shoestring"	220.07
	B	Bill Boland	Boland Mong	177.45
	U	Clay Lacy	"Miss Van Nuys" (North American P-51D)	387.34
1971	FI	Ray Cote	"Shoestring"	224.14
	AT-6	Bob Mitchem	"Miss Colorado"	205.85
	B	Bill Boland	"Prop Wash"	181.67
	U	Darryl Greenamyer	"Conquest I" (Grumman F8F-2 Bearcat)	413.99
1972	FI	Ray Cote	"Shoestring"	223.95
	AT-6	Mac McClain	"Miss Eufaula"	201.59
	B	Don Beck	"Sorceress"	189.72
	U	Gunther Balz	"Roto-Finish Spl." (North Amwerican P-51D)	416.16

National Championship Air Races cont.

Year	Class	Pilot	Plane	Speed
1973	FI	Ray Cote	"Shoestring"	231.26
	AT-6	Bill Turnbull	"Old Ironsides"	206.60
	B	Sid White	"Sundancer"	194.95
	U	Lyle Shelton	"7 1/4% Spl." (Grumman F8F-2 Bearcat)	428.16
1974	FI	Ray Cote	"Shoestring"	235.42
	AT-6	Pat Palmer	"Gotcha"	211.35
	B	Sid White	"Sundancer"	198.17
	U	Ken Burnstine	"Miss Suzi Q" (North American P-51D)	381.48
1975	FI	Ray Cote	"Shoestring"	227.46
	AT-6	Pat Palmer	"Gotcha"	207.19
	B	Don Beck	"Sorceress"	198.99
	U	Lyle Shelton	"Aircraft Cyl. Spl." (Grumman F8F-2 Bearcat)	429.92
1976	FI	Vince DeLuca	"Lil' Quickie"	228.75
	AT-6	Pat Palmer	"Gotcha"	210.68
	B	Don Beck	"Sorceress"	202.15
	U	Lefty Gardner	"Thunderbird" (North American P-51D)	379.61
1977	FI	John Parker	"Top Turkey"	226.12
	AT-6	Ralph Twombly	"Spooled Up"	209.66
	U	Darryl Greenamyer	"Red Baron" (North American RB-51)	430.70
1978	FI	no race due to weather		
	AT-6	Ralph Rina	"Miss Everything"	205.71
	U	Steve Hinton	"Red Baron" (North American RB-51)	415.46
1979	FI	John Parker	"Top Turkey"	240.09
	U	John Crocker	"Sumthin' Else"	422.30
1980	FI	John Parker	"American Special"	249.07
	B	Pat Hines	"Sundancer"	206.62
	U	Mac McClain	"Jeannie" (North American P-51D)	433.01
1981	FI	Ray Cote	"Shoestring"	232.13
	AT-6	John Mosby	"Miss Behavin"	222.78
	B	Pat Hines	"Sundancer"	209.44
	U	Skip Holm	"Jeannie" (North American P-51D)	431.29
1982	FI	Jon Sharp	"Aero Magic"	224.52
	AT-6	Ralph Twombly	"Miss Behavin"	214.90
	B Sport	Don Fairbanks	"White Knight"	172.73
	B Racing	Pat Hines	"Sundancer"	223.88
	U	Ron Hevle	"Dago Red" (North American P-51D)	405.09
1983	FI	Chuck Wentworth	"Flexi-Flyer"	239.02
	AT-6	Richard Sykes	"The Mystery Ship"	225.94
	B Sport	Don Fairbanks	"White Knight"	172.73
	B Racing	Pat Hines	"Sundancer"	217.60
	U	Neil Anderson	"Dreadnought" (Hawker Sea Fury)	425.24
1984	FI	Ray Cote	"Judy"	236.07
	AT-6	Ralph Rina	"Miss Everything"	217.26
	B	Don Beck	"Miss Tahoe"	189.97
	U	Skip Holm	"Stiletto" (North American P-51D)	437.62
1985	FI	Ray Cote	"Judy"	229.09
	AT-6	Randy Difani	"Thunderbolt"	213.89
	B	Don Beck	"Miss Lake Tahoe"	195.62
	U	Steve Hinton	"Supercorsair" (Grumman F4U Corsair)	438.19
1986	FI	Jon Sharp	"Aero Magic"	229.61
	AT-6	Eddie Van Fossen	"Miss TNT"	223.45
	B	Alan Preston	"Sitting Duck"	192.66
	U	Rick Brickert	"Dreadnought" (Hawker Sea Fury)	434.48
1987	FI	Alan Preston	"Sitting Duck"	232.98
	AT-6	Eddie Van Fossen	"Miss TNT"	226.36
	B	Tom Aberle	"Long Gone Mong"	196.47
	U	Bill Destefani	"Strega" (North American P-51D)	452.55

National Championship Air Races cont.

1988	FI	Alan Preston	"Sitting Duck"	240.74
	AT-6	Eddie Van Fossen	"Miss TNT"	229.75
	B	Alan Preston	"Top Cat"	205.91
	U	Lyle Shelton	"Rare Bear" (Grumman F8F Bearcat)	456.82
1989	FI	Ray Cote	"Alley Cat"	231.25
	AT-6	Tom Dwelle	"Tinkertoy"	222.32
	B	Tom Aberle	"Wanna Play II"	196.14
	U	Lyle Shelton	"Rare Bear" (Grumman F8F Bearcat)	450.91
1990	FI	Jim Miller	"Texas Gem"	237.40
	AT-6	Tom Dwelle	"Tinkertoy"	229.26
	B	Dan Mortensen	"Amsoil Pacific Flyer"	192.27
	U	Lyle Shelton	"Rare Bear" (Grumman F8F Bearcat)	468.62
1991	FI	Jon Sharp	"Nemesis"	245.26
	AT-6	Eddie Van Fossen	"Miss TNT"	227.02
	B	Ken Ueno	"Sanurari"	195.27
	U	Lyle Shelton	"Rare Bear" (Grumman F8F Bearcat)	481.61(r)
1992	FI	Jon Sharp	"Nemesis"	238.17
	AT-6	Eddie Van Fossen	"Miss TNT"	234.78(r)
	B	Jim Smith	"Glass Slipper"	193.89
	U	Bill Destefani	"Strega" (North American P-51D)	450.83
1993	FI	Jon Sharp	"Nemesis"	246.85
	AT-6	Eddie Van Fossen	"Miss TNT"	226.89
	B	Nelson	"Full Tilt Boogie"	208.47(r)
	U	Bill Destefani	"Strega" (North American P-51D)	455.38
1994	FI	Jon Sharp	"Nemesis"	248.91
	AT-6	Eddie Van Fossen	"Miss TNT"	224.70
	B	Earl Allen	"Class Action"	203.31
	U*	John Penney	"Rare Bear" (Grumman F8F Bearcat)	424.41
	U**	Allen Preston	"Miss America" (North American P-51D)	413.77

** "Super Gold Shootout" among top three from preliminary race ** Gold race*

1995	FI	Jon Sharp	"Nemesis"	249.904(r)
	AT-6	Charles Hutchins	"Mystical Power"	231.43
	B	Patti Johnson	"Full Tilt Boogie"	202.12
	U	Bill Destefani	"Strega" (North American P-51D)	467.03
1996	FI	Jon Sharp	"Nemesis"	238.95
	AT-6	Sherman Smoot	"Bad Company"	221.67
	T-28	John Herlihy		263.31
	B	Patti Johnson	"Full Tilt Boogie"	212.81
	U	Bill Destefani	"Strega" (North American P-51D)	467.95

Notable Air Racing Pilots

Pilot (nationality)	Race record	Aircraft
Acosta, Bert (US)	1-1921 Pulitzer, 3-1920	Curtiss CR-1
Balz, Gunther (US)	1-1972 Reno Unlimited, 3-1971 Alton	P-51, F8F-1
Barnes, Florence "Pancho" (US) (1902-75)	Women's speed record 1930	Travel Air S
Bayles, Lowell (US) (d. 1931)	1-1931 Thompson, 2-All-American	Gee Bee Z, X
Biard, Henry (GB)	1-1922 Schneider, 3-1923	Supermarine Sea Lion II
Boothman, John (GB)	1-1931 Schneider	Supermarine S.6B
Brickert, Rick (US) (d. 1993)	1-1986 Reno Unlimited	"Dreadnought"
Cleland, Cook (US)	1-Thompson Trophy 1947, 1949	F2G-1 Corsair
Cochran, Jacqueline (US) (1906-80)	1-1938 Bendix, 2-1946,3-1948	Beech D17, Seversky
Cote, Ray (US)	1-Reno FI 1968-75, 1981, 1984-85, 1989	"Shoestring"
Cuddihy, George (US)	4-1925, 5-1926 Schneider, 1-1926 National Air Races	Boeing FB-3 Curtiss R3C-2
Curtiss, Glenn (US) (1878-1930)	1--first Gordon Bennett in 1909	Curtiss
Davis, Doug (US) (1899-1934)	1-1929 National Air Races, 1934 Bendix	Travel Air R, Wedell-Williams
Destafani, Bill (US)	1-Reno Unlimited 1987, 1992-93, 1995-96	"Strega"
Detroyat, Michel (F)	1-1936 Thompson Trophy, Greve Trophy	Caudron 460
Doolittle, James (US) (1897-1993)	1-1925 Schneider, 1931 Bendix, 1932 Thompson Trophy	Curtiss R3C-2, Laird S. Solution Gee Bee R-1
DeBernardi, Mario (I)	1-1926 Schneider	Macchi M-39
Earhart, Amelia (US) (1897-1937)	3-1929 Women's Derby, 5-1935-36 Bendix	Lockheed Vega
Fuller, Frank (US)	1-1937 and 1939 Bendix, 2-1938	Seversky
Garros, Roland (F) (d. 1918)	2-1911 Paris-Rome, European Circuit, Schneider Trophy, first pilot to cross Mediterranean	Morane-Saulnier
Gehlbach, Lee (US)	4-1932 Bendix, 5-1932 Thompson, 2-1933 Thompson	Wedell-Williams Gee Bee R-1, R-2
Gilbert, Eugène (F)	1-1913 Deutsch de la Meurthe	Deperdussin
Goebel, Arthur (US)	1-1927 Dole Derby, 5-1931 Bendix	Travel Air, Lockheed Altair
Graham-White, Claude (GB)	1-1913 Bennett Cup	Bleriot XI
Greenamyer, Darryl (US)	1-Reno Unlimited 1965-69, 1971, 1977	F8F-2 Bearcat
Haizlip, James (US)	1-1931 Thompson, 2-1930	Travel Air R, Wedell-Williams
Hall, Bob (US)	Designer of Gee Bees, 4-1931 Thompson Trophy, 6-1932	Gee Bee Y, Hall "Bulldog"
Hinton, Steve (US)	1-Reno Unlimited 1978	"Red Baron"
Holm, Skip (US)	1-Reno Unlimited 1981, 1984	"Jeannie," "Stiletto"
Holman, Charles (US) (d. 1931)	1-1930 Thompson Trophy	Laird Solution
Howard, Ben O. (US) (1904-)	1-1935 Bendix in his *Mr. Mulligan* design, which also won the 1935 Thompson Trophy flown by Harold Neumann	DGA-3 "Pete," DGA-4 "Mike," DGA-5 "Ike" DGA-6
Janello, Guido (I)	1-1919 Schneider	Savoia S.13
Johnston, Tex (US)	1-1946 Thompson Trophy	Bell P-39Q
Kestle, Richard (US)	1-1969-70 Harold's Club, 3-1967, 2-1968	P-51D
Kling, Rudy (US)	4-1936 Thompson Trophy, 1-1937,	Folkerts
Lacy, Clay (US)	1-Reno Unlimited 1970, 1-1971 Alton	P-51D

Pilot (nationality)	Race record	Aircraft
LaVier, Tony (US)	1-1938 Greve Trophy, 2-1939 and 1946 Thompson, 5-1947, 4-1947 Goodyear	Rider R-4, P-38, Cosmic Wind
Liggett, Roy (US) (d. 1933)	1932-33: 11 starts 1-1,4-2, 3-3	Cessna CR-2
Livingston (US)	1931: 65 starts, 41 wins, 19 seconds 1933: 5 starts, 5 wins	Monocoupe Cessna CR-2A
Mantz, Paul (US) (1904-65)	3-1938-39 Bendix, 1-1946-48	Orion 9C, P-51C
Neumann, Harold (US)	1-1935 Thompson Trophy, Greve Trophy, 2-1936 Greve	Howard DGA-6, DGA-3
Ortman, Earl	2-1936-37 Bendix and Thompson, 2-1938 Thompson, 3-1939 Thompson	Rider R-3
Preston, Alan (US)	1-1986 Reno Biplane, 1-1988 Reno FI and Biplane, 1-1994 Unlimited	"Miss America," "Sitting Duck," "Top Cat,"
Prévost, Marcel (F)	1-1913 Gordon Bennett, 1913 Schneider	Deperdussin
Rittenhouse, David (US)	1-1923 Schneider	Curtiss CR-3
Sadi-Lecointe, Joseph (F)	1-1920 Gordon Bennett, 1919 Deutsch de la Meurthe, 1924-25 Beaumont Cup	Nieuport 29V
Sharp, Jon (US)	1--1986,, 1991-96 Reno FI	
Shelton, Lyle (US)	1-Reno Unlimited 1973, 1975, 1988-91	"Rare Bear"
Slovak, Mira (US)	1-Reno Unlimited 1964	"Miss Smirnoff"
Thaden, Louise (US) (1905-79)	1-1929 Women's Derby, 1936 Bendix	Travel Air , Beech 17R
Turner, Roscoe (US) (1895-1970)	1-1933 Bendix, 1934 Thompson, 2-1935 Bendix, 1-1938-39 Thompson	Wedell-Williams
Van Fossen, Eddie (US)	1-Reno AT-6 1986-88, 1991-94	"Miss TNT"
Verdines, Jules (F)	1-1912 Gordon Bennett Cup, first man to fly faster than 100 mph	Deperdussin
Waghorn, H.R.D . (GB)	1-1929 Schneider, 3rd 1923	Supermarine S.6
Wedell, James R. (US) (1899-1934)	2-1931-32 Thompson Trophy, 2-1933 Bendix, 1-1933 Thompson; Wedell designs won the Bendix 1932-34 and the Thompson 1933-34	Wedell-Williams
Weiner, Ellis (US)	1-1965, 1968 Harold's Club, 2-1967,	
Weyman, Charles (US)	1-1911 Gordon Bennett, 3-1913 Schneider, 5-1914,	Nieuport
Williams, Alford (US)	1-1923 Pulitzer, 2-1925, 4-1922, 2-1925	R-1 Navy Racer, Curtiss R2C-1, RC3-1
Wittman, Steve (US) (1905-96)	2-1935 Thompson Trophy, 5-1937 and 1939, 2-1937 Greve Trophy, 2-1948 Goodyear, 3-1949 Goodyear	"Bonzo," "Chief Oshkosh," P-39
Worthen, J.A. (US)	2-1934 Bendix, 3-1934 Thompson	Wedell-Williams

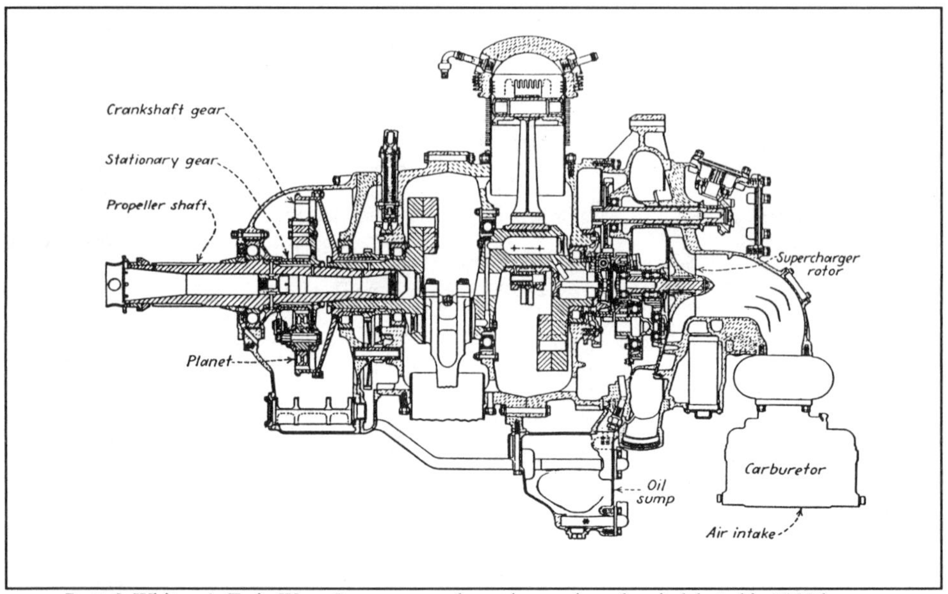

Pratt & Whitney's Twin Wasp Jr. was a popular racing engine when it debuted in 1932 because it could provide up to 1,000 horsepower with a low-drag diameter of 44 in.

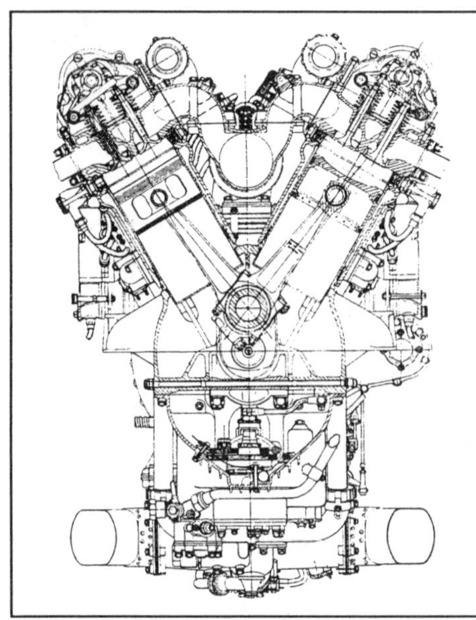

The Rolls-Royce Kestrel, inspired by the Curtiss D-12, led to the racing engine which gave Britain permanent possession of the Schneider Trophy, and ultimately to the Merlin.

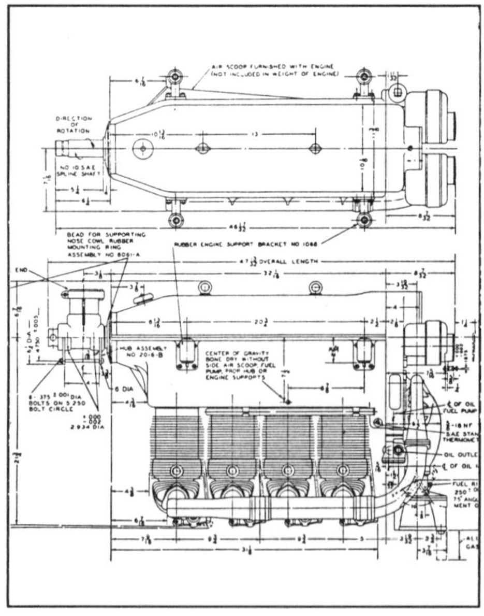

From the small-engine school of design, the Menasco 4 and 6-cylinder engines were full-race screamers that delivered up to 315 hp (at 3500 rpm) while weighing a mere 415 lb.

X. MILITARY AVIATION HISTORY

The crew of the "Memphis Belle" with its namesake, Margaret Polk (center). The famous Boeing B-17 was the first U.S. bomber in World War II to survive 25 raids.

AIRCRAFT OF THE USAAF/USAF by category, 1909-1990

Designation	Name-variants	Manufacturer	Year	Notes
Model B	Wright Flyer	Wright	1909	2-place
Model D		Curtiss	1911	Tricycle gear
Model H		Burgess	1912	Enclosed fuselage
Model G		Curtiss	1913	2-place side-by-side
JN-1		Curtiss	1914	Tandem 2-place
JN-4	Jenny	Curtiss	1916	
SJ		Standard	1917	3 front wheels
S-4		Thomas-Morse	1917	1 machine gun
Avro 540K		Avro	1917	2 wheels + front ski
Sopwith 1A-2		Sopwith Works	1917	
Breguet 14	Corps d' Armee	Breguet	1917	
Salmson 2A-2		Salmson	1917	
Spad-VII		S.P.A.D.	1917	
Nieuport 11	Bebe	Nieuport	1917	
Nieuport 12		Nieuport	1917	tandem side-by-side
Nieuport 17		Nieuport	1917	
HD-1		Hanriot-Macchi	1917	
Nieuport 28		Nieuport	1918	
Spad-XIII		S.P.A.D.	1918	
Sopwith	Camel	Sopwith Works	1918	
SE-5	Scout	Royal Aircraft	1918	2 guns, 4 bombs
Lusac-11	Le Pere	Packard	1918	Turbo-supercharger

Trainers 1919-present

Key: BT=basic trainers; PT=primary trainer; TA=training, air-cooled; TW training, water-cooled; Y=limited production prototype

Designation	Name-variants	Manufacturer	Year	Notes
Primary Trainers				
TA-3	Chummy	Dayton-Wright	1921	2-place side-by-side
TW-3	Trusty	Dayton Wr.-Consolidated	1921	Watercooled TA-3
TW-5		Huff-Deland	1923	Became AT-1
PT-1	Trusty	Consolidated	1924	Refined TW-3
PT-3	Trusty	Consolidated	1928	Uncowled engine
YPT-6	Husky Jr.	Fleet	1930	
YPT-9	Cloudboy	Stearman	1930	
PT-11		Consolidated	1932	
PT-13	Kaydet	Stearman	1936	280 hp
YPT-14		Waco	1940	
YPT-15		St. Louis	1940	
YPT-16	PT-20	Ryan	1939	1st monoplane tr.
PT-17	Kaydet	Boeing Stearman	1940	220 hp
PT-18	Kaydet	Boeing Stearman	1940	
PT-19	Cornell	Fairchild	1941	
PT-20	PT-16	Ryan	1940	
PT-21	Recruit	Ryan	1941	No wheel pants
PT-22	Recruit	Ryan	1942	
PT-23	Cornell	Fairchild	1942	
YPT-25		Ryan	1942	Bonded wood
PT-26	Cornell	Fairchild	1942	Enclosed cockpit
PT-27	Kaydet	Boeing Stearman	1942	Lend-Lease trainer
Basic Trainers				
BT-1		Douglas	1930	Last w/Liberty eng.
BT-2		Douglas	1931	Conv. from O-32
YBT-7		Consolidated	1932	
BT-8		Seversky	1934	Enc. cockpit, pants
BT-9		North American	1936	2-gun, camera

Designation	Name-variants	Manufacturer	Year	Notes
Basic Trainers cont.				
BT-12	Sophomore	Fleetwings	1938	Stainless steel
BT-13	Valiant	Vultee	1939	Enc., retractable
BT-14	Yale	North American	1941	BT-9 w/ metal skin
BT-15	Valiant	Vultee	1941	

Advanced Trainers

Key: AT=advanced trainer; BC=basic combat; T=trainer (from 1948)

Designation	Name-variants	Manufacturer	Year	Notes
AT-1		Huff-Deland	1925	Tandem
AT-4	Hawk	Curtiss	1927	Single seat
AT-5	Hawk	Curtiss	1927	
BC-1		North American	1938	Enc., retractable
AT-6	Texan	North American	1940	Conv. from BC-1
AT-7	Navigator	Beech	1940	Nav. trainer
AT-8	Bobcat	Cessna	1941	Transition trainer
AT-9	Jeep	Curtiss	1940	Adv. transition tr.
AT-10	Wichita	Beech	1941	All wood
AT-11	Kansan	Beech	1941	Bomb trainer
AT-12	Guardsman	Republic	1941	Enlarged P-35A
AT-16	Harvard	Noorduyn	1942	Canadian AT-6
AT-17	Bobcat	Cessna	1942	
AT-18	Hudson	Lockheed	1942	Bomb crew trainer
AT-19	Reliant	Stinson-Vultee	1943	
AT-20	Avro Anson	Federal	1943	Canadian-built
AT-21	Gunner	Fairchild	1943	Adv. gunner trainer
AT-22	Liberator	Consolidated	1943	Adv. flt eng. trainer
AT-23	Marauder	Martin	1943	Target tug
AT-24	Mitchell	North American	1943	Transition/crew tr.
T-6	Texan	North American	1948	Tr./Korean spotter
T-7	Navigator	Beech	1948	Adv. nav. trainer
T-11	Kansan	Beech	1948	Adv. bomb/gun tr.
T-19	Cornell	Fairchild	1948	
TB-25	Mitchell	North American	1948	
T-28	Trojan	North American	1950	2-seat, 2-gun
T-29	Flying Classroom	Consolidated	1949	
T-33	Silver Star/Thunderbird	Lockheed	1949	First jet trainer
T-34	Mentor	Beech	1949	
T-37	Tweety Bird	Cessna	1955	
T-38	Talon	Northrop	1961	1st supersonic tr.
T-39	Sabreliner	North American	1960	
T-41	Mescalero	Cessna	1964	Military 172
T-43	737	Boeing	1973	
T-45C	Expediter	Beech	1948	
TF-80C	Shooting Star	Lockheed	1948	Stretch F-80

Obersvation/Liaison 1922-present

Key: AO=artillery observation; CO=corps observation; L=liason; O=observation; X=experimental prototype; Y limited production prototype

Designation	Name-variants	Manufacturer	Year	Notes
CO-4		Fokker	1922	
AO-1		Fokker-Atlantic	1923	
O-1	Falcon	Curtiss	1925	2-set, 4-gun
O-2		Douglas	1924	2-seat, 3-gun
O-5	DWC	Douglas	1924	World Cruiser

Designation	Name-variants	Manufacturer	Year	Notes
Observation/Liaison cont.				
YO-6		Thomas-Morse	1925	All-metal
O-17	Courier	Consolidated	1928	
O-19		Thomas-Morse	1928	All-metal biplane
YO-27		Fokker	1931	Enc. monoplane
O-31		Douglas	1931	Butterfly-wing
O-32		Douglas	1931	
YO-35		Douglas	1932	Retractable gear
O-47		North American	1937	Mid-wing, 3-place
O-49	Vigilant (L-1)	Stinson	1940	
O-52	Owl	Curtiss	1940	Retractable gear
O-57	Grasshopper (L-2)	Taylorcraft	1941	
O-58	Defender (L-3)	Aeronca	1941	
O-59	Cub (L-4)	Piper	1941	
YO-60		Kellett	1942	Helicopter
O-62	Sentinel (L-5)	Stinson-Vultee	1942	
XO-63	Cadet (L-6)	Interstate	1942	XL-6
L-9	Voyager	Stinson-Vultee	1942	
L-13		Consolidated	1947	Triphib, fold wing
L-19	Bird Dog	Cessna		
L-20	Beaver	deHavilland	1952	
L-21	Super Cub	Piper	1951	
L-26	Commander	Aero Design	1956	Presidental
L-27	Administrator	Cessna	1957	Model 310
L-28	Super Courier	Helio-GAC	1958	
O-1	Bird Dog	Cessna	1963	
O-2	Skymaster	Cessna	1967	COIN
OV-10	Bronco	Rockwell	1968	
Amphibian				

Key: A=amphibian; OA=observation ampibian; SA=search amphibian

Designation	Name-variants	Manufacturer	Year	Notes
OA-1		Loening	1925	
OA-2		Loening	1929	
OA-3	Dolphin	Douglas	1933	Monoplane
OA-4	Dolphin	Douglas	1934	
YOA-8		Sikorsky	1937	
OA-9	Goose	Grumman	1938	
OA-10	Catalina	Consolidated Vultee	1942	S&R/Patrol
OA-13	Goose	Grumman	1942	
OA-14	Widgeon	Grumman	1942	
A-9	Goose	Grumman	1948	
A-10	Catalina	Consolidated Vultee	1948	
SA-16	Albatross	Grumman	1949	Korean triphibian

Photography/Reconnaissance 1930-present

Key: F=photo reconnaissance; SR=strategic reconnaissance

Designation	Name-variants	Manufacturer	Year	Notes
F-1		Fairchild	1930	Folding wing
F-2	Expediter	Beech	1941	
F-3	Havoc	Douglas	1940	Bomb bay camera
F-4	Lightning	Lockheed	1942	Four K-17 cameras
F-5	Lightning	Lockheed	1942	
F-6	Mustang	North American	1942	
F-7	Liberator	Consolidated	1942	Nine cameras

Designation	Name-variants	Manufacturer	Year	Notes
Photography/Reconniassance cont.				
F-8	Mosquito	deHavilland	1943	
F-9	Flying Fortress	Boeing	1942	
F-10	Mitchell	North American	1941	
F-13	Superfortress	Boeing	1944	
F-14	Shooting Star	Lockheed	1944	
F-15	Reporter	Northrop	1946	
SR-71	Blackbird	Lockheed	1964	2,190 mph

Utility 1952-present

Key: HU=search & rescue utility; U=utility, general

Designation	Name-variants	Manufacturer	Year	Notes
U-2	Gray Ghost/Shady Lady	Lockheed	1955	10-hr. endurance
U-3	Administrator/ Blue Bird	Cessna	1958	
U-4	Commander	Aero Design	1960	VIP transport
U-5	Twin Courier	Helio-GAC	1964	
U-6	Beaver	deHavilland	1962	
U-7	Super Cub	Piper	1962	
U-10	Super Courier	Helio-GAC	1961	
HU-16	Albatross	Grumman	1962	S&R amphibian
U-17	Skywagon	Cessna	1965	Model 185
AU-23	Peacemaker	Fairchild Hiller	1971	Utility attack
AU-24	Stallion	Helio-GAC	1972	

Racers 1920-1925

Designation	Name-variants	Manufacturer	Year	Notes
R-1	VCP-R	Engineering Div.	1921	Won 1920 Pulitzer
R-2	MB-6	Thomas-Morse	1921	For 1921 Pulitzer
R-3		Eng. Div.-Sperry	1922	Won 1924 Pulitzer
R-5	MB-9	Thomas-Morse	1922	For 1922 Pulitzer
R-6		Curtiss	1922	1922 Pulitzer 1-2
R-8		Curtiss	1923	Navy R2C-1
R3C-1	Land Racer	Curtiss	1925	Won 1925 Pulitzer
R3C-2	Sea Racer	Curtiss	1925	1st 1925 Schneider

Rotary Wing 1935-present

Key: G=rotary wing, autogiro; H=helicopter, army; HH=search & rescue healicopter; R=rotary wing, helicopter; UH=utility helicopter; Y=limited production prototype

Designation	Name-variants	Manufacturer	Year	Notes
YG-1	(Gyroplane)	Kellett	1934	
XR-2	(Gyroplane)	Kellett	1941	
R-4		Vought-Sikorsky	1942	Coastal patrol
R-5		Vought-Sikorsky	1944	Air Rescue Serv.
R-6		Sikorsky	1944	Improved R-4
YR-12		Bell	1947	Model 47
YR-13	Sioux	Bell	1947	
H-4	(R-4)	Vought-Sikorsky	1948	Redesignation R-4
H-5	(R-5)	Vought-Sikorsky	1948	
H-6	(R-6)	Sikorsky	1948	
YH-12	(YR-12)	Bell	1948	
H-13	Sioux (YR-13)	Bell	1948	
H-19	Chickasaw	Sikorsky	1949	
H-21	Workhorse	Piasecki	1952	Double rotor
H-43	Huskie	Kaman	1958	

Designation	Name-variants	Manufacturer	Year	Notes
Rotary Wing cont.				
HH-1	Iroquois	Bell	1964	
HH-3	Jolly Green Giant	Sikorsky	1963	
UH-13	Sioux	Bell	1962	
HH-19	Chicasaw	Sikorsky	1962	
HH-21	Workhorse	Piasecki-Vertol	1962	Tandem rotor
HH-43	Huskie	Kaman	1962	
HH-53	Sea Stallion	Sikorsky	1967	

Attack 1918-1942

> Key: A=attack; DB=day bombardment; DH=deHavilland; GA=ground attack;
> Y=limited production prototype

Designation	Name-variants	Manufacturer	Year	Notes
DH-4	Flaming Coffin	Dayton-Wright	1918	
DB-1		Gallaudet	1921	Day Bomber
GA-1		Boeing	1921	Pusher triplane
GA-2		Boeing	1922	Armor-plated
A-3	Falcon	Curtiss	1927	
YA-8	Shrike	Curtiss	1932	Trailing edge flaps
YA-11		Consolidated	1934	All-metal enclosed
A-12	Shrike	Curtiss	1933	
A-17	Nomad	Northrop-Douglas	1936	
YA-18	Shrike	Curtiss	1937	Twin attack
YA-19		Vultee	1939	Foreign aid model
A-20	Havoc	Douglas	1941	
A-24	Dauntless	Douglas	1941	Adapted from Navy
A-25	Shrike-Helldiver	Curtiss	1942	Adapted from Navy
A-26	Invader	Douglas	1943	Shoulder wing
A-27	(AT-6)	North American	1941	775 hp
A-28	Hudson	Lockheed	1941	Britain/Australia
A-29	Hudson	Lockheed	1941	
A-30	Baltimore	Martin	1942	Mid-wing
A-31	Vengence	Vultee	1941	
A-33	Nomad	Douglas	1942	
A-35	Vengance	Vultee	1942	England/Brazil
A-36	Mustang	North American	1942	

Pursuit-Fighter 1919-1925

> Key: PA=pursuit, air-cooled; PG=pursuit, ground attack; PN=pursuit, night;
> PW=pursuit, water-cooled

Designation	Name-variants	Manufacturer	Year	Notes
Orenco-D		Curtiss	1919	
MB-3		Morse-Boeing	1920	
PN-1		Curtiss	1921	Pursuit-Night
PA-1		Loening	1922	
PG-1		Aeromarine	1923	
PW-1		Engineering Div.	1921	
PW-7	D.XI	Fokker	1923	
PW-8		Curtiss	1923	Transcontinental
PW-9		Boeing	1924	
TP-1		Engineering Div.	1924	

Pursuit 1925-1948

Key: FM=fighter, multi-place; P=pursuit; PB=pursuit, bi-place; X=experimental prototype; Y=limited production prototype				
Designation	*Name-variants*	*Manufacturer*	*Year*	*Notes*
P-1	Hawk	Curtiss	1925	
P-2	Hawk	Curtiss	1926	Turbosupercharged
P-3	Hawk	Curtiss	1927	
P-5	Hawk	Curtiss	1927	
P-6	Hawk	Curtiss	1928	Nat. Air Races 1-2
P-12		Boeing	1929	Mail carrier
YP-16		Berliner-Joyce	1932	
PB-1		Berliner-Joyce	1935	
P-26	Peashooter	Boeing	1932	All-metal
P-30		Consolidated	1934	
PB-2		Consolidated	1935	
P-35		Seversky	1937	Retractable gear
P-35A	Guardsman	Republic	1940	
P-36	Hawk	Curtiss	1937	
Hawk-75	Hawk-Mohawk	Curtiss	1937	
YP-37	Hawk	Curtiss	1939	
YFM-1	Airacuda	Bell	1939	Five-place twin
P-38	Lightning	Lockheed	1941	
P-39	Airacobra	Bell	1941	
P-40	Warhawk	Curtiss	1941	
P-43	Lancer	Republic	1940	Flown in China
Mk. V	Spitfire	Supermarine	1940	
P-47	Thunderbolt	Republic	1942	
P-51	Mustang	North American	1943	
P-59	Airacomet	Bell	1944	First jet
P-61	Black Widow	Northrop	1943	
P-63	Kingcobra	Bell	1943	
YP-64	(AT-6)	North American	1940	Export figher
P-66	Vanguard	Vultee	1941	Chinese-flown
P-70	Night Havoc	Douglas	1942	A-20 conversion
P-80	Shooting Star	Lockheed	1945	Twin jet
XP-81		Convair	1945	Jet/prop twin
P-82	Twin Mustang	North American	1945	
XP-83		Bell	1945	Long-range P-59
XP-87		Curtiss	1948	Four-jet

Fighters 1948-1959

Key: F=Fighter; Q=target drone; RF=reconnaissance fighter; Y=limited production prototype				
Designation	*Name-variants*	*Manufacturer*	*Year*	*Notes*
F-24	Dauntless	Douglas	1948	A-24 redesignation
F-38	Lightning	Lockheed	1948	P-38 redesignation
F-40	Warhawk	Curtiss	1948	P-40 redesignation
F-47	Thunderbolt	Republic	1948	P-47 redesignation
F-51	Mustang	North American	1948	P-51 redesignation
F-59	Airacomet	Bell	1948	P-59 redesignation
RF-61	Black Widow	Northrop	1948	P-61 redesignation
QF-63	Kingcobra	Bell	1948	Manned target
F-80	Shooting Star	Lockheed	1948	P-80 redesignation
F-82	Twin Mustang	North American	1948	P-82 redesignation
F-84	Thunderjet	Republic	1948	

Designation	Name-variants	Manufacturer	Year	Notes
Fighters cont.				
F-84F	Thunderstreak	Republic	1952	Swept wing
RF-84F	Thunderflash	Republic	1954	B-36 mother ship
XF-85	Goblin	McDonnell	1948	
F-86	Sabre Jet	North American	1948	
XF-88		McDonnell	1948	Became F-101
F-89	Scorpion	Northrop	1950	
XF-90		Lockheed	1949	Penetration fighter
XF-91		Republic	1949	
XF-92		Consolidated-Vultee	1949	Delta wing
XF-93	(F-86C)	North American		(F-86C)
F-94	Starfire	Lockheed	1949	Afterburner
YF-95	(F-86D)	North American		(F-86D)
XF-96	(F-86F)	Republic		(F-84F)
XF-97	(F-94C)	Lockheed		(F-94C)
XF-98	(Falcon missle)	Hughes		
XF-99	(BOMARC missle)	Boeing		
F-100	Super Sabre	North American	1953	Supersonic cruise
F-101	Voodoo	McDonnell	1954	
F-102	Delta Dagger	Convair	1955	
XF-103	cancelled	Republic		
F-104	Starfighter	Lockheed	1956	Mach 2
F-105	Thunderchief	Republic	1957	
F-106	Delta Dart	Convair	1959	
XF-107	(F-100B variant)	North American		
XF-108	cancelled	North American		
XF-109	V/STOL	Bell		
XF-110	(F-4C)	McDonnell		

Fighters 1963-present

Designation	Name-variants	Manufacturer	Year	Notes
F-4	Phantom	McDonnell	1963	
F-5	Freedom Fighter/ Tiger II	Northrop	1963	
YF-12	Blackbird	Lockheed	1964	
F-111		General Dynamics	1965	Variable wing
F-15	Eagle	McDonnell-Douglas	1974	
F-16	Fighting Falcon	General Dynamics	1976	
F-117	Stealth	Lockheed	1980	
YF-22		Lockheed	1990	
YF-23		Northrop-MDD	1990	

Attack 1963-present

Designation	Name-variants	Manufacturer	Year	Notes
A-1	Skyraider	Douglas	1963	
A-7	Corsair II	LTV	1968	
A-10	Thunderbolt II	Fairchild	1975	
A-37	Dragonfly	Cessna	1967	

Bombers 1918-present

Key: FB=fighter-bomber; NBL=Barling bomber; HB=heavy bombardment; KB=tanker-bomber; LB=light bombardment; MB=Martin bomber; NBS=night bombardment, short distance; RB=reconnaisance bomber; X=experimental prototype; Y=limited production prototype

Designation	Name-variants	Manufacturer	Year	Notes
MB-1	Martin Bomber	Martin	1918	
MB-2	Martin Bomber	Martin-Curtiss	1920	
NBS-1		Martin-Curtiss	1920	

Designation	Name-variants	Manufacturer	Year	Notes
Bombers cont.				
NBL-1	Barling Bomber	Witteman-Lewis	1923	
XHB-1	Cyclops	Huff-Deland	1926	Single-engine
LB-1		Huff-Deland	1926	
LB-5	Pirate	H-D-Keystone	1927	
LB-6		Keystone	1928	
B-2	Condor	Curtiss	1928	
B-3		Keystone	1929	
YB-7		Douglas	1932	Retractable gear
YB-9		Boeing	1932	Mid-wing
B-10	Martin Bomber	Martin	1934	All-metal
B-12	Martin Bomber	Martin	1934	Float-fitted
XBLR-1	(XB-15)	Boeing	1935	
XB-15		Boeing	1937	10-man crew
B-17	Flying Fortress	Boeing	1940	
B-18	Bolo	Douglas	1937	
XBLR-2		Douglas	1938	
XB-19		Douglas	1941	200 ft. wing
B-23	Dragon	Douglas	1939	Tail gun turret
B-24	Liberator	Consolidated	1941	
B-25	Mitchell	North American	1941	
B-26	Marauder	Martin	1941	
B-26	(A-26) Invader	Douglas	1948	
B-29	Superfortress	Boeing	1944	
KB-29	Superfortress	Boeing	1948	Refueling tanker
B-32	Dominator	Consolidated-Vultee	1944	
B-34	Ventura	Lockheed	1942	
YB-35	Flying Wing	Northrop	1946	
B-36	Peacemaker	Consolidated-Vultee	1947	10-engined
B-37	Ventura	Lockheed	1943	
XB-42	Mixmaster	Douglas	1944	
XB-43		Douglas	1946	Twin jet
B-45	Tornado	North American	1948	
XB-46		Consolidated	1947	
B-47	Stratojet	Boeing	1951	Swept wings
KB-47	Stratojet	Boeing	1953	
XB-48		Martin	1947	
YB-49	Flying Wing	Northrop	1947	Eight jets
B-50	Superfortress	Boeing	1947	Re-engined B-29
KB-50	Superfortress	Boeing	1952	
XB-51		Martin	1949	
B-52	Stratofortress	Boeing	1955	
XB-53	cancelled	Convair		
XB-55	cancelled	Boeing		
XB-56	cancelled	Boeing		
B-57	Intruder (Canberra)	Martin	1953	
RB-57D	Intruder	Martin	1954	
RB-57F	Intruder	General Dynamics	1964	
B-58	Hustler	Convair	1956	Supersonic
YB-60	(B-36)	Consolidated-Vultee	1952	Swept-wing B-36
B-61	Matador	Martin	1951	Pilotless missle
B-66	Destroyer	Douglas	1956	High-wing twin jet
RB-66	Destroyer	Douglas	1954	
XB-70	Valkyrie	North American	1964	Triplesonic
B-1		Rockwell	1974	Variable-wing
B-26K	Counter-Invader	On Mark	1964	COIN
FB-111		General Dynamics	1969	
B-2	Stealth	Northrop	1989	

Cargo Transports 1919-present

> Key: AC=attack-cargo transport; C=cargo transport; KC=tanker; LC=liason cargo transrport; T=transport; UC=utility cargo transport; VC=VIP transport; X=experimental prototype; Y=limited production prototype

Designation	Name-variants	Manufacturer	Year	Notes
T-1		Martin	1919	10-passenger
T-2		Fokker	1923	
C-1		Douglas	1925	
C-2	Fokker Trimotor	Fokker	1926	Byrd North Pole
C-4	Ford Trimotor	Ford	1929	Byrd South Pole
C-6		Sikorsky	1929	Amphibian
C-8		Fairchild	1930	
YC-14		Fokker	1931	Parasol wing
C-21	Dolphin	Douglas	1932	Amphibian
C-26	Dolphin	Douglas	1932	Border Patrol
C-27	Airbus	Bellanca	1932	
C-29	Dolphin	Douglas	1933	Conv. from C-26
C-30	Condor	Curtiss	1933	Byrd Antarctic
C-32	DC-2	Douglas	1936	14-seat
C-33	DC-2	Douglas	1936	
YC-34	DC-2	Douglas	1936	C-32 variant
XC-35	Electra	Lockheed	1936	Pressurized
UC-36	Electra	Lockheed	1937	Amelia Earhart
UC-37	Electra	Lockheed	1937	
C-38	DC-2 1/2	Douglas	1937	
C-39	DC-2 1/2	Douglas	1939	
UC-40	Electra	Lockheed	1938	12A
UC-43	Traveler	Beech	1941	Staggerwing
UC-45	Expediter	Beech	1940	AT-7/AT-11
C-46	Commando	Curtiss	1942	
C-47	Skytrain (DC-3)	Douglas	1942	
C-48	DC-3	Douglas	1941	
C-49	DC-3	Douglas	1941	
C-50	DC-3	Douglas	1941	
C-51	DC-3	Douglas	1941	
C-52	DC-3	Douglas	1941	
C-53	Skytrooper (DC-3)	Douglas	1941	
C-54	Skymaster (DC-4)	Douglas	1942	
C-56	Lodestar	Lockheed	1941	
C-57	Lodestar	Lockheed	1941	
C-58	Bolo	Douglas	1942	
C-59	Lodestar	Lockheed	1942	British C-57
C-60	Lodestar	Lockheed	1942	
UC-61	Forwarder	Fairchild	1941	
C-63	Hudson	Lockheed	1941	
C-64	Norseman	Noorduyn	1942	
C-66	Lodestar	Lockheed	1942	
C-67	Dragon	Douglas	1942	B-23 conversion
C-68	DC-3A	Douglas	1942	
C-69	Constellation	Lockheed	1943	
UC-70	Nightingale	Howard	1942	
UC-71	Executive	Spartan	1942	
UC-72		Waco	1942	Cabin biplane
C-73	Globemaster	Douglas	1945	
C-75	Stratoliner	Boeing	1942	Pressurized
UC-78	Bobcat	Cessna	1941	

Designation	Name-variants	Manufacturer	Year	Notes
Cargo Transports cont.				
UC-81	Reliant	Stinson-Vultee	1942	
C-82	Packet	Fairchild	1945	Twin booms
UC-83	Coupe	Piper	1943	
C-84	DC-3B	Douglas	1942	
UC-86	Forwarder	Fairchild	1942	
C-87	Liberator	Consolidated	1942	
UC-95	Grasshopper	Taylorcraft	1942	
UC-96		Fairchild	1942	High folding wing
C-97	Stratofreighter	Boeing	1949	B-29/B-50 variant
KC-97	Stratofreighter	Boeing	1951	Tanker
XC-99	B-36	Consolidated	1949	400-seat
UC-100	Gamma	Northrop	1942	
UC-101	Vega	Lockheed	1942	
XC-105	XB-15	Boeing	1943	
YC-108	Flying Fortress	Boeing	1941	MacArthur VIP
C-109	Liberator	Consolidated	1943	"Hump" fuel ferry
C-111	Super Electra	Lockheed	1944	
YC-116	Skymaster II	Douglas	1946	
C-117	Skytrain II	Douglas	1945	
C-118	Liftmaster	Douglas	1946	118-seat
C-119	Flying Boxcar	Fairchild	1947	
C-121	Super Constellation	Lockheed	1949	
YC-122	Cargo Glider	Chase	1948	All-metal glider
XC-123	Avitruc	Chase	1949	
XC-123A		Chase	1951	Four-jet
C-123B	Provider	Chase-Fairchild-Stroukoff	1955	
C-124	Globemaster II	Douglas	1950	
C-125	Raider	Northrop	1950	
LC-126		Cessna	1949	
C-130	Hercules	Lockheed	1956	
C-131	Samaritan	Convair	1954	Airborne hospital
C-133	Cargomaster	Douglas	1957	
YC-134	Pantobase	Stroukoff	1958	STOL C-123
C-135	Stratolifter	Boeing	1961	
KC-135	Stratotanker	Boeing	1957	
VC-137	707	Boeing	1959	Air Force One
C-140	Jetstar	Lockheed	1961	
C-141A	Starlifter	Lockheed	1965	
YC-141B	Starlifter	Lockheed	1977	
XC-142	Tilt-wing	LTV-Hiller-Ryan	1964	Tilt-wing transport
C-5	Galaxy	Lockheed	1969	
VC-6	King Air	Beech	1965	
C-7	Caribou	deHavilland	1967	
C-8	Buffalo	Bell-deHavilland	1967	Air cushion exp.
C-9	Nightingale	McDonnell-Douglas	1968	
KC-10	Extender	McDonnell-Douglas	1980	
C-12	Super King Air	Beech	1975	
YC-14	AMST-STOL	Boeing	1976	
YC-15	AMST-STOL	McDonnell-Douglas	1975	
AC-47	DC-3	Douglas	1965	Gunship
AC-119	C-119 gunship	Fairchild-Hiller	1968	
AC-130	C-130 gunship	Lockheed-LTV	1970	

X-Planes

Designation	Name-variants	Manufacturer	Year	Notes
X-1	Rocketship (XS-1)	Bell	1946	First supersonic
X-2	Transonic	Bell	1952	Mach 3
X-3	Stiletto	Douglas	1952	Subsonic

Designation	Name-variants	Manufacturer	Year	Notes
X-planes cont.				
X-4	Bantam	Northrop	1948	Flying wing twin
X-5	Variable-Sweep	Bell	1951	
X-13	Vertijet	Ryan	1955	Vertical op.
X-14	Thrust-Diverter	Bell	1956	
X-15	Hypersonic	North American	1959	4,534 mph
X-18	Tilt-Wing	Hiller	1959	Convertiplane
X-19	Tilt-Prop	Curtiss-Wright	1963	VSTOL
X-21	Laminar-Flow	Northrop	1963	
X-22	Tilt-Duct	Bell	1966	
X-24	Lifting Body	Martin-Marietta	1967	Wingless re-entry
X-25	Gyro-Glidedr	Bensen	1978	Gyrocopter
X-31A		Rockwell-MBB	1990	

U.S. MILITARY AIRCRAFT 1908-1945 BY MANUFACTURER

Specifications and performance of each type varies among several models; figures are generally for the final variant. Models are listed in ascending order of weight. Production numbers reflect approximation of those built for the military. Manufacturer listing includes location and date of founding.

Type (year flown)	Max. wt., lb.	Powerplant	Speed, mph	No. built
Aeronca, Aeronautical Products, Inc., Middletown, OH, 1928				
L-3 Grasshopper	1,260	C. O-170	87	1,487
Beech, Beech Aircraft Corp., Wichita, KS, 1932				
UC-43 Traveller (34)	4,250	PW R-985	200	207
C-45 Expediter (37)	7,500	(2) PW R-985	225	2.645
F-2 (40)	7,725	(2) PW R-985	225	56
AT-7 (SNB-2) (41)	7,850	(2) PW R-985	224	1,037
AT-10 (41)	6,465	(2) L. R-680	190	1,771
AT-11 (41)	8,195	(2) PW R-985	215	1,582
UC-43 (42)	4,800	PW R-985	195	325
Bell Aircraft Corp., Buffalo, NY, 1935				
P-39 Airacobra (37)	8,052	A. V-1710	385	9,558
P-63 Kingcobra (42)	8,442	A. V-1710	410	3,300
P-59 Airacomet (44)	13,000	(2) GE J-31	400+	50
Berliner-Joyce				
P-16 (29)	4,209	Curtiss V-1570	170	25
Boeing Aircraft Company, Seattle, WA, 1916				
GA-1 (20)	9,740	(2) Liberty 12	115	10
GA-2 (21)	9,150	E.D.	98	2
PW-9 (24)	3,030	Curtiss D-12	159	110
Stearman 75 (35)	2,717	L. R-680/C. R-670	124	10,000
C-73 Model 247 (33)	13,650	(2) PW R-1340	20	27
C-75 Stratoliner(38)	45,000	(4) W. GR-1820	246	10
B-17 Flying Fortress (36)	60,000	(4) W. R-1820-97	295	12,731
C-98	82,500	(4) W. R-2600	193	
B-29 Superfortress (42)	135,000	(4) W. R-3350-23	357	4,281
C-98 Clipper (42)	84,000	4 W. R-2800	190	4
Brewster Aeronautical Corp., Long Island City, NY, 1932				
SBA/SBN (36)	6,759	W. R-1820-22	254	30
F2A Buffalo (37)	7,159	W. R-1820-40	321	
SB2A Bucaneer (41)	14,289	W. R-2600-8	274	771
Burgess				
Model F (11)	1,379	Sturdevant D-4	43	1
Model H (12)	1,600	Renault V-8	57	7
I-Scout (13)	2,038	Sturdevant D-4	59	1
J-Scout (13)	1,533	Sturdevant D-4	45	1
Twin Hydro (17)	5,380	(2) Sturdevant 5A	78	1
Cessna Aircraft Co., Inc., Wichita, KS, 1927				
AT-8 Bobcat (41)	5,100	(2) L. R-680	185	33
AT-17 Bobcat (42)	5,303	(2) J. R-775	176	1,199
UC-77 (42)	5,700	(2) W. R-975	160	11
UC-78 Bobcat (42)	5,700	(2) J. R-755	175	3,437
UC-94 (42)	2,450	W. R-500	162	3
C-106 (43)	14,800	(2) PW R-1340	200	2

Type (year flown)	Max. wt., lb.	Powerplant	Speed, mph	No. built
Chance Vought Div. United Aircraft Corp., Stratford, CT				
OS2U Kingfisher (38)	4,980	PW R-985	170	1,306
SB2U Vindicator (38)	9,421	PW R-1535-02	243	219
F4U Corsair (40)	14,670	PW R-2800-18W	446	12,571
TBY Sea Wolf (41)	18,488	PW R-2800-20	306	180
Consolidated Vultee Aircraft Corp., San Diego, CA				
TW-3 (23)	2,407	Wright E	103	20
A-11 (33)	5,490	Curtiss V-1570	225	
BT-13/15 Valiant (39)	4,360	PW R-985	164	11,000
P-66 Vanguard (39)	7,384	PW R-1830-33	340	144
A-31/A-35 Vengeance	16,400	W. R-2600-13	279	1,931
PBY Catalina (35)	34,000	(2) PW R-1830-92	196	4,000
B-24 Liberator (39)	60,000	(4) PW R-1830-65	297	18,475
A-31 (41)	13,659	Wright R-2600	275	100
A-35 (42)	15,600	Wright R-2600	273	931
C-87 Liberator	56,000	(4) PW R-1830-43	300	276
PB4Y Privateer (43)	65,000	(4) PW R-1830-94	250	73
PB2Y-3 Coronado(37)	66,000	(4) PW R-1830-88	194	257
B-32 Dominator (42)	100,000	(4) W. R-3350-23	357	115
Curtiss, Wright and Curtiss-Wright Corp., St. Louis, MO; Buffalo, NY, 1910				
Wright A (08)	1,360	Wright 4	40	1
Wright D (11)	1,387	Curtiss E-4	50	1
Wright B (11)	1,216	Wright 4	42	2
E (11)	1,262	Curtiss S-6	44	3
Wright C (12)	1,431	Wright 6-60	48	7
Flying Boat (13)	1,760	Curtiss L or O	59	3
Wright D Scout (13)	1,270	Wright 6-60	67	2
G Scout (13)	1,290	Curtiss O	52	2
J (14)	1,345	Curtiss OXX	84	2
N (14)	1,360	Curtiss OX	82	1
Wright F (15)	2,100	Austro-Daimler 6	60	1
N-8 (16)	1,750	Curtiss OX-2	78	4
R-2 (16)	2,800	Curtiss VX	86	12
JN-4 (17-18)	1,850	Curtiss OX-5	75-93	6,163
Twin JN (17)	3,150	(2) Curtiss OXX-2	85	8
R-4 (17)	3,242	Curtiss V2-3	90	53
S-3 (17)	1,320	Curtiss OXX-2	112	4
N-9 (17)	2,410	Curtiss OXX-3	70	14
L-2 (17)	1,060	Curtiss OXX-2	115	4
R-3 (17)	3,945	Curtiss V2-3	83	18
18 (19)	3,000	Kirkham K-12	162	4
R-4L (18)	3,280	Liberty	105	6
Navy R6 (18)	3,945	Curtiss V2-3	83	10
Eagle (20)	9,460	(3) Liberty	124	3
R-6 (22)	2,120	Curtiss D-12	236	2
Wright PW-8 (24)	3,151	Curtiss D-12	161	25
R-8 Navy (24)	2,151	Curtiss D-12	266	1
R3C-1 (25)	2,150	Curtiss V-1400	249	1
O-1 Falcon (25)	4,488	Curtiss V-1150	145	104
P-1 Hawk (25)	2,950	Curtiss V-1150	155	93
P-2 Hawk (25)	3,255	Curtiss V-1400	180	5
O-11 (27)	4,560	Liberty V-1650	146	66

Type (year flown)	Max. wt., lb.	Powerplant	Speed, mph	No. built
Curtiss cont.				
A-3 Shrike (27)	4,378	Curtiss D-12	141	154
AT-4 (27)	2,375	Wright V-720	126	35
B-2 Condor (28)	16,500	(2) Curtiss V-1570	133	12
P-3A (28)	2,788	PW R-1340	153	5
P-5 (28)	3,360	Curtiss V-1150	159	4
P-6 Hawk (29)	3,150	Curtiss V-1570	180	18
P-11 (29)	3,310	Curtiss H-1640	170	3
O-39 (32)	4,725	Curtiss V-1570	173	10
A-12 Shrike (33)	5,745	W. R-1820	177	46
SNC	3,626	W. R-975-E3	201	305
CW-21 Demon (39)	4,500	W. R-1820-G5		315
O-52 Owl	5,364	PW R-1340-51	220	203
SOC Seagull (34)	5,437	PW R-1340-22	165	302
SO3C Seamew (39)	5,729	R. SGV-770-8	172	795
P-36 Hawk (36)	5,800	PW R-1830-17	311	240
AT-9 Jeep (41)	6,000	(2) L. R-680-9	197	790
SC Seahawk (44)	6,320	W. R-1820-62	313	576
SBC Helldiver (33)	7,141	W. R-1820-34	237	900
P-40 Warhawk (38)	8,720	A. V-1710-81	378	12,440
P-40 Warhawk (38)	8,720	P. V-1650-1		1,300
SB2C Helldiver (40)	14,760	W. R-2600-8	281	7,000
C-76 Caravan (43)	28,000	(2) PW R-1830-92	200	14
C-46 Commando (40)	56,000	(2) PW R-2800-51	269	3,355

Dayton-Wright

Douglas Aircraft Co., Santa Monica, CA, 1920

Type (year flown)	Max. wt., lb.	Powerplant	Speed, mph	No. built
D-WC (23)	9,162	Liberty 12	92	5
DT-2 (24)	6,505	Liberty 12	103	2
C-1 (25)	6,483	Liberty 12	123	26
O-2 (25)	4,753	Liberty V-1650	128	241
O-32A (30)	4,018	PW R-1340	130	30
O-38 (30)	4,650	PW R-1690	158	156
A-33	8,949	W. GR-1820-G205A (R)	265	121
SBD Dauntless	10,700	W. R-1820-66 (R)	255	4,558
TBD Devasator (35)	10,914	PW R-1830-64 (R)	206	129
O-46A (35)	6,135	PW R-1535	201	90
C-33 (36)	18,500	(2) W. R-1820	202	18
C-39 (38)	18,500	(2) W. R-1820	210	35
C-41 (38)	25,000	(2) PW R-1830	225	1
BTD Destroyer (43)	19,000	W. R-3350-14 (R)	344	28
A-20 Havoc (38)	20,000	(2) W. R-2600-29 (R)	325	7,385
B-23 Dragon (39)	26,500	(2) W. R-2600-3 (R)	282	38
B-18 (35)	27,673	(2) W. R-1820-53 (R)	215	350
A-26/B-26 Invader	32,000	(2) PW R-2800-71 (R)	345	2,450
C-47 Skytrain (35)	26,000	(2) PW R-1830-92 (R)	229	13,000
C-54 Skymaster (38)	73,000	(4) PW R-2000 (R)	274	953

Fairchild Aircraft Div. Fairchild Engine and Airplane Corp., Hagerstown, MD, 1935

Type (year flown)	Max. wt., lb.	Powerplant	Speed, mph	No. built
F-1A (31)	5,300	PW R-1340	140	6
PT-19 Cornell (40)	2,545	R. L-440-3 (I)	132	4,879
PT-23 (42)		C. R-670		1,125
PT-26 (42)	2,741	R. L-440	128	670
24W-41 Argus	2,562	Warner R-500 (R)	132	
AT-21 Gunner (42)	11,288	(2) R. V-770-15 (V)	225	175

Type (year flown)	Max. wt., lb.	Powerplant	Speed, mph	No. built
Fairchild cont.				
UC-86 (42)	2,550	R. 6-410	134	9
UC-88 (42)	4,000	W. R-760	170	2
C-82 (42)	42,000	(2) PW R-2800	223	200
Fleet Aviation Ltd., Fort Erie, Ont.				
PT 26A/B (42)	2,750	R. L-440	129	1,057
Fleetwings Div. Kaiser Cargo, Inc., Bristol, PA, 1929				
BT-12 (42)	4,410	PW R-985 (R)	195	25
General Motors, Fisher Body Div., Cleveland, OH				
P-75 Eagle (43)	18,210	A. V-3420-23 (W)	420	13
Grumman Aircraft Engineering Corp., Bethpage, NY, 1929				
F3F (35)	4,116	PW R-1535-84 (R)	231	162
J2F Duck	6,170	W. R-1820-20 (R)	180	60
F4F Wildcat (37)	7,002	PW R-1830-76 (R)	330	7,898
JRF Goose (37)	8,000	(2) PW R-985 (R)	201	345
F6F Hellcat (42)	12,441	PW R-2800-10 (R)	375	12,275
F8F Bearcat (44)	12,947	PW R-2800-34W (R)	421	1,263
TBF/TBM Avenger(41)	18,250	W. R-2600-20 (R)	267	9,939
F7F Tigercat (43)	25,720	(2) PW R-2800-34W	435	363
Lockheed Aircraft Corp., Burbank, CA, 1932				
P-80 Shooting Star	14,500	GE J33-A-11 (TJ)	558	
P-38 Lightning (39)	17,500	(2) A. V-1710 (V)	414	10,423
C-60 Lodestar	18,500	(2) W. R-1820-87 (R)	266	161
A-28/A-29 Hudson	20,000	(2) W. GR-1820 (R)	255	2,822
B-34 Ventura (41)	26,000	(2) W. GR-2800 (R)	312	463
B-37 Ventura		(2) W. R-2600	300	1,627
C-69 Constellation	72,000	(4) W. R-3350-35 (R)	330	22
Martin, Glenn L. Martin Co., Baltimore, MD, 1909				
A-30 Baltimore (41)	22,600	(2) W. GR-2600-A5B	305	1,575
B-26 Marauder (40)	37,000	(2) PW R-2800-4	282	5,157
PBM Mariner (39)	51,330	(2) W. R-2600-22 (R)	211	1,184
JRM Mars (43)	140,000	(4) W. R-3350-18 (R)	207	6
North American Aviation, Inc., Inglewood, CA, 1928				
AT-6 Texan	5,300	PW R-1340 (R)	208	4,361
A-36 Mustang	10,000	A. V-1710-87 (V)	310	500
F-6 Mustang				482
P-51 Mustang (40)	11,600	P. V-1650-7 (V)	437	14,319
B-25 Mitchell (39)	27,100	(2) W. R-2600-9 (R)	315	11,655
Northrop Aircraft, Inc., Hawthorne, CA, 1939				
P-61 Black Widow (42)	38,000	(2) PW R-2800-65 (R)	366	742
Piper Aircraft Corp., Lock Haven, PA, 1937				
L-4 Grasshopper	1,220	C. O-170-3 (O)	85	5,413
Republic Aviation Corp., Farmingdale, NY				
P-35 (Seversky) (35)	6,723	PW R-1830-45 (R)	310	136
P-43 Lancer	7,810	PW R-1830-35 (R)	349	282
P-47 Thunderbolt (41)	13,500	PW R-2800-21 (R)	433	15,660

Type (year flown)	Max. wt., lb.	Powerplant	Speed, mph	No. built
Ryan Aeronautical Co., San Diego, CA, 1933				
PT-16/20 (39)		M. L-365-1 (I)	125	30
PT-21		K. R-440-3 (R)	132	100
PT-22 Recruit (41)	1,860	K. R-540-1 (R)	131	1,023
FR-1 Fireball (44)	10,595	W.R-1820-72W/GE J31-GE-3	426	66
Sikorksy Div. United Aircraft Corp., Bridgeport, CT, 1943				
R-4 (42)	2,535	Warner R-550-1 (R)	75	132
R-5 (43)	4,825	PW R-985 (R)	106	131
R-6 (43)		L. O-435 (O)		228
JRS-1	19,096	(2) PW R-1690-52 (R)	190	15
Spartan Aircraft Co., Tulsa, OK, 1928				
NP-1	2,775	L. R-680-8 (R)	108	
Stinson Div. Consolidated Vultee Aircraft Corp., Wayne, MI				
L-5 Sentinel	2,020	L. O-435-1 (O)	130	3,284
L-1 Vigilant	3,400	L. R-680-9 (R)	122	324
AT-19 Reliant	4,000	L. R-680 (R)	141	500
Taylorcraft Aviation Corp., Alliance, OH, 1936				
L-2 Grasshopper	1,300	C. O-170-3 (O)	88	1,911
Timm Aircraft Corp., Van Nuys, CA				
N2T Tutor 410	2,725	C. R-670-4 (R)	144	262

Engine manufacturers: A=Allison; C=Continental; GE General Electric; J=Jacobs; K=Kinner; L=Lycoming; M=Menasco; P=Packard; PW=Pratt & Whitney; R=Ranger; W=Wright
Engine Type: I=Inline; O=Horizontally opposed; R=Radial; V=V-type

BRITISH AIRCRAFT OF WORLD WAR II

Specifications and performance of each type varies among several models; figures are generally for the final variant, Manufacturer listing includes location and date of founding. Production numbers are approximations.

Type (year flown)	Max wt., lb.	Powerplant (type)	Speed mph	No. built
Airspeed Ltd., Portsmouth, Hants., 1934				
AS.5 Courier (1933)	4000	AS Cheetah V (R)	165	16
AS.6 Envoy (1934)	6300	AS Cheetah IX (R)	210	50
AS.10 Oxford (1937)	8000	2 P&W R-985 (R)	202	8586
A.S.30 Queen Wasp (1937)	3500	AS Cheetah IX (R)	172	5
A.S.39 Fleet Shadower		4 Pobjoy Niagara V		
AS.45 Cambridge (1941)		B. Mercury VII (R)	237	2
AS.57 Ambassador	45000	2 B.Centaurus 57 (R)		300+
Armstrong Whitworth, Sir W.G. Armstrong Whitworth Aircraft, Ltd., Coventry, 1921				
A.W..27A Ensign	55000	4 W. GR-1820 (R)	210	
A.W..38 Whitley (1936)	33500	2 RR Merlin X (V)	230	1814
A.W..41 Albermarle(1939)	36500	2 B.Hercules XI (R)	265	598
Avro, A.V. Roe & Co., Ltd., Manchester, 1909				
621 Tutor/Sea Tutor	2458	AS Lynx IVC (R)	122	795
626 Prefect (1931)	2750	AS Lynx IVC (R)	112	178
652A Anson (1935)	8000	2 AS Cheetah IX (R)	188	11085
679 Manchester (1939)	56000	2 RR Vulture (X)	265	202
683 Lancaster (1941)	70000	4 RR Merlin XXIV (V)	287	7377
685 York (1942)	68597	4 RR Merlin XX (V)	298	257
688 Tudor I	76000	4 RR Merlin 100 (V)	290	
964 Lincoln (1944)	75000	4 RR Merlin 85 (V)	295	373
Blackburn Aircraft, Ltd., Brough, E. Yorkshire, 1910				
B.6 Shark (1933)	8050	AS Tiger VI (R)	150	363
B.20 flying boat (1940)	35000	2 RR Vulture (V)	268	1
B.24 Skua (1937)	8228	B. Perseus XII (R)	225	190
B.25 Roc (1938)	7950	B. Perseus XII (R)	223	136
B.26 Botha (1038)	18450	2 B. Perseus XA (R)	249	580
B.37 Firebrand (1942)	17500	B. Centaurus IX (R)	340	85
T.F.Mk IV (1943)	16227	B. Centaurus IX (R)	355	102
Boulton Paul Aircraft, Ltd., Wolverhampton, 1934				
Defiant (1937)	8424	RR Merlin XX (V)	313	1065
Bristol Aeroplane Co. Ltd., Filton, Bristol, 1910				
105 Bulldog (1927)	3490	B. Jupiter VII (R)	174	312
130 Bombay (1935)	20000	2 B. Pegasus XXII (R)	192	50
Blenheim (1936)	14400	2 B. Mercury XV (R)	266	1930
152 Beaufort (1938)	21230	2 B. Taurus (R)	260	1200
156 Beaufighter (1939)	25200	2 B. Hercules XVIII	303	5500
163 Buckingham (1943)	38050	2 B. Centaurus (R)	330	123
164 Brigand (1944)	39000	2 B. Centaurus 57 (R)	358	147
166 Buckmaster (1944)	33700	2 B. Centaurus VII (R)	352	110
British Taylorcraft, Thurmaston, Leicester				
Auster series	1850	Auster I--Cirrus Minor (I)		
		Auster III--DH Gipsy Major (I)		
		Auster IV-V Lyc. O-290-2 (O)	130	1500

Type (year flown)	Max wt., lb.	Powerplant (type)	Speed mph	No. built
de Havilland Aircraft Co., Ltd., Hatfield, Herts., 1920				
D.H.82 Tiger Moth (1931)	1770	DH Gipsy Major (I)	109	7759
D.H.89 Dominie (1934)	5500	2 DH Gipsy Queen (I)	157	728
D.H.91 Albatross (1937)	29500	4 DH Gipsy Twelve (V)	225	7
D.H.95 Flamingo (1938)	18000	2 B. Perseus (R)	243	15
D.H.98 Mosquito (1940)	22300	2 RR Merlin 25 (V)	362	7781
D.H.103 Hornet (1944)	20900	2 RR Merlin 133 (V)	472	204
Fairey Aviation Co., Ltd., Hayes, Middlesex, 1916				
Fox (1925)	4117	Curtiss D-12 (V)	156	300+
Seafox (1936)	5420	N. Rapier VI (H)		124
Gordon (1931)	5906	AS Panther IIA (R)	145	178
Swordfish (1934)	7510	B. Pegasus XXX (R)	138	2391
Battle (1937)	10792	RR Merlin I (V)	257	185
Albacore (1938)	10460	B. Taurus XII (R)	161	800
Fulmar (1940)	10700	RR Merlin VIII (V)	247	602
Barracuda (1940)	14250	RR Merlin 32 (V)	240	1718
Firefly (1941)	14020	RR Griffon IIB (V)	316	1631
Gloster Aircraft Co., Ltd., Hucclecote, Gloster, 1917				
SS.19B Gauntlet	3970	B. Mercury VIS.2 (R)	230	228
SS.37 Gladiator (1934)	4864	B. Mercury IX (R)	257	768
E.28/39 (1941)	3748	Power Jets W.2/500(TJ)	466	2
G.41 Meteor (1942)	13795	2 RR W.2B/23C Welland (TJ)	415	304
Handley Page, Ltd., Cricklewood, London, 1914				
H.P.52 Hampden (1936)	18756	2 B.Pegasus XVII	254	1432
H.P.52 Hereford (1937)	17800	2 N.Dagger VIII (H)	265	152
H.P.54 Harrow (1936)	23000	2 B.Pegasus XX (R)	200	100
H.P.57 Halifax (1939)	65000	4 B.Hercules XVI (R)	282	6200
Hawker Aircraft, Ltd. Kingston-on-Thames, Surrey, 1933				
Hart (1928)	4554	RR Kestrel IB (V) or Kestrel X (V)	184	1000
Osprey (1930)	4950	RR Kestrel V (V)	176	
Demon	4464	RR Kestrel IIS (V)	182	305
Audax (1931)	4386	RR Kestrel IB (V)	170	624
Hartebeeste (1935)	4787	RR Kestrel VFP (V)	176	65
Hardy (1934)	5005	RR Kestrel IB (V) or Kestrel X (V)	161	47
Hind (1934)	5298	RR Kestrel V (V)	186	452
Hector (1936)	4910	N. Dagger III MS (H)	187	178
Fury I/II	3609	RR Kestrel VI (V)	223	258
Nimrod (1931)	4059	RR Kestrel VFP (V)	193	84
Henley (1937)	8480	RR Merlin II (V)	272	
Hurricane (1935)	7300	RR Merlin XX (V)	342	15866
Sea Hurricane	8100	RR Merlin XX (V)	342	
Typhoon (1940)	11400	N. Sabre IIA (H)	374	3200
Tempest (1942)	13540	N. Sabre IIA (H)	426	1797
Tornado (1939)	10668	RR Vulture V (X)	398	2
Sea Fury (1944)	12500	B. Centaurus 18 (R)	460	667
Miles Aircraft, Ltd. Reading, Berkshire, 1935				
M.3B Falcon Six (1934)	2650	DH Gipsy Six (I)	180	36
M.11 Whitney Straight	1896	DH Gipsy Major (I)	145	

Type (year flown)	Max wt., lb.	Powerplant (type)	Speed mph	No. built
Miles cont.				
M.14 Magister (1937)	1900	DH Gipsy Major I (I)	132	1329
M.16 Mentor (1938)	2710	DH Gipsy Six I (I)	156	45
M.17 Monarch (1938)	2150	DH Gipsy Major (I)	140	11
M.20 (1940)	7758	RR Merlin XX (V)	350	2
M.25 Martinet (1942)	6750	B. Mercury XX (R)	240	1700
M.28 Mercury (1941)	2500	Cirrus Major III (I)	159	6
M.33 Monitor (1944)	21075	2 W. R-2600-31 (R)	330	20
M.38 Messenger (1942)	1900	DH Gipsy Major (I)	116	92
Master (1939)	5573	B. Mercury XX (R)	242	3295
Percival Aircraft, Ltd., Luton, Bedfordshire, 1932				
Proctor (1939)	3500	DH Gipsy Queen II (I)	160	200
Q.6 Petrel (1937)	5500	2 DH Gipsy Six (I)	195	7
Saro, Saunders-Roe, Ltd., Isle of Wight				
A.27 London (1934)	18400	2 B. Pegasus X (R)	155	29
S.36 Lerwick (1938)	33200	2 B. Hercules II (R)	216	21
Short Bros., Ltd., Rochester, Kent, 1898				
S.19 Singapore	27500	4 RR Kestrel (V)	145	37
S.25 Sunderland (1937)	65000	4 PW R-1830-90B (R)	213	739
S.45 Seaford (1944)	75000	4 B. Hercules XIX	242	8
S.26 'G' Class (1939)	74500	4 B. Hercules IVC	209	3
S.29 Stirling (1939)	70000	4 B. Hercules XVI	270	2369
Supermarine (1912), Vickers-Armstrong, Ltd., Southampton, 1928				
Stranraer (1935)	19000	2 B. Pegasus X (R)	165	63
Walrus (1933)	7200	B. Pegasus VI (R)	135	740
Sea Otter (1943)	10000	B. Mercury XXX (R)	150	290
Spitfire (1936)	10280	RR Griffon 65 (V)	448	20351
Seafire	7100	RR Merlin 55 (V)	352	3024
Spiteful/Seafang(1944)	10200	RR Griffon 65 (V)	475	18
Vickers, Vickers-Armstrongs, Ltd., Weybridge, 1938				
Vildebeest (1928)	8500	B. Perseus VIII (R)	156	194
Vincent	8100	B. Pegasus IIM3 (R)	142	171
Valentia	19500	2 B. Pegasus IIM3 (R)	120	28
Wellesley (1937)	11100	B. Pegasus XX (R)	228	176
Wellington (1936)	29500	2 B. Hercules XI (R)	235	11461
Warwick (1939)	51250	2 B. Centaurus VI(R)	262	750
Windsor (1943)	54000	4 RR Merlin 65 (V)	317	3
Westland Aircraft, Ltd., Yeovil, Somerset, 1935				
Wapiti (1927)	5400	B. Jupiter VIII (R)	135	544
Wallace	5750	B. Pegasus IV (R)	158	
Lysander (1936)	6318	B. Mercury XX (R)	212	1650
Whirlwind (1938)	11388	2 RR Peregrine (V)	360	114
Welkin (1942)	17500	2 RR Merlin 76 (V)	387	67

Engine manufacturers: AS=Armstrong Siddeley; B=Bristol; DH=de Havilland; N=Napier; P&W=Pratt & Whitney; RR=Rolls Royce; W=Wright

Engine types: H=H-type; I=Inline; O=Horizontally opposed; R=Radial; TJ=Turbo-jet; V=V-type; X=X-type

GERMAN AIRCRAFT OF WORLD WAR II

Type (year flown)	Max wt., lb.	Powerplant (type)	Speed mph	No. built
Arado Flugzewugwerke G.m.b.H., Babelsberg Bei Berlin, 1925				
Ar 96 (1938)	3307	As 10C-3 (V)	171	11500
Ar 196 (1937)	8223	BMW 132K (R)	193	536
Ar 232 (1941)	44092	4 BMW 323R-2 (R)	191	8
Ar 234 Blitz (1943)	21715	2 BMW 004B (TJ)	461	224
Blohm und Voss, Hamburg, 1933				
Bv 138 (1937)	31967	2 J. Jumo 205D (VO)	171	263
Ha 139 Nordwind	38581	4 J Jumo 205C (VO)	196	3
Bv 222 Wiking (1940)	100531	6 BMW 323R-2 (R)	184	13
Bücker Flugzeugbau G.m.b.H., Rangsdorf Bei Berlin, 1933				
Bu 131 Jungmann (1934)	1474	Hirth HM 504 (I)	115	
Bu 133 Jungmeister	1290	Siemens Sh 14A-4 ®	134	
Bu 180 Student	1188	Walter Mikron II (I)	99	
Bu 181 Bestmann	1650	Hirth HM 504 (I)	121	
Bu 182 Kornett	1122	Buker M 700 (I)	127	
Dornier-Werke G.m.b.H., Friedrichshafen, 1922				
Do 18/24/26 (1935)	23810	2 J. Jumo 205C (VO)	165	
Do 217 (1938)	29101	2 DB 603A (V) or		
		2 BMW 801M (R)	320	1905
Do 17 (1934)	18872	2 Bramo Fafnir 323P	224	1200
Do 335 Pfeil (1943)	20966	2 DB 603A-2 (V)	455	36
Fieseler, Gerhard Fiesler Werke G.m.b.H., Kassel, 1930				
Fi 156 Storch (1936)	2923	As 10C (V)	109	2549
Fi 103 Reichenberg	4806	As 109-104	401	
Flettner				
Fl 282 Kolibri helicopter (1941)	2205	Bramo SH.14A (R)	93	24
Focke-Achgelis & Co., G.m.b.H., Hoyenkamp				
Fa 223 Drache helictoper (1939)	9480	Bramo 323Q-3 (R)	114	
Focke-Wulf Flugzeugbrau G.m.b.H., Flughafen, Bremen, Johannisthal, 1924				
Fw 61 (1936)	2100	Bramo Sh.14A (R)	62	
Ta 152B		Jumo 213E (V)	428	
Ta 154 A-2 (1943)	19480	2 Jumo 211R (V)	394	
Ta 183 (1945)		Jumo 004 (TJ)	590	1
Fw 189 Uhu (1938)	8708	2 As 410A-1 (V)	217	842
Fw 190 (1939)	8378	BMW 801D-2 (R)	416	20200
Fw 200 Condor (1937)	50045	4 Bramo 323R-2 (R)	207	276
Heinkel, Ernst Heinkel A.G., Marienehe, 1922				
He 112 (1935)	4960	J. Jumo 210Ea (V)	317	74
He 219 Uhu (1942)	33730	2 DB 603E (V)	416	268
He 115 (1937)	22928	2 BMW 132K (R)	203	90
He 111 (1935)	24912	2 J. Jumo 211D-1 (R)	258	6000
He 177 Greif (1939)	59966	2 DB 610 (R)	303	1174
He 162 (1944)	6184	BMW 003E-1 (TJ)	562	280

Type (year flown)	Max wt., lb.	Powerplant (type)	Speed mph	No. built
Henschel Flugzeug-Werke A.G., Schonefeld, 1933				
Hs 123 (1935)	4894	BMW 132Dc (R)	211	
Hs 129 (1939)	9259	2 GR 14M 04/05 (R)	253	848
Junkers Flugzeug und Motorenwerke, A.G., Dessau, 1910				
Ju 87 Stuka (1935)	14550	J. Jumo 211J-1 (V)	255	5750
Ju 88 (1936)	28880	(2) BMW 801D (R)	342	14676
Ju 52/3m (1931)	24317	3 BMW 132T (R)	190	4850
Ju 352	43000	3 BMW 323 R-2 (R)	150	
Ju 388 Stortebeker	30700	2 BMW 801 TJ	414	
Ju 290 (1941)	43,000	3 BMW 323 ®	150	
Messerschmitt A.G., Augsberg, 1926				
Bf 109 (1935)	5523	DB 601Aa (V)	357	33000
Me 210/410 Hornisse	23500	2 DB 603A (V)	388	619/1160
Bf 110 (1936)	13289	2 DB 601A-1 (V)	336	6105
Me 262 Schwalbe (1941)	14101	2 J. Jumo 004B (TJ)	541	1430
Me 163 Komet (1941)	9502	Walter HWK 109-509A-1 liquid rocket	597	350
Me 323 (1941)	95901	6 GR 14N 48/49 (R)	137	198
Siebel Flugzeugwerke A.G., Halle				
Si 204	12324	2 As 411A-1 (V)	229	

Engine <u>manufacturers</u>: As=Argus; Bramo=Siemens-Halske; DB=Daimler- Benz; GR=Gnome Rhone; J=Junkers;

Engine <u>types</u>: I=Inline; O=Horizontally opposed; R=Radial; TJ=Turbojet; V=V-type; VO=vertically opposed

JAPANESE AIRCRAFT OF WORLD WAR II

(Allied code names in quotes)

Type	Max wt., lb.	Powerplant (type)	Speed mph	No. built
Aichi Tokei Denki Kabushiki Caisha, Nagoya, 1899				
E16A Zuiun "Paul"	8360	M. Kinsei 54 (R)	278	
D3A "Val"	8378	M. Kinsei 44 (R)	242	1294
B7A1 Ryusei "Grace"	10780	N. Homare 11 (R)	350	
E13A1 "Jake"		M. Kinsei 43 (R)		
Kawanishi Kokuki Kabushiki Kaisha, Kobe, 1928				
N1K Shiden-Kai "Rex"	8818	N. Homare 21 (R)	369	1440
H8K "Emily" (1941)	71650	4 M. Kasei 22 (R)	290	161
H6K "Mavis"	37479	4 M. Sinsei 46 (R)	211	
Kawasaki Kokuku Kogyo Kabushiki Kaisha, Kobe				
Ki-61 Hein "Tony" (1941)	6504	K. Ha-40 (V)	368	2734
Ki-45 "Nick" (1939)	12125	2 M. Ha-102 (R)	340	1687
Type 99 "Lily"		2 N. Ha 115 (R)	285	
Mitsubishi Jukogyo Kabushiki Kaisha, Tokyo, 1917				
A5M (1935) "Claude"	3684	N. Kotobuki 41 (R)	270	782
A6M Zero-Sen (1939)	5313	N. Sakae 12 (R)	332	10937
J2M "Jack"		M. Kasei 23 (R)	400+	
K2M Raiden (1942)	7573	M. Kasei 23a (R)	380	475
Ki-21 "Sally" (1936)	21407	2 M. Ha-101 (R)	302	2564

Type (year flown)	Max wt., lb.	Powerplant (type)	Speed mph	No. built
Mitsubishi cont				
G4M "Betty" (1939)	33069	2 M. Kasei 25 (R)	272	2479
Ki-67 Hiryu "Peggy"	30346	2 M. Ha-104 (R)	334	727
Ki-46 "Dinah"	14330	2 M. Ha-112 (R)	391	
Ki-49 Donryu (1939)	23545	2 N. Ha-109-II	304	828
Ki-57 "Topsy"	18300	2 M. Ha 5 (R)	266	
Nakajima Hikoki Kabushiki Kaisha, Tokyo, 1917				
Ki-27 (1936)	3638	N. Ha.1b (R)	286	3386
Ki-43 Hayabusa				
"Oscar" (1939)	6746	M. Ha-112 (R)	342	5751
Ki-44-I Shoki "Tojo"	5512	N. Ha-41 (R)	360	1233
Ki-84 Hayate "Frank"	7940	N. Ha-45-21 (R)	427	3470
J1N "Irving" (1941)	15212	2 N. Sakae 21 (R)	315	477
B5N "Kate" (1937)	9039	N. Saeke 11 (R)	235	1200
B6N Tenzan "Jill"	12456	M. Kasei 25 (R)	299	1268
C6N Saiun "Myrt"	11596	N. Homare 21 (R)	395	498
Ki-49 Donryu	23545	2 N. Ha-109 (R)	304	
Tachikawa Hikoki Kabushiki Kaisha, Tokyo, 1937				
Ki-54 "Hickory" (1940)	8995	2 N. Ha-13, 515 (R)	228	1200
Yokosuka Navy Depot				
D4Y Suisei "Judy" (1940)	9597	Aichi Atsuta 32 (V)	366	2319
P1Y Ginga "Frances"	29762	2 N. Homare 11 (R)	345	
MXY-7 Ohka piloted bomb (1944)	4718	Type 4 Model 20 rocket	534	755

<u>Engine</u> <u>manufacturers</u>: K=Kawasaki; M=Mitsubishi; N=Nakajima. <u>Engine</u> <u>type</u>: R=Radial; V=V-type

ITALIAN AIRCRAFT OF WORLD WAR II

	Max wt., lb.	Powerplant (type)	Speed mph	No. built
Cant, Cantieri Riuniti Dell'Adriatico, Monfalcone, Trieste. 1923				
Z.506B Airone	28008	3 AR 126 RC 34 (R)	217	
Z.1107	28260	3 P.XI RC 40 (R)	280	
Caproni, Soceita Italiana Caproni, Milan				
Ca 313	12450	2 IF Betta R.C.35(V)	248	
Fiat, Aeronautica d'Italia S.A., Turin, 1916				
C.R.42 Falco (1939)	5033	Fiat A.74R.1C 38 (R)	267	143
G.50 Freccia (1937)	5512	Fiat A.74 RC 38 (R)	302	783
B.R.20 Cicogna	23038	2 Fiat A.80 RC 41	267	
Macchi, Aeronautica Macchi, Varese				
C.200 Saetta (1937)	5132	Fiat A.74 RC 38 (R)	313	379
C.202 Folgore (1940)	6459	AR R.A.1000 RC 41(V)	370	1492
Reggiane, Officine Meccaniche Reggiane S.A., Reggio Emilia				
Re 2000 Falco (1939)	6349	P.XI RC 40 (R)	329	173
Re 2001	7000	DB 601 (V)	348	
Savoia-Marchetti, Societa Anonima Industrie Meccaniche Aeronautiche Navali, Rome				
S.M.79 Sparviero	25133	3 P.XI RC 40 (R)	295	

<u>Engine</u> <u>manufacturers</u>: AR=Alfa-Romeo; DB=Daimler-Benz; IF=Isotta- Fraschini; P=Piaggio

RUSSIAN AIRCRAFT OF WORLD WAR II

Design Bureau *Type (year flown)*	*Max wt., lb.*	*Powerplant (type)*	*Speed mph*	*No. built*
Beriev				
MBR-2	9359	Mikulin AM-34N (V)	171	
Ilyushin				
Il-2 Shturmovik (1939)	12147	Mikulin AM-38F (V)	251	35000
Il-4	22046	2 M-88B (R)	255	
Lavochkin				
LA 5/7 (1942)	7408	Shvetsov M-82FN (R)	402	15200
LAAG-3	7040	M-105P (V)	348	
Mikoyan-Gurevich				
MiG-3 (1940)	7242	Mikulin AM-35A (V)	398	3350
Petlyakov				
Pe-2	18783	2 Klimov VK-107A (V)	407	22500
Pe-8	73469	4 Mikulin AM-35A (V)	272	
Polikarpov				
I-16 (1933) Ishak	4519	Shvetsov M-62 (R)	326	20000
Sukhoi				
SU-2	9645	M-88B (R)	283	
Yakovlev				
Yak-1	5137	Klimov VK-105PA (V)	373	
Yak-3	4864	Klimov VK-107A (V)	448	
Yak-9 (1942)	6867	Klimov M-105 PF (V)	435	

POLISH AIRCRAFT OF WORLD WAR II

PZL, Panstwowe Zaklady Lotnicze, Waresaw				
P.11 (1931)	3505	PZL Bristol Mercury	230	80
P.37	18739	2 Bristol Pegasus	273	

FRENCH AIRCRAFT OF WORLD WAR II

(Prior to June 17, 1940 capitulation)

Bloch 152 (1937)	6058	GR 14N-49 (R)	316	140
Dewoitine D.520 (1938)	6135	HS 12Y-45 (V)	326	905
Potez P.63 (1936)	9773	2 GR 14M (R)	264	

Engine manfuacturers: GR=Gnome-Rhone; HS=Hispano-Suiza;

DUTCH AIRCRAFT OF WORLD WAR II

Fokker				
D.XXI	4519	Bristol Mercury VII	286	
G.1	10582	2 Bristol Merc. VIII	295	

SWEDISH AIRCRAFT OF WORLD WAR II

SAAB, Svenska Aeroplan A.B., Linkoping, 1937

SAAB-21 (1943)	9149	DB 605B (V)	398	594
SAAB-18	17946	2 PW R-1830 (R)	289	

AIRCRAFT ENGINES OF WORLD WAR II

* S=Supercharged; N=non-supercharged, ambient air; TJ=Turbojet; GT=Gas Turbine; #=thrust

Model series (yr. introduced)	Type, no. cyl.	Inlet air (S/A)*	cooling	cu.in disp.	wt.,lb. (dry)	takeoff hp
GREAT BRITAIN						
Armstrong Siddeley, Coventry						
Lynx (1920)	R-7	N	A	756	512	215
Cheetah (1930)	R-7	S	A	834	635	295- 420
Cougar	R-9	S	A	1176	1050	690
Panther (1929)	R-14	S	A	1829	980	525
Leopard (1928)	R-14	S	A	2969	1637	800
ASX	TJ				1900	2000#
Bristol, Filton, Bristol						
Jupiter (1920)	R-9	S	A			
Mercury (1931)	R-9	S	A	1520	1065	725-905
Peresus (1933)	R-9	S	A	1520-1635	1105-1380	830-1175
Taurus	R-14	S	A	1550	1335	1085
Pegasus (1932)	R-9	S	A	1753	1050-1180	920-1050
Hercules (1938)	R-14	S	A	2360	1870-1970	1590-1725
Centaurus (1938)	R-18	S	A	3270	2695	2520
Theseus	GT				2310	--
Cirrus, Brough, E. Yorkshire						
Minor I (1935)	Iv-I-4	N	A	220	238	90
Minor II	Iv-I-4	N	A	238	248	100
Major II (1935)	Iv-I-4	N	A	384	338	150
Major III	Iv-I-4	N	A	384	338	155
de Havilland, Hatfield, Herts.						
Gipsy Major I (1932)	Iv-I-4	N	A	374	305	130
Gipsy Six I (1932)	Iv-I-6	N	A	561	468	200
H-1 Goblin	TJ				1500	3400#
Metropolitan-Vickers, Manchester						
F.2/4	TJ				1750	3500#
Napier, Acton, London						
Sabre (1940)	H-24	S	L	2240	2500	2300
Sabre VII	H-24	S	L	2240	2540	3000
Power Jets						
W1	TJ				700	850#
W2/850	TJ				950	2485#
Rolls-Royce, Derby						
Merlin (1936)	V-12	S	L	1647	1375-1650	990-1635
Griffon (1940)	V-12	S	L	2240	2090	2340

Model series (yr. introduced)	Type, no. cyl.	Inlet air (S/A)*	cooling	cu.in disp.	wt.,lb. (dry)	takeoff hp
Ross-Royce cont.						
Vulture (1939)	X-24	S	L	2592		1845
WR1	TJ				1100	2000#
Derwent	TJ				975	2000#
Nene	TJ				1600	4000#

GERMANY

Argus Motorengesellschaft M.B.H., Berlin

Model series (yr. introduced)	Type, no. cyl.	Inlet air (S/A)*	cooling	cu.in disp.	wt.,lb. (dry)	takeoff hp
As 410 (1939)	Iv-V-12	S	A	732	695	465
As 10 (1931)	Iv-V-8	S	A	771	423	220

BMW Bayerische Motorenwerke A.G., Munich

Model series (yr. introduced)	Type, no. cyl.	Inlet air (S/A)*	cooling	cu.in disp.	wt.,lb. (dry)	takeoff hp
Bramo 323	R-9	S	A	1637	1210	900
132	R-9	S	A	1690	1155	865
801 (1940)	R-14	S	A	2560	2321-2702	2000
801 TJ/TQ	R-14	T	A	2560	3542	2270
803	R-28	S	L	5094	2950	3900
003 A-1	TJ				1252	1760#

Daimler-Benz Aktiengesellschaft, Stuttgart

Model series (yr. introduced)	Type, no. cyl.	Inlet air (S/A)*	cooling	cu.in disp.	wt.,lb. (dry)	takeoff hp
600 (1938)	Iv-V-12	S	L	2069	1510	1000
601 (1938)	Iv-V-12	S	L	2069	1540	1360
605	Iv-V-12	S	L	2178	1663	1435-2000
603 (1941)	Iv-V-12	S	L	2715	2002-2145	1280-1900
610 (1940)	Iv-W-24	S	L	5438	3476	2450

Hirth, Heinkel-Hirth, Stuttgart

Model series (yr. introduced)	Type, no. cyl.	Inlet air (S/A)*	cooling	cu.in disp.	wt.,lb. (dry)	takeoff hp
HM 504	Iv-I-4	N	A	243	245	105
HM 506	Iv-I-6	N	A	365	328	160
HM 508	Iv-V-8	S	A	486	458	280
HM512 (1938)	Iv-V-12	S	A	729	593	400
He S 011	TJ				2090	2860#

Junkers Flugzeug und Motorenwerke A.G., Dessau

Model series (yr. introduced)	Type, no. cyl.	Inlet air (S/A)*	cooling	cu.in disp.	wt.,lb. (dry)	takeoff hp
Jumo 205 (1932)	O-6	S	L	1014	1144	700
Jumo 207	O-6	S	L	1014	1903	1000
Jumo 210 (1936)	Iv-V-12	S	L	1202	968	700
Jumo 211 (1936)	Iv-V-12	S	L	2136	1408	1500
Jumo 213 (1942)	Iv-V-12	S	L	2136	2024	1776
Jumo 222	R-24	S	L	2837	2464	2500
Jumo 004B	TJ				1585	1890#

JAPAN

Mitsubishi Jukogyo Kabushiki Kaisha, Nagoya

Model series (yr. introduced)	Type, no. cyl.	Inlet air (S/A)*	cooling	cu.in disp.	wt.,lb. (dry)	takeoff hp
Kinsei (1937)	R-14	S	A	1970	1200	1000
Ha-104/211 (1942)	R-18	S	A	2546		2200
Kasei (1937)	R-14	S	A	2576		1300

Nakajima Hikoki Kabushiki Kaisha, Tokyo

Model series (yr. introduced)	Type, no. cyl.	Inlet air (S/A)*	cooling	cu.in disp.	wt.,lb. (dry)	takeoff hp
Homare (1940)	R-18	S	A	2195		1850
Kotobuki (1935)	R-9	S	A			550

Model series (yr. introduced)	Type, no. cyl.	Inlet air (S/A)*	cool- ing	cu.in disp.	wt.,lb. (dry)	takeoff hp
ITALY						
Alfa Romeo, Milan						
135 R/C 34 (1938)	R-18	S	A	2940	2100	1500
Fiat, Turin						
A 74 R/C 38	R-14	S	A	1970	1257	870
A 80 R/C 41(1937)	R-14	S	A	2788		1000
Isotta-Fraschini, Milan						
Betta R.C.35	Iv-V-12	S	L			650
Piaggio, Genoa						
Stella XI R/C 40 (1937)	R-14	S	A	2354	1444	1000
UNITED STATES						
Allison, Indianapolis						
V-1410 Liberty (1922)	V-12	N	A	1410		420
V-1710 (1935)	V-12	S	L	1710	1385-1620	1200-1475
V-3420 (1940)	W-24	S	L	3420	2600	2600
Continental, Muskegon, MI						
O-170 (A-65) (1939)	O-4	N	A	171	175	65
R-545 (1931)	R-7	N	A	545		165
R-670 (1935)	R-7	N	A	668	465	220-250
Franklin, Syracuse, NY						
4AC-150 (1942)	O-4	N	A	150		65
4AC-176 1942)	O-4	N	A	176		75
4AC-199 (1940)	O-4	N	A	199	224	90
6AC-298 (1940)	O-6	N	A	298	295	130
6ACV-405 (1941)	O-6	N	A	405		230
6AC-425 (1944)	O-6	N	A	425		215
8ACGSA-538(1942)	O-8	N	A	538		300
12ACGSA-806 (1942)	O-12	N	A	806		450
Guiberson, Dallas, TX						
A-1020 Diesel	R-7	N	A	1021	653	310
Jacobs, Pottstown, PA						
R-755 L-4 (1934)	R-7	N	A	757	505	225
R-830 L-5 (1941)	R-7	N	A	830		330
R-915 L-6 (1940)	R-7	N	A	914	555	330
Kinner, Glendale, CA						
R-370 K-5 (1927)	R-5	N	A	372	304	100
R-440 B-5 (1930)	R-5	N	A	441	312	125
R-540 R-5-2 (1934)	R-5	N	A	540	345	160
R-720 C-5 (1931)	R-5	N	A	720		210
Lycoming, Williamsport, PA						
O-145 (1942)	O-4	N	A	145		65
O-235 (1940)	O-4	N	A	233	239	104

Model series (yr. introduced)	Type, no. cyl.	Inlet air (S/A)*	cool-ing	cu.in disp.	wt.,lb. (dry)	takeoff hp
Lycoming cont.						
O-290 (1941)	O-4	N	A	290	241	130
O-435 (1941)	O-6	N	A	434	350	190
R-680 (1929)	R-9	N	A	680	515	300
O-1230 (1941)	O-12	N	L	1230		1200
H-2470 (1941)	H-24		L	2430		2300
Menasco						
L-365 (C-4) (1931)	Iv-I-4	N	A	363	296	125
IV-2040 (1942)	Iv-V-12	N	L	2040	2000	
H-4070 (1943)	H-24	N	L	4070		3400
Packard, Detroit, MI						
DR-980 (1932)	R-9	S	A	980		225
V-1650 Merlin	V-12	S	L	1650	1320	1300
Pratt & Whitney, East Hartford, CT						
R-985 Wasp Jr. (1929)	R-9	S	A	985	668	450
R-1340 Wasp (1926)	R-9	S	A	1344	864	600
R-1535 Twin Wasp Jr. (1932)	R-14	S	A	1535	1162	625-825
R-1690 Hornet A (1926)	R-9	S	A	1690	800	525
R-1830 Twin Wasp (1932)	R-14	S	A	1830	1460	1200
R-1860 Hornet B (1929)	R-9	S	A	1860		575
R-2000 Twin Wasp D (1941)	R-14	S	A	2000	1585	1200
R-2180 Twin Hornet (1938)	R-14	S	A	2180		1250
H-2600 (1943)	H-24	S	L			2200
R-2800 Double Wasp (1939)	R-18	S	A	2804	2350	2000
R-4360 Wasp Major (1943)	R-28	S	A	4360	3600	3500
Ranger, Farmingdale, NY						
6-390 (1942)	Iv-I-6	N	A	390		120
6-410 (1942)	Iv-I-6	N	A	410		165
6-440-C	Iv-I-6	N	A	441	376	200
V-770	Iv-V-12	S	A	773	730	520
Warner, Detroit, MI						
R-420 Scarab (1930)	R-7	N	A	420		125
Super Scarab R-500 (1938)	R-7	N	A	499	341	165
Super Scarab 185	R-7	N	A	555	344	185
Wright, New York						
R-760 Whirlwind 7	R-7	S	A	756	570	235-285
R-790 (1925)	R-9	S	A	790		225
R-975 Whirlwind 9 (1929)	R-9	S	A	973	675	365-412
V-1460 (1926)	V-12	N	A	1460		550
R-1510 Whirlwind 14 (1934)	R-14	S	A	1510		725
V-1560 (1930)	V-12	N	A	1560		600
H-1640 Chieftan (1928)	R-12	S	A	1640		600
R-1670 (1934)	R-14	S	A	1670		850
R-1820 Cyclone 9 (1927)	R-9	S	A	1823	1000-1320	770-1200
R-2160 Tornado (1942)	R-42	S	A	2160		2350
R-2600 Cyclone 14 (1937)	R-14	S	A	2603	1935-2045	1600-1900
R-3350 Duplex Cyclone (40)	R-18	S	A	3347	2670	3000
R-4090 (1943)	R-22	S*	A	4090		3000

Model series (yr. introduced)	Type, no. cyl.	Inlet air (S/A)*	cool- ing	cu.in disp.	wt.,lb. (dry)	takeoff hp
USSR, POLAND, FRANCE						
Gnome Rhone, Gennevilliers						
14N (1936)	R-14	S	A	2360	1300	1085
M.14 (1936)	R-14	S	A	1158	922	810
Hispano-Suiza Fabrica de Automoviles, S.A., Madrid						
12y-45 (1933)	V-12	S	L	2196	1014	860
Klimov						
VK-107A	V-12	S	L	2075	1655	1600
Mikulin						
AM-38	V-12	S	L	2850	1892	1600
M-88						1100
Shvetsov						
ASh 62 (1938)	R-9	S	A	1818	1184	1000
ASh 82	R-14	S	A	2513	1947	1675

A Chronology of
AIRCRAFT PISTON POWERPLANTS

Horsepower and weight specifications reflect models at introdution; year approximate.
Engine Type: B=Barrel; Ds=Diesel; Ro=Rotary; *=two-stroke. Cooling: A=Air; L=Liquid
Cylinder arrangement: F=Fan-type; H=H-type; HO=Horizontally Opposed; I=Inline; Iv=Inverted;
R=Radial; T=T-type; U=U-type; V=V-type; VO=Vertically Opposed; W=W-type; X-X-type

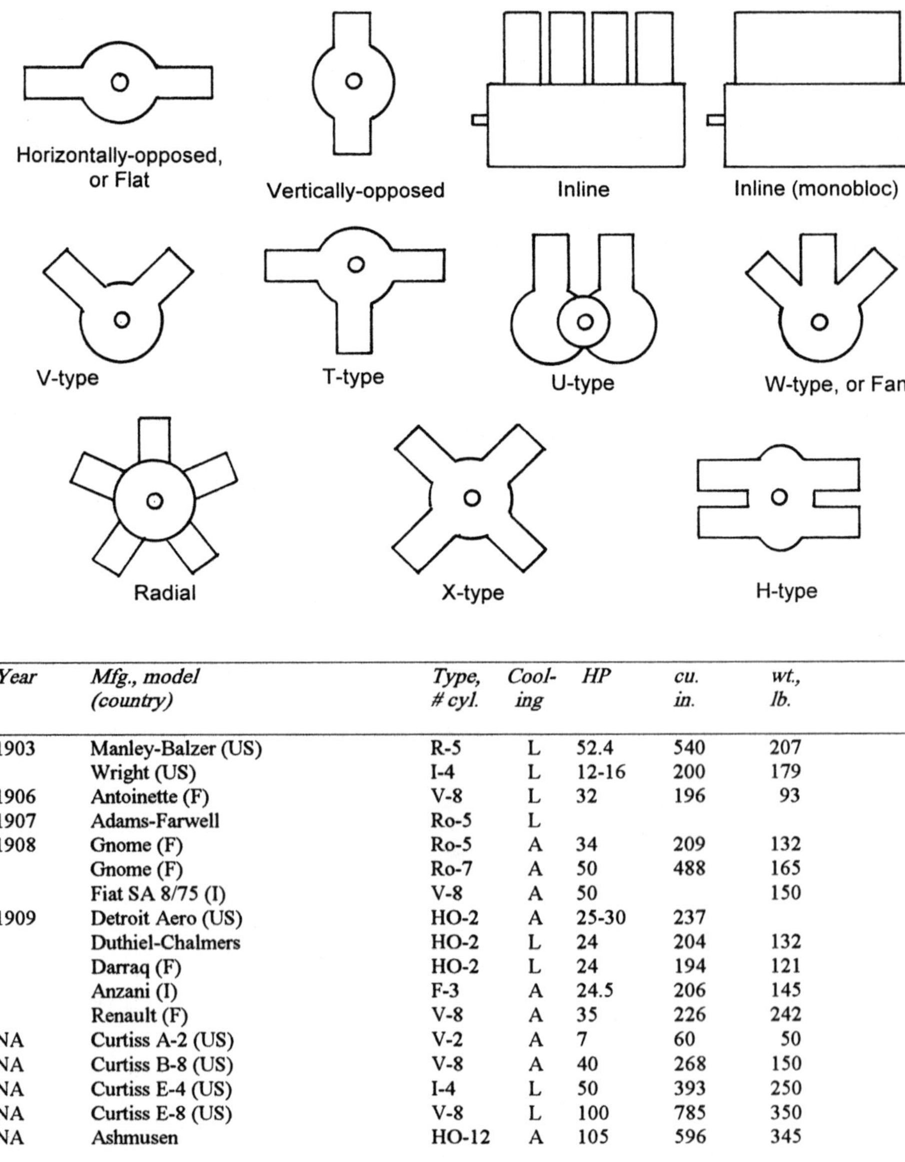

Year	Mfg., model (country)	Type, # cyl.	Cooling	HP	cu. in.	wt., lb.
1903	Manley-Balzer (US)	R-5	L	52.4	540	207
	Wright (US)	I-4	L	12-16	200	179
1906	Antoinette (F)	V-8	L	32	196	93
1907	Adams-Farwell	Ro-5	L			
1908	Gnome (F)	Ro-5	A	34	209	132
	Gnome (F)	Ro-7	A	50	488	165
	Fiat SA 8/75 (I)	V-8	A	50		150
1909	Detroit Aero (US)	HO-2	A	25-30	237	
	Duthiel-Chalmers	HO-2	L	24	204	132
	Darraq (F)	HO-2	L	24	194	121
	Anzani (I)	F-3	A	24.5	206	145
	Renault (F)	V-8	A	35	226	242
NA	Curtiss A-2 (US)	V-2	A	7	60	50
NA	Curtiss B-8 (US)	V-8	A	40	268	150
NA	Curtiss E-4 (US)	I-4	L	50	393	250
NA	Curtiss E-8 (US)	V-8	L	100	785	350
NA	Ashmusen	HO-12	A	105	596	345

Year	Mfg., model (country)	Type, # cyl.	Cooling	HP	cu. in.	wt., lb.
1910	Elbridge Featherweight (US)	I-3*	L	30-45	227	150
	Kirkham BG-6	I-6	L	70-90	449	420
	E.N.V.	V-8	L	39	249	150
	Clement-Bayard (F)	R-7	L	50	387	154
	Clement-Bayard (F)	HO-2	L	30	181	110
	Curtiss OX-5 (US)	V-8	L	90	503	320
	Fiat A-10 (I)	I-4	L	100		
	Anzani (I)	R-3	A	30	190	121
	Gordon-Brille	R-4/8	L	102	1102	350
1911	Elbridge Aero Spl. (US)	I-4*	L	50-60	200	150
	Roberts	I-4*	L	50	318	170
	Roberts	I-6*	L	75	477	243
	Sturdevant D-4	I-4	L	48-55	553	400
	Isotta-Fraschini V-1(I)	I-4	L	90	583	682
	Anzani (I)	R-6	A	45	279	154
	Gnome Lambda (F)	Ro-7	A	80	773	207
	Macomber	Ro/B-7	A	50	1145	230
1912	Hall-Scott A-2 (US)	V-8	L	65	402	260
	Argus type II (D)	I-4	L	100	525	345
	Argus type II (D)	I-6	L	110	575	430
	Curtiss OXX-6 (US)	V-8	L	100	567	401
1913	Renault (F)	V-8	A	80	548	463
NA	Renault (F)	V-12	A	90	651	823
NA	RAF 1 (GB)	V-8	A	90	537	
	Green C-4 (US)	I-4	L	32-35	253	160
	Austro Daimler (D)	I-6	L	120	850	575
	Salmson M-7 (F)	R-7	L	90	765	375
	Gnome Lambda-Lambda (F)	Ro-14	A	160	1446	396
	Edleweiss	Ro-6	A			
	ABC Gnat (GB)	HO-2	A	45	139	115
NA	Green E-6 (US)	I-6	L	100-120	447	855
NA	NEC (GB)	I-4*	A	70	251	290
NA	Aeromarine L-6 (GB)	I-6	L	130-145	553	400
1914	Hall-Scott A-5 (US)	I-6	L	125	825	525
	Beardmore AR-3 (GB)	I-6	L	135	850	600
	Gnome Delta (F)	Ro-9	A	110	993	297
	Salmson M-9 (F)	R-9	L	140	868	465
	Smith Static Radial	R-10	A	150	875	372
	RAF 4 (GB)	V-12	A	140	806	637
	Statax	Ro/B-5	A	40	288	200
1915	Rolls-Royce Hawk (GB)	I-6	L	75	453	405
	Rolls-Royce Eagle (GB)	V-12	L	360	1240	900
	Benz Bz 3a (D)	I-6	L	180	1070	605
	Hispano-Suiza (Es)	V-8	L	180	718	467
	Hispano-Suiza (Es)	I-6	L	200	942	695
	Christofferson Aero (US)	I-6	L	120	638	
1916	Fiat A-12 (I)	I-6	L	268	1326	920
	Renault RE 8 Gd (F)	V-8	L	220	899	474
	Hall-Scott A-7 (US)	I-4	L	100	550	410
	Beardmore AR-3 (GB)	I-6	L	160	1025	
	Rolls-Royce Falcon (GB)	V-12	L	205	867	
	Maybach (D)	I-6	L	200	1228	
	Le Rhone type J (F)	Ro-9	A	120	920	323
	Clerget type 9 (F)	Ro-9	A	110	931	295
	Oberursel (D)	Ro-9	A	110	995	

Year	Mfg., model (country)	Type, # cyl.	Cool-ing	HP	cu. in.	wt., lb.
1916 cont.						
	Salmson Z-9 (F)	R-9	L	250	1146	473
	ABC Wasp (GB)	R-7	A	170	667	290
	Lawrance A-3 (US)	HO-2	A	28		
	RAD 4D (GB)	V-12	A	240	886	695
1917	Duesenberg A-44 (US)	I-4	L	125	361	365
	Renault RE 8 Gd (F)	V-12	L	300	1347	794
	Fiat A-14 (I)	V-12	L	725	3488	1740
	Isotta-Fraschini V6 (I)	I-6	L	275	1014	620
	Packard 905 (US)	V-12	L	750	2500	1210
	Curtiss K-12 (US)	V-12	L	400	1145	679
	BHP (GB)	I-6	L	230	1148	680
	Mercedes type 1468 (D)	I-8	L	287	1201	900
	Mercedes type 1468 (D)	I-6	L	180	901	700
	Benz Bz 4 (D)	I-6	L	230	1148	848
	Basse und Selve (D)	I-6	L	269-302	1381	885
	Mercedes D-IVa (D)	I-6	L	260	1326	936
	Fiat A-18 (I)	R-9	L	320	1094	
	Clerget type 9B (F)	Ro-9	A	130	992	381
	Bentley BR1 (GB)	Ro-9	A	150	1053	405
	Siemens-Halske (D)	Ro-11	A	160	1431	428
	ABC Dragonfly (GB)	R-9	A	320	1389	656
1918	BMW IIIA (D)	I-6	L	180	1163	644
	Isotta-Fraschini V-8 (I)	I-6	L	300	1101	595
	Liberty L-4 (US)	I-4	L	102	412	398
	Liberty L-6 (US)	I-6	L	215	825	550
	Liberty L-8 (US)	V-8	L	270	1099	575
	Liberty L-12 (US)	V-12	L	400	1649	786
	Bugatti (F)	U-16	L	410	488	814
	Hispano-Suiza 42 (Es)	V-12	L	300	1127	528
	Curtiss K-6 (US)	I-6	L	150	573	417
	Napier Lion (GB)	W-12	L	450	1462	858
	Austro Daimler (D)	I-6	L	200	917	729
	Benz Bz3 BV (D)	V-8	L	200	941	683
	Cosmos Jupiter (GB)	R-9	A	450	1752	700
	Gnome type N (F)	Ro-9	A	160	970	290
	Bentley BR2 (GB)	Ro-9	A	230	1522	500
1919	Fiat A-15 (I)	V-12	L	430	1240	800
	Armstrong-Siddeley Jaguar (GB)	R-14	A	400	1512	850
	Lawrance L-3/L-5 (US)	R-3	A	65	223	175
	Mercedes F7502 (D)	HO-2	A	20	54	106
	ADC Airsix (GB)	I-6	A	300	1024	620
1920	Armstrong-Siddeley Lynx (GB)	R-7	A	215	756	512
	Wright R-1 (US)	R-9	A	350	1454	
1921	BMW IV (D)	I-6	L	250	1938	
	Farman 8 VI (F)	Iv-W-18	L	350	977	704
1922	Lorraine 12 Ed (F)	W-12	L	450	1490	935
	Allison Liberty (US)	V-12	L	420	1410	
	Isotta-Fraschini Asso 200 (I)	I-6	L	250	901	616
	Curtiss D-12 (US)	V-12	L	350	1145	704
	Lawrance J-1 (US)	R-9	A	200	787	476
	Anzani (I)	R-6	A	70-80	396	215
	Walter NZ60 (CZ)	R-5	A	75	316	225
1923	Junkers L-5 (D)	I-6	L	280-310	1399	711
	Isotta-Fraschini V10 (I)	I-6	L	350	1202	726

Year	Mfg., model (country)	Type, # cyl.	Cool-ing	HP	cu. in.	wt., lb.
1923 cont.						
	Lawrance J-2 (US)	R-9	A		924	
	Siemens Sh 5 (D)	R-7	A	77	403	
	Morehouse M-42	HO-2	A	28	80	85
1924	ABC Scorpion Mk 1 (GB)	HO-2	A	24	73	90
	Bristol Cherub I (GB)	HO-2	A	34	67	81
	Farman 12 WE (F)	V-12	L	500	1555	1034
1925	Fiat A-20 (I)	V-12	L	430	1143	733
	Fiat A-22 (I)	V-12	L	750	1476	1023
	Fiat A-25 (I)	V-12	L	1000	3322	1874
	Isotta-Fraschini Asso 500 (I)	V-12	L	500	1691	957
	Attendu	I-2	L	85	310	500
	Wright J-5 (US)	R-9	A	220	788	510
	Wright Wimoon R-1200 (US)	R-9	A	350	1176	640
	Salmson AD9 (F)	R-9	A	45	182	150
	Bristol Lucifer IV (GB)	R-3	A	130	584	325
	Siemens SH 11 (D)	R-7	A	96	403	326
	Fairchild-Caminez (US)	R-4	A	135	447	360
	Hermann	Ro-7	A	53	251	150
	Cirrus Mk 1 (GB)	I-4	A	65	274	268
	ABC Hornet (GB)	HO-4	A	82	243	219
	Wright-Morehouse W-M80 (US)	HO-2	A	28	80	85
1926	Renault 12MC (F)	V-12	L	800	2652	1342
	Wright V-1460 (US)	V-12	A	550	1460	
	Farman 18 W.I. (F)	Iv-W-18	L	730	1304	701
	Pratt & Whitney Wasp (US)	R-9	A	425	1344	650
	Pratt & Whitney Hornet (US)	R-9	A	525	1690	800
	Armstrong-Siddeley Genet (GB)	R-5	A	82	251	203
	Armstrong-Siddeley Mongoose (GB)	R-5	A	150	540	340
	Tips & Smith Super Rhone	R-9	A	125	667	315
	Fiat A-50 (I)	R-7	A	95	402	273
	Cirrus Mk II (GB)	I-4	A	84	302	280
1927	Wright Cyclone R-1820 (US)	R-9	A	575	1820	974
	Salmson AC5 (F)	R-5	A	60	312	242
	Salmson AC7 (F)	R-7	A	95	437	286
	Wolseley AR9 Aries (GB)	R-9	A	203	589	452
	Renard Typ. 100 (F)	R-5	A	145	482	275
	Curtiss Challenger (US)	R-6	A	185	603	420
	Detroit Aircat (US)	R-5	A	80	234	230
	Kinner K-5 (US)	R-5	A	100	372	275
	Rolls-Royce Kestrel (GB)	V-12	L	490	1296	865
	deHavilland Gypsy I (GB)	I-4	A	100	318	285
	Allison V-1410 (US)	Iv-V-12	A	410	1410	1010
1928	Junkers L-55 (D)	V-12	L	500-600	2798	1254
	Mercedes F-2 (D)	V-12	L	1000	3294	1654
	Argus AS VI (D)	V-12	L	700	2652	1168
	Curtiss Conqueror (US)	V-12	L	600	1570	770
	Hispano-Suiza 6Mb (Es)	I-6	L	300	831	550
	Lorraine Elder (F)	V-12	L	1050	2743	1397
	Maybach VL-12 (gaseous fuel) (D)	V-12	L	570	2029	2530
	Rolls-Royce Condor (GB)	V-12Ds	L	500	2138	1504
	Packard DR-980 (US)	R-9Ds	A	225	982	510
	Gnome-Rhone Titan 5Kds (F)	R-5	A	260	843	473
	Gnome-Rhone Titan Major 7Ksd (F)	R-7	A	370	1180	787
	Gnome-Rhone Mistral Major 9Kfr (F)	R-9	A	770	1517	1111

Year	Mfg., model (country)	Type, # cyl.	Cooling	HP	cu. in.	wt., lb.
1928 cont.						
	Gnome-Rhone Mistral Major 14Kbs (F)	R-14	A	705	2360	
	Armstrong-Siddeley Leopard (GB)	R-14	A	800	2969	1637
	Wright R-1640 (US)	R-12	A	600	1640	
	Brownback/Anzani (US)	R-6	A	85	379	265
	Pobjoy P-type (GB)	R-7	A	65	151	119
	Walter Porlaris I (CZ)	R-3	A	50	190	158
	Walter Regulus (CZ)	R-5	A	230	697	463
	Farini Algol	R-7	A	100	279	253
	Continental A-70 (US)	R-7	A	170	544	415
	Wright J-6-7 (US)	R-7	A	225	756	425
	Velie M-5/LeBlond 5D (US)	R-5	A	65	251	240
	Szekeley SR-3 (US)	R-3	A	40	191	148
	Cirrus Hermes (GB)	Iv-I-4	A	115	347	300
	Renault 4Pb (F)	I-4	A	90	355	286
	Michel AM 14 Type I	I-4	A	90	381	304
	Michel AM 14 Type II	I-4	A	120	449	356
	Argus As. 8 (D)	Iv-I-4	A	110	386	248
	Colombo S-53	I-4	A	94	374	226
	Isotta-Fraschini Asso Caccia (I)	V-12	A	480	1257	838
	Isotta-Fraschini 80T (I)	I-6	A	240	370	
1929	Junkers L-88 (D)	V-12	L	700-850	2798	
	Fiat A.S. 5 (I)	V-12	L	1050	1589	783
	Isotta-Fraschini Asso 750 (I)	W-18	L	820	2871	1460
	Packard A-2500 (US)	V-12	L	750	2500	1210
	Lorraine Courlis (F)	W-12	L	600	1960	935
	Rolls-Royce R (GB)	V-12	L	2500	2240	1630
NA	Isotta-Fraschini Asso 1000 (I)	W-18	L	1100	3501	1782
	Wright J-6-9 Whirlwind (US)	R-9	A	300	975	550
	Bristol Jupiter VI FS (GB)	R-9	A	465	1752	775
	Bristol Titan IIIF (GB)	R-5	A	240	924	775
	Armstrong-Siddeley Panther (GB)	R-14	A	525	1829	980
	Pratt & Whitney Wasp Jr. R-985 (US)	R-9	A	450	985	653
	Pratt & Whitney Hornet B (US)	R-9	A	575	1860	
	Wichita Blue Streak (US)	R-5/10*	A	125	250	125
	Hurricane	R-8*	A	150	450	225
	Irwin Meteor	R-4*	A	20	84	58
	Potez 6Ac (F)	R-6	A	100	396	
	Potez 3B (F)	R-3	A	60	198	172
	Hispano-Suiza 5Q (ES)	R-5	A	150	541	407
	Elizalde Dragon V	R-5	A	200	524	378
	Lycoming R-680 (US)	R-9	A	225	680	505
	Jacobs L-3 (US)	R-3	A	55	190	189
	LeBlond 7D (US)	R-7	A	90	351	295
	Lambert R-266 (US)	R-5	A	90	266	224
	Warner Scarab (US)	R-7	A	125	372	275
	Szekeley SR-3L (US)	R-3	A	30	191	152
	Axelson A7R	R-7	A	150	612	430
	Kimball Beetle	R-7	A	135	585	380
	Fairchild Ranger 6-370 (US)	Iv-I-6	A	122	374	325
	Chevrolair D-4 (US)	Iv-I-4	A	120	352	345
	Heath B-4	Iv-I-4	A	30	83	119
	Menasco 4-A (US)	Iv-I-4	A	90	326	265
1930	Farman 18T (F)	Iv-T-18	L	1200	1498	1062
	Beardmore Tornado (GB)	I-8Ds	L	585	5132	4179

Year	Mfg., model (country)	Type, # cyl.	Cooling	HP	cu. in.	wt., lb.
1930 cont.						
	Wright Cyclone F R-1820 (US)	R-9	A	575	1820	974
	Armstrong-Siddeley Cheetah (GB)	R-7	A	295	834	556
	Renault 7A (F)	R-7	A	120-130	403	286
	Avia R-7 (Cz)	R-7	A	140	430	352
	LeBlond 5DF (US)	R-5	A	85	266	219
	LeBlond 7DF (US)	R-7	A	110	372	275
	Kinner B-5 (US)	R-5	A	125	441	295
	Comet 7D	R-7	A	130	612	
	McClatchie Panther	R-7	A	150	612	439
	Dawn	R-5/10	A		156	501
	deHavilland Gipsy I (GB)	I-4	A	120	347	298
	Napier Rapier (GB)	H-16	A	305	539	720
	Regnier R.4 (F)	Iv-I-4	A	120	375	302
	Chaise AV2	Iv-I-4	A	100	386	336
	Curtiss V-1460 (US)	Iv-V-12	A	525	1460	925
	Dayton Bear (US)	I-4	A	100	445	375
	Dayton Aero Four (US)	I-4	A	50	201	173
	Ranger 6-390 (US)	Iv-I-6	A	120	386	345
	Michigan Rover R-267 (US)	Iv-I-4	A	75	267	232
	Wright V-1560 (US)	V-12	A	600	1560	
	Colombo S-63	I-6	A	145	521	344
	Aeronca E-107 (US)	HO-2	A	30	107	114
1931	Delage CIDRS (F)	Iv-V-12	L	400	488	814
	Delage CIDRS (F)	Iv-V-12	L	450	732	968
	Pietenpol	I-4	L	35	201	244
	Bristol Mercury (GB)	R-9	A	560	1520	950
	Continental R-545 (US)	R-7	A	165	545	
	Pobjoy R-type (GB)	R-7	A	173	135	
	Fiat A-53 (I)	R-7	A	120	443	324
	Kinner C-5 (US)	R-5	A	210	715	420
	Szekeley SR-3 0 (US)	R-3	A	45	191	152
	Aeromarine AR 3	R-3	A	55	160	145
	Tank V-502 (US)	V-8	A	115	502	
	Tank V-470 (US)	V-8	A	115	471	398
	Menasco B-4 (US)	Iv-I-4	A	95	326	295
	Menasco C-4 (US)	Iv-I-4	A	125	363	296
	Menasco B-6 (US)	Iv-I-6	A	160	486	398
	Argus As. 10 (D)	Iv-V-8	A	220	771	423
	Hirth G-M 60 (D)	Iv-I-4	A	70	211	181
	Fiat A.60 (I)	Iv-I-4	A	140	400	302
	Aeronca E-113 (US)	HO-2	A	40	114	113
	Continental A-40 (US)	HO-4	A	37	115	144
1932	Farman 12 W.I. (F)	Iv-V-12	L	550	1466	902
	Farman 12 Brs (F)	Iv-V-12	L	460	471	565
	Lorraine Petrel (F)	V-12	L	500	1803	1045
	Daimler-Benz LOF 6 (D)	V-12Ds	L	1200	3299	4400
	Junkers Jumo 205 (D)	I-6/12	L	600	1012	1147
	DKW (D)	Iv-I-2*	L	21	34	82
	Fiat A.S.6 (I)	V-24	L	2800	3178	2050
	Packard DR-980 (US)	R-9 Ds	A	225	980	
	Bristol Pegasus (GB)	R-9	A	590	1753	
	Pratt & Whitney Tw. Wasp Jr. R-1535	R-14	A	625	1535	1162
	Pratt & Whitney Tw. Wasp R-1830	R-14	A	825	1830	1467
	Siemens Sh 14 (D)	R-7	A	113	436	308

Year	Mfg., model (country)	Type, # cyl.	Cool-ing	HP	cu. in.	wt., lb.
1932 cont.						
	Jacobs LA-2 (US)	R-7	A	195	589	400
	Warner Super Scarab 145 (US)	R-7	A	145	499	305
	Von Festenberg-Packisch	R-4*	A	40		136
	Napier E 97 Javelin (GB)	Iv-I-4	A	170	501	410
	Ranger V-770 (US)	Iv-V-12	A	250	772	565
	Hirth H-M 150 (D)	Iv-V-8	A	170	400	330
	Walter Junior (Cz)	Iv-V-4	A	105	355	292
	CNA C-X	X-12	A	240		
	Brownback Tiger Kitten 30	Iv-I-2*	A	30	77	75
	Grade	Iv-I-4*	A	30	81	66
	Chaise 4E	Iv-V-4	A	40	122	186
	Long Harlequin	HO-2	A	30	96	90
	Argus As 16 (D)	HO-2	A	40	251	156
1933	Hispano-Suiza 12Ycrs (Es)	V-12	L	860	2196	1014
	Bristol Peresus (GB)	R-9	A	580	1520	1026
	Curtiss-Wright R-1510 (US)	R-14	A	765	1512	1000
	Fiat A 58C (I)	R-14	A	740	2360	1157
	Potez 9Ba (F)	R-9	A	250	485	324
	Walter Bora II (Cz)	R-9	A	210	569	363
	Fiat A-54 (I)	R-7	A	140	443	331
	Renault 4Pdi Bengali (F)	Iv-I-6	A	145	386	386
	Regnier R.6 (F)	Iv-I-6	A	210	484	460
	Argus As.16 (D)	Iv-I-6	A	230	538	352
	Hirth H-M B (D)	Iv-V-8	A	245	485	539
	Walter Junior Major (Cz)	I-4	A	120	375	312
1934	Maybach GO 56 (D)	Iv-V-12	L	410	1973	4620
	Deschamps (B)	Iv-V-12Ds	L	1600	3052	2400
	Wright R-1670 (US)	R-14	A	850	1670	
	Siemens SAM 322 H2 (D)	R-9	A	715	1636	1080
	Siemens Sh 20b (D)	R-9	A	540	1921	925
	Lorraine P 5B (F)	R-5	A	125	524	354
	Jacobs L-4 R-755 (US)	R-7	A	225	757	415
	Kinner R-5-2 (US)	R-5	A	160	490	305
	Bristol Aquila (GB)	R-9	A	500	950	830
	Walter Major Six (Cz)	Iv-I-6	A	200	563	385
	CNA C-4	I-4	A	90	110	230
	Renault Special (F)	Iv-I-6	A	325	485	485
	Napier Dagger (GB)	H-24	A	890	1027	1390
	Sarolea Epervier	HO-2	A	27.5	56	109
1935	Allison V-1710	V-12	L	1200	1710	1385
	Gnome-Rhone M.14 (F)	R-14	A	810	1158	922
	Nakajima Kotobuki (J)	R-9	A	550		
	Alfa Romeo 125 R/c (I)	R-9	A	680	1747	1045
	Piaggio P-VII c35 (I)	R-7	A	500	1178	716
	Piaggio Stella VII C/15 (I)	R-7	A	400	1178	
	Continental R-670 (US)	R-7	A	220	668	465
	Piaggio Stella IX R/C 40 (I)	R-9	A	600	1515	981
	Security	R-5	A	125	441	355
	Cirrus Major (GB)	Iv-I-4	A	135	383	310
	Cirrus Minor (GB)	Iv-I-4	A	90	220	205
	Menasco 6CS-4 (US)	Iv-I-4	A	260	545	550
	Walter Minor (Cz)	Iv-I-4	A	85	242	211
	Walter Mikron (Cz)	Iv-I-4	A	60	142	135
	Scott Flying Squirrel (US)	Iv-I-2*	A	16	42	85

Year	Mfg., model (country)	Type, # cyl.	Cooling	HP	cu. in.	wt., lb.
1935 cont.						
	Aubier et Dunne (F)	Iv-I-2*	A	17	33	
	Train	I-2	A	20	61	77
	Mengin Type B	HO-2	A	25	76	77
	Mengin Type C	HO-2	A	38	95	82
	DKW (D)	I-2*	L	21	36	82
1936	Isotta-Fraschini XI. R (I)	V-12	L	700	1963	1135
	Junkers Jumo 210 (D)	V-12	L	700	1202	968
	Junkers Jumo 211 (D)	V-12	L	1100	2136	1356
	Rolls-Royce Merlin (GB)	V-12	L	1030	1650	1320
	Coatalen	V-12	L	550	2196	1210
	Junkers Jumo 206 (D)	I-6/12	L	1050	1526	
	Alvis Leonides (GB)	R-9	A	415	950	653
	Alvis Alcides (GB)	R-18	A	1700	3310	1050
	Alvis Pelides (GB)	R-14	A	1060	1158	1475
	Alvis Maeondies (GB)	R-14	A	700	1158	920
	Hispano-Suiza 14A (Es)	R-14	A	1000	2816	1408
	Gnome-Rhone 14N (F)	R-14	A	1085	2360	1300
	Nakajima Hikari (J)	R-9	A	710		
	Siemens 323R (D)	R-9	A	1000	1636	1199
NA	Nakajima Sakae (J)	R-14	A	1020	1700	1175
	Siemens SH 14A-4 (D)	R-7	A	160	469	298
	Farina T-58	R-5	A	142	488	317
	Jacobs L-5 (US)	R-7	A	285	831	415
	Soviet M-11D (R)	R-5	A	125	518	344
	Avia (P)	Iv-I-4	A	64	181	180
	Regnier 4J (F)	Iv-I-4	A	60	170	165
	Regnier 4L (F)	Iv-I-4	A	90	384	297
	AVA type 4A00	HO-4*	A	28	66	82
	Aspin	HO-4	A	114	202	200
1937	Piaggio P XI R/C 30 (I)	R-14	A	950	2293	1366
	Curtiss-Wright Cyclone G (US)	R-9	A	1000	1823	
	Curtiss-Wright R-2600 (US)	R-14	A	1500	2603	950
	Mitsubishi Zusei (J)	R-14	A	865	1716	1200
	Mitsubishi Kinsei (J)	R-14	A	1280	1980	1200
	Mitsubishi Kasei (J)	R-14	A	1300	2576	
	Fiat A 80 R/C 41 (I)	R-14	A	1000	2788	
	Piaggio Stella P XI R/C 40 (I)	R-14	A	1000	2354	1444
	Piaggio Stella P XI R/C 35 (I)	R-14	A	700		
	Pobjoy Niagara V (GB)	R-7	A	125	192	175
	Cirrus New Major (GB)	Iv-I-4	A	155	383	338
	deHavilland Gipsy Twelve (GB)	Iv-V-12	A	525	1121	1058
	Zlin Toma 4 (Cz)	Iv-I-4	A	106	244	209
	Walter Saggita (Cz)	Iv-V-12	A	370	727	715
	Deeble Duplex	Iv-I-6	A	300	447	918
	Sarolea Albatross	HO-2	A	35	67	95
	Praga D (Cz)	HO-4	A	79	174	142
	Dieke	HO-2*	A	18	46	62
	Fahlin	I-6	L	80-90		
	Jalbert	I-6Ds	L	180	772	
1938	Mercedes-Benz D-B 600 (D)	Iv-V-12	L	1000	2069	1510
	Mercedes-Benz D-B 601 (D)	Iv-V-12	L	1360	2069	1540
	Pratt & Whitney X-1800 (US)	H-24	L	2000	2599	2400
	Pratt & Whitney H-3130 (US)	H-24	L	2650	3730	3350
	Arrow	V-8	L	82	221	402

Year	Mfg., model (country)	Type, # cyl.	Cool- ing	HP	cu. in.	wt., lb.
1938 cont.						
	Waterman (US)	I-6	L	100		
	Bristol Taurus XII (GB)	R-14	A	1130	1550	1300
	Bristol Centaurus (GB)	R-18	A	2520	3270	2695
	Pratt & Whitney Twin Hornet A (US)	R-14	A	1150	2180	1635
	Nakajima Mamoru (J)	R-14	A	1870		
	Alfa Romeo 135 R/C 34 Tornado (I)	R-18	A	1500	2940	2100
	Soviet Shvetsov ASh 62 (R)	R-9	A	1000	1818	1184
	Warner Super Scarab 165 (US)	R-7	A	165	499	333
	Gadoux	Ro-B-12	L		900	1555
	Mawen	Ro-7	A		50	109
	Cirrus Midget (GB)	Iv-I-4	A		55	155
	deHavilland Gipsy Minor (GB)	Iv-I-4	A	90	222	216
	Renault 12R.01 (F)	Iv-V-12	A	450	1157	940
	Skymotor Model 70	Iv-I-4	A	60	201	186
	Hirth H-M 12B (D)	Iv-V-12	A	450	727	606
	Walter Minor Twelve (Cz)	Iv-V-12	A	370	727	715
	Shackleton Model C	Iv-I-4*	A	60	129	120
	Argus As 410 (D)	Iv-V-12	A	450	732	695
	Rolls-Royce Exe (GB)	X-24	A	1200	1324	
	Franklin 4AC150 (US)	HO-4	A	50	150	163
	Lycoming O-145 (US)	HO-4	A	50	145	152
	Continental A-50 (US)	HO-4	A	50	171	170
	Menasco M-50 (US)	HO-4	A	50	144	156
	Mengin type GHM	HO-2	A	45	116	99
	Walter Atom (Cz)	HO-2	A	28	66	88
	Prage B2 (Cz)	HO-2	A	46	116	106
	Jawa (Cz)	HO-2	A	35	61	86
1939	Bristol Taurus XII (GB)	R-14	A	1130	1550	1300
	Bristol Hercules (GB)	R-14	A	1290	2364	1680
	Alvis Leonides (GB)	R-9	A	415	905	653
	Pratt & Whitney Double Wasp R-2800	R-18	A	2000	2804	2350
	Regnier 12C-01 (F)	Iv-V-12	A	450	915	
	Renault 4P 03 (F)	Iv-I-4	A	100		
	Ranger 6-410 (US)	Iv-I-6	A	165	411	350
	Akron Funk (US)	I-4	L	63	201	260
	Rolls-Royce Vulture (GB)	X-24	L	1845	2592	
1940	Mercedes-Benz D-B 610 (D)	Iv-W-24	L	2450	5438	3476
	Lycoming O-1230 (US)	HO-12	L	1200	1234	1325
	Rolls-Royce Griffon II (GB)	V-12	L	1735	2240	2090
	Napier Sabre (GB)	H-24	L	2200	2240	2500
	Allison V-3420 (US)	W-24	L	2600	3420	2655
	Jacobs R-915 L-6 (US)	R-7	A	330	914	555
	Curtiss-Wright R-3350 (US)	R-18	A	2800	3350	2779
	Curtiss-Wright Cyclone 7 (US)	R-7	A	800	1300	919
	Nakajima Homare (J)	R-18	A	1850	2195	
	BMW 802 (D)	R-18	A	1675	3294	
	BMW 801 (D)	R-14	A	1580	2562	2702
	Fiat A 82 R/C 42 (I)	R-18	A	1250	2788	1600
NA	Fiat A 74 R/C 38 (I)	R-14	A	870	1907	1257
NA	Soviet Shvetsov ASh 21 (R)	R-9	A	630		
NA	Soviet Shvetsov ASh 82 (R)	R-14	A	1675	2513	1947
NA	Soviet Shvetsov ASh 73 (R)	R-18	A	2000		
	Guiberson A-1020 (US)	R-7 Ds	A	185	1021	653
	Phillips V-8	Iv-V-8	A	280	664	

Year	Mfg., model (country)	Type, # cyl.	Cool-ing	HP	cu. in.	wt., lb.
1940 cont.						
	Hirth H-M 4 (D)	Iv-I-4	A	90	242	238
	Hirth H-M 6 (D)	Iv-I-6	A	135	364	330
	Lycoming O-235 (US)	HO-4	A	100	233	207
	Franklin 6AC298 (US)	HO-6	A	130	298	295
1941	Mercedes-Benz D-B 603 (D)	Iv-V-12	L	2830	2715	2002
	Allison V-1710-G6 (US)	V-12	L	1250	1710	1595
	Ford XV-1650 (US)	V-12	L	2000		
NA	Mikulin AM-38 (R)	V-12	L	1600	2850	1892
NA	VK-105 (R)	V-12	L	1100	2147	1270
NA	VK-107 (R)	V-12	L	1600	2075	1655
	Chrysler IV 2220 (US)	V-16	L	1800	2220	
	Studebaker H-9350 (US)	H-24	L	5000	9350	
	Lycoming H-2470 (US)	H-24	L	2300	2430	
	Lycoming XR 7755 (US)	R-36	L	5000	7755	6050
	Pratt & Whitney Twin Wasp D (US)	R-14	A	1200	2000	1585
	Nakajima Ha-177 (J)	R-14	A	2400	2715	
	Jackobs R-830 L-5 (US)	R-7	A	330	830	
	Ranger 6-440 (US)	Iv-I-6	A	175	440	350
	Franklin 4AC-199 (US)	HO-4	A	90	199	224
	Lycoming O-290 (US)	HO-4	A	125	289	244
	Lycoming O-350 (US)	HO-6	A	180	352	289
	Lycoming O-435 (US)	HO-6	A	190	434	364
	Franklin 6ACV-405 (US)	HO-6	A	230	405	
1942	Mercedes-Benz D-B 605 (D)	Iv-V-12	L	1475	2179	1663
	Junkers Jumo 213 (D)	Iv-V-12	L	1776	2136	2024
	Menasco IV-2040	Iv-V-12	L	2040	2000	
	Lycoming O-145 (US)	HO-4	A	65	145	
	Franklin 4AC-176 (US)	HO-4	A	75	176	
	Franklin 8ACGSA-538 (US)	HO-8	A`	300	538	
	Franklin 12ACGSA-806 (US)	HO-12	A	450	806	
	Ranger 6-390 (US)	Iv-I-6	A	120	390	
	Ranger 6-410 (US)	Iv-I-6	A	165	410	
	Ranger 6-440-C (US)	Iv-I-6	A	200	441	376
	Ranger V-770 (US)	Iv-V-12	A	520	773	730
	Wright Tornado R-2160 (US)	R-42	A	2350	2160	
1943	Pratt & Whitney Wasp Major R-4360	R-28	A	2800	4360	3600
	Wright R-4090 (US)	R-22	A	3000	4090	
	Mercedes DB 613 (D)	Iv-V-24	L	3800	5430	4312
	Menasco H-4070 (US)	H-24	L	3400	4070	
	Pratt & Whitney H-2600 (US)	H-24	L	2200		
	Junkers Jumo 205 (D)	VO-6 Ds	L		1014	
1944	Rolls-Royce Eagle II (GB)	H-24	L	3500	2808	3900
	Kinner O-552 (US)	HO-6	A	250	552	
1946	Franklin 6AC425 (US)	HO-6	A	215	425	
	Continental C-115 (US)	HO-6	A	115	282	257
	Continental C-140 (US)	HO-6	A	140	282	298
1947	Hispano-Suiza 12Z-11Y (Es)	V-12	L	1500	2196	1364
	Fedden Flat Six	HO-6	A	185	325	310
	Franklin 4A4-B3 (US)	HO-4	A	100	225	228
	Franklin 6A4-B3 (US)	HO-6	A	150	335	365
	Franklin 6A8-215-B8F(US)	HO-6	A	215	500	485
	Continental A-100 (US)	HO-6	A	100	256	223
1948	PZL WN-1 (P)	HO-4	A	65	176	128
	Lycoming GSO-580 (US)	HO-8	A	375	578	560

Year	Mfg., model (country)	Type, # cyl.	Cool- ing	HP	cu. in.	wt., lb.
1948 cont.						
	SNECMA-Regnier 4L-04 (F)	Iv-I-4	A	147	384	341
1950	Napier Nomad (GB)	HO-12 Ds	L	3750	2510	4200
1952	Mathis	HO-4	A	92	211	251
1953	Polish WN 6 BR 2 (P)	R-5	A	185		
	Agusta G.A.40 (I)	HO-2	A	42	91	104
	Lycoming O-320 (US)	HO-4	A	150	320	272
	Franklin 425 (US)	HO-6	A	300	425	352
1954	Continental O-315 (US)	HO-4	A	150	315	261
1955	Potez 4E (F)	HO-4	A	105	209	
	Lycoming GO-480 (US)	HO-6	A	340	480	492
	Continental GSO-526 (US)	HO-6	A	340	526	540
1956	Praga Doris (CZ)	HO-6	A	220	437	436
	Agusta G.A.70 (I)	HO-4	A	86	140	150
1957	Lycoming O-340 (US)	HO-4	A	170	340	275
	Lycoming O-360 (US)	HO-4	A	180	361	282
	Continental O-300 (US)	HO-6	A	145	301	277
1959	Continental O-200 (US)	HO-4	A	100	201	188
	Lycoming O-540 (US)	HO-6	A	250	542	396
	Meteor G90 CA	R-4*	A	90	165	121
1960	Hindustan P.E.90H	HO-4	A	90	192	182
1961	M 110H (Cz)	HO-4	A	120	201	
	M 108H (Cz)	HO-6	A	270		
	NP-1 (Cz)	HO-4	A	68	166	
	WN-6 (Cz)	HO-6	A	220	421	346
	Potez 6E (F)	HO-6	A	155	314	
1962	Nelson H-63-CP	HO-4*	A	48	63	68
1964	Lycoming O-720 (US)	HO-8	A	400	722	597
	Franklin 2A120 (US)	HO-2	A	60	118	130
	Franklin 4A235 (US)	HO-4	A	130	235	240
1965	ENMA Flecha	HO-4	A	93	211	323
1967	Franklin 6V350 (US)	HO-6	A	235	350	329
1969	Polish PZL A-1 (P)	R-9	A	260	616	434
1972	Continental Tiara 6-285 (US)	HO-6	A	285	406	382
1976	Machen Merlyn (US)	I-3 Ds	L	650	210	580
1979	Continental 520 (US)	HO-6	A	325	520	380
	Continental 368 (US)	HO-4	A	180	346	260
	Continental Voyager 200 (US)	HO-4	L	110	201	220
	Continental Voyager 300 (US)	HO-6	L	170	301	291
	Continental 370 (US)	HO-4	L	190	370	280
	Continental TSIOL-370 (US)	HO-4	L	225	370	314
	Continental Voyager 550 (US)	HO-6	L	350	550	504
	Continental GT-550 (US)	HO-6	L	400	550	550
	AMI (US)	HO-2	A	53	100	100
	Arrow GT500 (I)	I-2*	A	65	30.5	79
	Arrow GT1000 (I)	I-2*	A	110	30.5	119
	Bakanov M-16 (R)	X-8	A	300		
	Emdair 112 (GB)	HO-2	A	85	112	128
	F.I 509 V2 (F)	V-2	A*	52	31	46
	Limbach L2000 (D)	HO-4	A	87	148	181
	Sauer ST 2500 H1S (D)	HO-4	A	92	151	174
	CAM Turbo 90 (C)	I-3	L	90	61	144
	Fuji EC51PL (J)	I-3	L	75	30.5	99
	Greth (US)	V-8	L	447	406	384
	Kawasaki Invader 440 (J)	Iv-I-2*	L	65	27	55
	RotorWay RW152 (US)	HO-4	L	152	162	170
	Westermayer W5/33 (D)	O-4	L	100	201	225

U.S. AIR FORCE BASES

Name & location	*Namesake*
Albrook, Panama	Lt. Frank Albrook, k. Chanute Field 1924
Andersen, Guam	Brig. Gen. James Andersen, k. Marshall Is. 1945
Andrews, MD	Lt. Gen Frank Andrews, k. Keflavik 1943
Bakalar, IN	Lt. John Bakalar, k. France 1943
Barksdale, LA	2nd Lt. Gene Barksdale, k. Wright Field 1926
Bergstrom, TX	Capt. John Bergstrom, k. Clark Field 1941
Biggs, TX	Lt. James Biggs, k. France 1918
Bolling, DC	Col. Ray Bolling, k. France 1918
Brookley, AL	Capt. Wendell Brookley, k. Bolling Field 1934
Brooks, TX	Cadet Sydney Brooks, k. Kelly Field 1917
Cannon, NM	Gen. John Cannon, d. 1955
Carswell, TX	Maj. Horace Carswell, k. S. China Sea 1944
Castle, CA	Brig. Gen. Fred Castle, k. Belgium 1944
Chennault, LA	Maj. Gen. Claire Chennault, d. 1958
Clark, Philippines	Maj. Harold Clark, k. Panama Canal Zone 1919
Connally, TX	Col. James Connally, k. Japan 1945
Davis-Monthan, AZ	2nd Lt. Sam Davis, k. Florida 1921; 2 Lt. Oscar Monthan, k. Hawaii 1924
Dow, ME	2nd Lt. James Dow, k. Queensboro, NY 1940
Dyess, TX	Lt. Col. Bill Dyess, k. Burbank, CA 1943
Edwards, CA	Capt. Glen Edwards, k. Muroc Field 1948
Elgin, IL	Lt. Col. Fred Elgin, k. Langley Field 1937
Eielson, AK	Col. Carl Eielson, k. Teller, AK 1929
Ellington, TX	2nd Lt. Eric Ellington, k. San Diego 1913
Ellsworth, SD	Brig. Gen Richard Ellsworth, k. Newfoundland 1953
Elmendorf, AK	Capt. Hugh Elmendorf, k. Patterson Field 1933
England, LA	Lt. Col. John England, k. France 1954
Ent, CO	Maj. Gen. Uzal Ent, d. 1948
Fairchild, CA	Gen. Muir Fairchild, d. 1950
Forbes, KS	Maj. Dan Forbes, k. Muroc Field 1948
George, CA	Brig. Gen. Harold George, k. Australia 1942
Goodfellow, TX	2nd Lt. John Goodfellow, k. France 1918
Griffiss, NY	Lt. Col. Townsend Griffiss, MIA Russia 1942
Hamilton, CA	Lt. Lloyd Hamilton, k. Belgium 1918
Harmon, Nfland.	Capt. Ernest Harmon, k. Stamford, CT 1933
Hickham, HI	Lt. Col. Horace Hickham, k. Ft. Crockett 1934
Hill, UT	Maj. Pete Hill, k. Wright Field 1935
Holloman, NM	Col. George Holloman, k. Formosa 1946
Howard, Panama	Maj. Charles Howard, k. Bryans Mill, TX 1936
Hunter, GA	Maj. Gen. Frank Hunter
Johnson, Japan	Lt. Col. Gerald Johnson, k. Tokyo 1945
Keesler, MS	2nd Lt. Sam Keesler, k. France 1918
Kelly, TX	2nd Lt. George Kelly, k. Ft. Sam Houston 1911
Kincheloe, MI	Capt. Iven Kincheloe, k. Edwards AFB 1958
Kindley, Bermuda	Capt. Field Kindley, k. Kelly Field 1920
Kingsley, OR	2nd Lt. Dave Kingsley, k. Rumania 1944
Kirtland, NM	Col. Roy Kirtland, d. 1941
Lackland, TX	Brig. Gen. Frank Lackland, d. 1943
Larson, WA	Maj. Don Larson, k. Germany 1944
Laughlin, TX	Lt. Jack Laughlin, k. Java 1942
Loring, ME	Maj. Charles Loring, k. N. Korea 1952
Lowry, CO	Lt. Frank Lowry, k. France 1918
Luke, AZ	2 Lt. Frank Luke, k. France 1918
McChord, WA	Col. Bill McChord, k. Richmond, VA 1937
McClellan, CA	Maj. Hez McClellan, k. Centerville, OH 1936

Name & location	*Namesake*
McConnell, KS	Capt. Fred McConnell, k. Garden Plain, KS 1945
	2nd Lt. Tom McConnell, k. Solomon Is. 1943
McCoy, FL	Col. Mike McCoy, k. Orlando, FL 1957
MacDill, FL	Col. Les MacDill, k. DC 1938
McGuire, NJ	Maj. Tommy McGuire, k. Philippines 1945
Malmstrom, MT	Col. Einar Malmstrom, k. Great Falls, MT 1954
March, CA	2nd Lt. Peyton March, k. Kelly Field 1918
Mather, CA	2nd Lt. Carl Mather, k. Ellington Field 1918
Maxwell, AL	2nd Lt. Bill Maxwell, k. Philippines 1920
Mitchel, NY	Cadet John Mitchel, k. Lake Charles, LA 1918
Moody, GA	Maj. George Moody, k. Wichita, KS 1941
Nellis, NV	Lt. BIll Nellis, k. Luxembourg 1944
Norton, CA	Capt. Leland Norton, k. France 1944
Offutt, NE	Lt. Jarvis Offutt, k. France 1918
Olmstead, PA	2nd Lt. Bob Olmstead, k. Holland 1923
Otis, MA	Lt. Frank Otis, k. Hennepin, IL 1937
Paine, WA	2nd Lt. Top Pain, d. 1922
Patrick, FL	Maj. Gen. Mason Patrick, d. 1942
Patterson, OH	Lt. Frank Patterson, k. McCook Field 1918
Pease, NH	Capt. Harl Pease, k. New Britain 1942
Perrin, TX	Lt. Col. Elmer Perrin, k. Baltimore, MD 1941
Pope, NC	Lt. Harley Pope, k. Fayetteville, NC 1919
Randolph, TX	Capt. Bill Randolph, k. Gorman, TX 1928
Reese, TX	Lt. Gus Reese, k. Sardinia 1943
Richards-Gebaur, MO	Lt. John Richards, k. France 1918; Lt. Col. Art Gebaur, k. N. Korea 1952
Robins, GA	Brig. Gen. Warner Robins, d. 1940
Schilling, KS	Col. Dave Schilling, d. 1956
Scott, IL	Cpl. Frank Scott, k. College Park, MD 1912
Selfridge, MI	Lt. Tom Selfridge, Ft. Myer, VA 1908
Shaw, SC	2nd Lt. Ervin Shaw, k. France 1918
Sheppard, TX	Sen. Morris Sheppard, d. 1941
Stead, NV	2nd Lt. Croston Stead, k. Reno, NV 1949
Stewart. TM	Maj. Allan Stewart, k. Solomon Is. 1942
Tinker, OK	Maj. Gen. Clarence Tinker, k. Wake Is. 1942
Travis, CA	Brig. Gen. Bob Travis, k. Fairfield-Suisun 1950
Truax, WI	Lt. Tom Truax, k. San Francisco 1941
Turner, GA	2nd Lt. Sullins Turner, k. Langley Field 1940
Tyndall, FL	Lt. Frank Tyndall, k. Mooresville, NC 1930
Vance, OK	Lt. Col. Leon Vance, MIA Iceland 1944
Vandenberg, CA	Gen. Hoyt Vandenberg, d. 1954
Walker, NM	Brig. Gen. Ken Walker, k. New Britain 1942
Webb, TX	Lt. Jim Webb, k. Japan 1949
Westover, MA	Maj. Gen. Oscar Westover, k. Burbank, CA 1938
Wheeler, HI	Maj. Sheldon Wheeler, k. Hawaii 1921
Whiteman, MO	2nd Lt. George Whiteman, k. Hawaii 12/7/41
Williams, AZ	Lt. Charles Williams, k. Hawaii 1927
Wright, OH	Wilbur Wright, d. 1912; Orville Wright, d. 1948
Wurtsmith, MI	Maj. Gen. Paul Wurtsmith, k. Asheville, NC 1943

TOP AMERICAN ACES

World War I (15 or more victories)
Rickenbacker, Capt. Edward V.* (USAS)--26
Lambert, Capt. Wm. C. (RFC)--22
Luke, 2Lt. Frank, Jr.* (USAS)--21
Gilette, Capt. Fred W. (RFC)--20
Malone, Capt. John (RN)--20
Wilkenson, Maj. ALan M. (RFC)--19
Hale, Capt. Frank L. (RFC)--18
Iaccaci, Capt. Paul T.--18
Lufbery, Maj. Raoul G. (FAS/LE)--17
Kulberg, Lt. Harold A. (RFC)--16
Rose, Capt. Oren J. (RFC)--16
Warman, Lt. C.T. (RFC)--15

World War II (20 or more victories)
Bong, Maj. Richard T.* (AAF)--40
McGuire, Maj. Thomas B.* (AAF)--38
McCampbell, Capt. David (USN)--34
Gabreski, Col. Francis N. (AAF)--31
Johnson, Lt. Col. Robert S. (AAF)--28
Boyington, Maj. Gregory (USMC)--28
MacDonald, Col. Chas. H. (AAF)--27
Foss, Capt. Joseph J. (USMC)--26
Preddy, Maj. George E. (AAF)--25.83
Hanson, Lt. Robt. M. (USMC)--25
Meyer, Col. John C. (AAF)--24
Harris, Lt. Cecil E. (USN)--24
Valencia, Lt. Cmdr. Eugene (USN)--23
Wetmore, Capt. Ray S. (AAF)--22.59
Schilling, Col. David C. (AAF)--22.5
Johnson, Lt. Col. Gerald R. (AAF)--22

Kearby, Col. Neel E.* (AAF)--22
Mahurin, Lt. Col. Walker M. (AAF)--22
Robbins, Col. Jay T. (AAF)--22
Christensen, Capt. Fred J. (AAF)--21.5
Voll, Maj. John J. (AAF)--21
Walsh, Capt. Kenneth A. (USMC)--21
Lynch, Lt. Col. Thomas J. (AAF)--20
Westbrook, Lt. Col. Robert B. (AAF)--20
Aldrich, Capt. Donald N. (USMC)--20

Korea (10 or more victories)
McConnell, Capt. Joseph, Jr. (USAF)--16
Jabara, Lt. Col. James (USAF)--15
Fernandez, Capt. Manuel J., Jr. (USAF)--14.5
Davis, Maj. George A., Jr.* (USAF)--14
Baker, Col. Royal N. (USAF)--13
Blesse, Maj. Frederick C. (USAF)--10
Fischer, Capt. Harold E. (USAF)--10
Garrison, Lt. Col. Vermont (USAF)--10
Johnson, Col. James K. (USAF)--10
Moore, Capt. Lonnie R. (USAF)--10
Parr, Capt. Ralph S., Jr. (USAF)--10

Vietnam (5 or more victories)
DeBellevue, Capt. Charles D. (USAF)
Cunningham, Lt. Randy (USN)
Driscoll, Lt. William (USN)
Feinstein, Capt. Jeffrey S. (USAF)
Ritchie, Capt. Richard S. (USAF)

* denotes Medal of Honor recipient

XI. AVIATION ANNIVERSARIES

Amelia Earhart was a revelation and an inspiration, intelligent without being overbearing, gently spoken yet persuasive, successful, modest and self-reliant--all the things many women of the Thirties dreamed of being,

AVIATION ANNIVERSARIES
Firsts, First Flights and Records

JANUARY

1--World's first regular air passenger service--St. Petersburg to Tampa, FL, 1914.

2--First issue of *Flight* published, London, 1909.

3--**Cessna 310**, Wichita, KS, 1953.

4--Vought XSB2U-1 Vindicator, 1936

5--Unofficial climbing record of 13,000 feet per minute by Maj. Charles Yeager in Bell X-1 at Muroc, 1949.

6--750,000 people gather in Calcutta to watch India's first airplane flight, 1911.

7--**Lockheed XF-104**, Burbank, CA, 1954.

8--**Lockheed XP-80**, Muroc, CA, 1944.

9--**Lockheed L-49 Constellation**, 1943.

10--**Piper PA-28**, Vero Beach, FL, 1960.

January 3, 1953

11--**First Trans-Pacific solo flight**--Amelia Earhart, 1935.

12--**World altitude record**--1,209 m (3,966 ft.) by Louis Paulhan in Los Angeles, 1910.

13--**World speed record**--90.2 mph in a Deperdussin, France, 1912.

14--**Sikorsky XR-4 helicopter**, 1941.

15--**First bomb dropped from an airplane**--San Francisco, CA, 1911.

16--Benjamin Franklin suggests military use of balloons, 1784.

17--Zeppelin LZ 2 flies at Lake Constance, 1905.

18--**First nonstop around-the-world flight**--flight of B-52s, 45:19 hours, 1957.

19--**Transcontinental speed record**--Howard Hughes, 7:18:25 hours in Hughes H-1, 1937.

20--**Beech King Air 90**, Wichita, KS, 1964.

21--**Supersonic passenger service**, Concorde, 1976.

22--**Boeing 747** enters service on Pan Am's NY-London route, 1970.

23--**Bleriot XI**, 1909.

24--Three Italian S.M.79 flying boats leave for nonstop flight to Rio de Janeiro, 1938.

25--**Boeing 314** flying boat certified, 1939.

26--First organizational meeting of the **Experimental Aircraft Assn.**, Milwaukee, WI, 1953.

27--**Lockheed P-38**, 1939; **Jet Commander**, 1963.

28--**Challenger** explodes 72 seconds after liftoff, 1986.

29--Pescara's helicopter makes a ten-minute flight, Paris, 1924.

30--Orville Wright dies, Dayton, OH, 1948.

31--Boeing buys deHavilland Aircraft of Canada, 1986.

FEBRUARY

1--**General Dynamics' F-16** debuts at Edwards AFB, CA, 1974.

2--**Boeing 727**, 1963.

3--A Mil Mi-26 helicopter lifts a record 125,153.8 lbs to 6,562 ft, 1982.

4--Franz Reicelt makes a fatal jump from Eiffel Tower using crude parachute, 1912.

February cont.

5--**Travel Air Co.** founded by Walter Beech, Clyde Cessna and Lloyd Stearman, Wichita, KS, 1925.

6--First TWA scheduled international flight, NY to Paris, 1946.

7--Robert Timm and John Cook set an **endurance record** of 64 days, 22:19:05 hours in a Cessna 172 at Las Vegas, NV, 1959.

8--**Boeing 247**, 1933.

Transcontinental speed record of 607.8 mph set by Boeing B- 47, 1949.

9--Sopwith Pup, 1916.

10. Sq. Ldr. J.W. Gillan flies Hurricane 327 mi. in 48 min., averaging 408 mph, 1938

11--Vivian C. Walsh makes first flight in New Zealand, 1911.

12--**Grumman F4F-3 Wildcat**, 1939.

13--**Swearingen SJ30**, San Antonio, TX, 1991.

14--Burgess H biplane is flown a record 244.8 mi. in 4 hr., 43 min., San Diego, 1913.

15--**Douglas DC-6**, 1946.

16--**Lockheed F-104**, 1956.

17--Gabriel and Charles Voisin complete their first airplane, 1907.

18--Short flying boat *Caledonia* flies nonstop from England to Egypt, 2,222 mi., 1937.

19--**Cessna 188 Agwagon**, Wichita, KS, 1965; **Boeing 757**, Renton, WA, 1982.

20--John H. Glenn, Jr. is **first American to orbit** in space, 1962.

Henry Beaird, Jr. flies **Lear 24B** to 12,000 meters (39,370 ft.) in six minutes, 19.1 seconds, Wichita, KS, 1968.

21--Neil Armstrong climbs to 50,000 ft in Learjet 28 in just over 12 minutes, breaking five world records for business jets, 1979.

22--**US Navy's first woman pilot**--Lt. (j.g.) Barbara Allen, 1974.

23--**First flight in Canada**--J.A.D. McCurdy, 1909.

Dassault **Mystere** fighter, Istres, France, 1951.

24--**Hawker Typhoon**, 1940.

25--**Cessna 210**, Wichita, KS, 1957.

26-27--**Around-the-world speed record** of 36:08:34 hours set by Allen Paulson and crew in Gulfstream IV, 1988.

27--**First autopilot**--Sperry in Navy F5L, 1920.

First around-the-world nonstop flight--Boeing B-50, 94:22:01 hours, 1949.

28--Republic XP-84, 1946; Cessna 336 Skymaster, 1961.

February 27, 1988

MARCH

1--**First night flight**--Henry Farman, Chalons, France, 1910.

2--**Concorde**, Toulouse, France, 1969.

3--National Advisory Committee for Aeronautics (NACA) established, 1915.

4--Airship **Hindenburg**, 1936.

5--Sikorsky Aero Engineering, 1923. **Supermarine Spitfire**, 1936.

March cont.

6--The Bell Model 47 becomes the first **commercially certificated helicopter**, 1946.
    SR-71 sets **Transcontinental speed record**--2,404 miles in 1:07:53.69, averaging
    2,124.51 mph, 1990.

7--E.H. Barksdale and B.Q. Jones fly a DH-4B from Dayton, OH to New York City solely on
    instruments, 1924.

8--The **first certificated woman pilot**--Baroness de Laroche, 1910.

9--Capt. James E. Miller is first American air casualty in WWI, 1918.

10--Fairey Delta 2 first airplane to officially exceed 1,000 mph, Sussex, England, 1956.

11--U.S. Navy authorizes "not more than $50" for study of vertical lift aircraft, 1912.

12--First flight of *Red Wing*, built by U.S. Aerial Experimental Assn. ends with crash, 1908.

13--**First licensed woman pilot in Canada**--Eileen Vollick, 1928.

14--Pan American Airways formed, 1927.

15--Rolls-Royce is registered as a public company, 1906.

16--The **first successful liquid-fuel rocket**--by Dr. Robert Goddard, 1926.

17--**North American B-45**, 1947.

18--**First flight in Australia**--by American Eric Weiss (aka Harry Houdini), 1910.

18-20--Amelia Earhart and crew fly **around the world** from Oakland, CA in a Lockheed
    Electra, 1937.

20--*U.S.S. Langley*, the first U.S. aircraft carrier, is commissioned, 1922.

21--The **last DC-3** of 10,654 built is delivered to Sabena, 1947.

22--**Grumman XF3F-1**, 1935.

23--**World record parachute jump**--Lt. Arthur Hamilton from 24,400 ft, 1921.
    Flt. Sgt. Nicholas Alkemade jumps **without a parachute** from an RAF Lancaster,
    falls 18,000 ft without injury, 1944.

24--**Lockheed C-141B**, 1977.

25--**Vought XF8U-1 Crusader**, 1955.

26--**Cessna T-50**, Wichita, KS, 1939. **Convair B-36D**, 1949.

27--**Sperry gyro-stabilized autopilot** flies in a Navy F5L, 1920.

28--**First seaplane flight**--Henry Fabre, 1910.

29--Flight Safety Inc. begins operations in New York, 1951.

30--Morris Titterington demonstrates his earth inductor compass, New York, 1925.

31--Collier Trophy is awarded to Packard Motor Car Co. for developing diesel aircraft
    engine, 1932.

APRIL

1--**Beech Aircraft Co.** founded, 1932.
    Mitsubishi A6M1 "Zero," 1939.

2--Svenska Aeroplan Aktiebolaget (SAAB)
    founded, 1937.

3--**Airbus A310**, Toulouse, France, 1982.

4--**Airship USS Akron** crashes off Farnegat
    Light, NJ, 1933.

April 3, 1982

5--**First twin-jet flight**, Heinkel He-280, 1941.

6--**Bell P-39 Airacobra**, 1939.

April cont.

7--First mid-air airliner collision kills seven, France, 1922.

8--German Fw 200 anti-shipping Condors begin patrolling the North Atlantic, 1940

9--**Gulf Coast Air Line** begins flights between New Orleans and Pilottown, LA., 1923

10--Charles Lindbergh appointed chief pilot for Robertson Aircraft Co., St. Louis, MO, 1926.

11--First official flight of YF-12 (SR-71) **Blackbird**, 1962.

12--**First successful jet engine**--Frank Whittle, Rugby, England, 1937.
      Yuri Gagarin makes **first manned orbital flight**, 1961.

13--Maurice Prevost wins 1913 **Schneider Cup**, averaging 45.7 mph in a Deperdussin,
      Monaco.

14--**Link Trainer**, first electro-mechanical flight simulator, 1929.

15--**Boeing B-52**, 1952.

16--Aviatrix **Harriet Quimby** dies in Boston air crash, 1912.

17--**First solo flight around the world by a woman**--Jerrie Mock, Cessna 180, 29.5 days,
      Columbus, OH, 1964

18--Lt. Col. James Doolittle leads 16 B-25s from USS Hornet to bomb Tokyo, 1942.
      Piper PA-31T Cheyenne, San Antonio, TX, 1967.

19--**Macchi C.205**, 1942.

20--First seven astronauts are chosen for Project Mercury, 1959.

21--Baron Manfred von Richthofen, the "Red Baron" dies, replaced in command by
      Hermann Goering, 1918.

22--Synchronized machine gun fitted to aircraft, Anthony Fokker, 1915.

23--First BAe 146 delivered to **The Queen's Flight** on 50th anniversary, Hatfield,
      England, 1986.

24--**First aerial motion picture photography**--Wilbur Wright, 1909.

25. U.S. Navy makes first operational aircraft mission, Vera Cruz, Mexico, 1914.

26--**First supersonic dogfight**--N. Vietnamese MiG-21 is shot down by an U.S. F-4, 1966.

27--First non-stop flight from England to India, 1929.

28--Ryan **"Spirit of St. Louis"**, 1927.
      McDonnell Aircraft and Douglas Aircraft merge, 1967.

29--First Boeing fighter, the XPW-9, flies at Camp Lewis, WA, 1923.

30--Helicopters fly the last U.S. missions in Viet Nam airlifting personnel, 1975.

MAY

1--**First fully enclosed cabin airplane**--Avro Type F, 1912.

2--**First jet airline service**--BOAC, London to Johannesburg, 1952.

3--**First nonstop transcontinental flight**--Kelly and Macready, Fokker T-2, 1923.

4--Dassault Mystere 20 (**Falcon 20**), Bordeaux, France, 1963.

5--Alan Shepard becomes the **first American in space**, 1961.

6--**Hindenburg** is destroyed by fire at Lakehurst, NJ, 1937. **Republic XP-47B**, 1941.

7--**First pressurized aircraft**--Lockheed XC-35, 1937.

8--**Battle of the Coral Sea**--first U.S. victory
      over Japan, 1942.

9--**First Flight over North Pole**--piloted by Floyd Bennett, 1926.

May cont.

10--Hitler's deputy, Rudolph Hess, flies from Germany to Scotland on a claimed peace mission, 1941.

11--Charles Lindbergh sets non-stop 14:25 record San Diego to St. Louis in *Spirit of St. Louis*, 1927.

12--**Messerschmitt Bf 110**, 1936.

13--Igor Sikorsky's **VS-300 helicopter** first flies at Stratford, CT, 1937.

14--Charles Furnas flies with Wilbur Wright as the **first airplane passenger**, 1908.

15--**First U.S. Airmail service**--New York-Washington, 1918.

AOPA founded, Trenton, NJ, 1939.

The **first British jet**, the Gloster E.28/39, flies at Cranwell, 1941.

16-17--**First Transatlantic Flight**--Navy Curtiss NC-4, 1919.

18--**Douglas DC-7**, 1953.

19--Jacqueline Cochran becomes the **first woman to fly at supersonic speeds**, 1953.

20--Paul Tissander sets the **first recognized airplane speed record**--34.03 mph in a Wright, 1909.

20-21--**First solo transatlantic flight**--Charles Lindbergh, 1927.

First women's solo transatlantic flight--Amelia Earhart, 1932.

22--**Marine Corps Aviation Branch**, 1912.

23--Fairey Swordfish sinks a U-boat in the first successful use of an air-launch rocket against a submarine, 1942.

24--**Piper Comanche**, Lock Haven, PA, 1954.

25--**North American F-100 Super Sabre**, 1953.

26--**First north-south transcontinental nonstop flight**--Lt. H.G. Crocker, Houston to Gordon, Ontario, 1923.

27--German battleship *Bismark* disabled by Fairey Swordfish torpedo bombers and sunk by Royal Navy gunfire, 1941. McDonnell **F-4 Phantom** , 1958.

28--469.220 mph **prop-driven speed record** is set by Fritz Wendel in a Bf.109R at Augsburg, Germany, 1938.

29--**Grumman "Goose"**, 1937. **Chance-Vought F4U-1 Corsair**, 1940.

German pilot **Mathias Rust** lands a Cessna F172 in Moscow's Red Square, 1987.

30--**Cessna Citation III**, 1979.

31--Navy NC-4 flying boat completes **first transatlantic flight** to England, 1919.

JUNE

1--Focke-Wulf Fw190, 1939.

2--Glenn Curtiss flies a redesigned **Langley Aerodrome** in an effort to discredit Wrights, 1914.

3--**Gemini IX** spacecraft flown by Stafford and Cernan, rendezvous and space walk tests, 1966.

4--Montgolfier brothers first public balloon ascension, 1783.

Max Conrad lands Piper Comanche in New York after a 7,683-mi. flight from Casablanca, 1959.

4-6--Clarence Chamberlain and Charles Levine fly 3,904 miles non- stop New York to Eisleben, Germany in the Bellanca W.B.2 "Columbia," 1927.

June cont.

5--Montgolfier brothers fly an unmanned hot air balloon to 6,000 ft, Annonay, France, 1783.

6--Socony-Vacuum Oil (Mobil) produces 100-octane fuel, 1936.

7--**Supermarine S.5** racer, 1927.

8--Lt. John Wilson makes record parachute jump from 19,861 ft, 1920, San Antonio,TX.

9--Avro Lincoln, 1944.

10--MS-880 **Rallye**, Villacoublay, France, 1959.

11--**Ford 4-AT** trimotor, 1926.

12--**Cessna 172**, 1955. Bryan Allen makes first
human-powered flight across the English
Channel in *Gossamer Albatross*, 1979.

13--The first V-1 bomb lands in Britain, 1944.

14--**First east-to-west around-the-world flight,**
Allen Paulson, 45:25:10 hours in a Gulfstream IV, Paris, 1987.

June 11, 1926

14-15--**First non-stop transatlantic flight**--Alcock and Brown in Vickers Vimy, 1919.

15--**Arado Ar 234, the world's first jet bomber,** 1943.

16--BAe 1000, 1990.

17--**Delta Air Service** (later Delta Airlines) begins with a Travel Air from Dallas, TX, 1929.

17-18--Earhart-Stutz-Gordon make transatlantic flight in Fokker VII "Friendship," 1928.

19--World speed record of 623.738 mph, P-80 Shooting Star, 1947.

20--**Glenn Curtiss** is the first American to fly after the Wright Brothers, 1908.
Piper PA-24 Comanche, 1956.

21--**Helicopter world altitude record**--Jean Boulet in Aerospatiale SA 315 Lama,
40,280 ft, 1972.

22--**"Voyager"** flies at Chino, CA, 1984.

23--First Focke-Wulf, the A-16, 1924.

24--First report of "flying saucers", Mt. Rainier, WA, 1947.

25--**Northrop B-35 Flying Wing,** 1946. **Korean War** begins, 1950.

26--**World's first practical helicopter**--Focke, Germany, 1936. Grumman F6F-1 **Hellcat,**
1942. First Bell helicopter, the Model 30, flies at Gardenville, NY, 1943. **Berlin
Airlift** begins, 1948.

27--**First in-flight refueling** is demonstrated, 1923.

28--**Cessna 140,** Wichita, KS, 1945.

28-29--**First U.S.-Hawaii flight**--Lts. Hegenberger and Maitland in Fokker tri-motor, 1927.

30--**Lockheed C-5A Galaxy,** 1968.

JULY

1--First flight of **Douglas DC-1**, Santa Monica, CA, 1933.
Al and Fred Key land after a **record non-stop flight** of 653 hours, 34 minutes in a
Curtiss Robin, 1935.

2--Amelia Earhart is lost in the Pacific Ocean, 1937.

3--Lt. Theodore Ellyson, the first Navy pilot, makes the first night flight from water without
lights, 1911.

4--American Airlines inaugurates DC-3 service, 1936.

July cont.

5--Clyde Cessna sets an American speed record, flying 76 miles at 124.5 mph, 1917.

6--British R-34 completes **first transatlantic crossing by a rigid airship**, 1919.
McDonnell Aircraft founded in St. Louis, MO, 1939.

7--Coast-to-coast passenger transportation in 48 hours (by rail and air) is begun by
Transcontinental Air Transport, 1929.

8--First combat mission by B-17s, by RAF over Wilhelmshaven, Germany, 1941.

9-15--**"Voyager"** breaks world closed-course unrefueled endurance record with flight of
11,857.39 miles, 1986.

10--The **Battle of Britain** begins, 1940.

11--Laura Ingalls becomes first woman to fly solo east-west across U.S., 1935.

12--U.S. Navy Bureau of Aeronautics, 1921.

13--President Eisenhower becomes the first U.S. president to fly in a helicopter, 1957.

14--Russians Gromoff, Yumasheff and Danilin establish 6,262-mi non-stop distance
record, 1937.

15--Boeing founded, Seattle, WA, 1916. **Boeing 367-80 (707)**, Renton, WA, 1954.
Cessna 177/Cardinal, 1966.

15-22--Wiley post makes the **first round-the-
world solo flight**, Lockheed "Winnie
Mae," 1933.

16--The first **atomic bomb** is tested at White
Sands, NM, 1945.

17--**Northrop B-2** stealth bomber, 1989.

18--**Douglas "Wrong Way" Corrigan** lands in
Ireland flying solo in a Curtiss Robin from
New York after telling officials he was
headed to California, 1938.

July 17, 1989

19--Joe Walker pilots the North American X-15A to an unofficial record altitude of
347,800 ft, 1963.

20--Astronauts Neil Armstrong and "Buzz" Aldrin make **the first moon landing**, 1969.

21--Pilot Denise Moore becomes the first woman killed in an airplane, Chalons, France,
1911.

22--Donald W. Douglas forms Davis-Douglas Co., Los Angeles, 1920.

23--The Gossamer Condor makes **the first man-powered flight**, 1977.

24--British scientists successfully operate "radio direction finding" equipment (radar),
Suffolk, 1935.

25--Louis Bleriot makes **the first flight across the English Channel**, 1909.

26--**First plane-to-ground radio** transmissions, 1912.

27--**de Havilland Comet**, the first jet airliner, 1949. **McDonnell Douglas F-15**, 1972.

28--**Lockheed SR-71** sets two world records--2,193.16 mph by Capt. E.W. Joersz, and an
altitude of 85,069 ft. in horizontal flight by Capt. R.C. Helt, Beale AFB, CA, 1976.

29--**Piper PA-23 Apache**, 1953.

30--**Northrop F-5**, 1959.

31--**Learjet 23** receives type certificate, 1964.

AUGUST

1--The USAAF names its Mojave facility Muroc Flight Test Base, 1944.

2--The Sikorsky-designed four-ton, 92-ft span *Bolshoi Baltskii,* or *Grand,* the world's first four-engine airplane, makes a 1 hour, 54 minute flight in Russia, 1913.

3--First crop-spraying flight--Lt. John Macready, Troy, OH, 1921.

4--Lockheed U-2, 1955.

5--A women's speed record of 196.1 mph is set by **Pancho Barnes**, Los Angeles, 1920.

6--B-29 "Enola Gay" drops first atomic bomb, Hiroshima, 1945.

7--Grumman TBF Avenger, 1941.

8--The Nakajima Kikka, the **first Japanese jet aircraft**, makes first of two flights, 1945.
Convair B-36, Ft. Worth, TX, 1946.

9--Curtiss-Wright Corp., 1929.

10--Lawrence Sperry demonstrates a gyro stabilizer, 1913.

11--Bell XV-3 convertiplane, 1955.

12--First transatlantic balloon crossing--"Double Eagle II," 1978.

13--Gee Bee R-1, 1932.
deHavilland DH-125 business jet, 1962.

14--Grumman G-159 Gulfstream, 1958.

15--Wiley Post and Will Rogers die in Point Barrow, AK crash of Lockheed Orion Explorer, 1935.

16--Darryl Greenamyer breaks a 30-year-old German piston-driven speed record with a **488.98 mph** flight in Grumman Bearcat, 1969.

17--Art Goebel lands Travel Air *Wollaroc* in Honolulu to win the **Dole Air Derby**, 1927.

18--First mid-air space capsule recovery--Discoverer 14, 1960.

19--First flight of **North American B-25**, 1940.

20--Col. Horace Hanes sets a speed record of **822.135 mph** in F-100 Super Sabre, 1955.

21--Lyle Shelton sets world piston speed record of **529.273 mph** in Grumman F8F-2 "Rare Bear" at Las Vegas, NM, 1989.

22--Learjet Model 35, 1973.

23--Lockheed C-130 Hercules, 1954.
Man-powered **Gossamer Albatross** is flown 3 mi. by designer Paul MacCready, 1977.

24--British airship R.38 crashes off coast of Yorkshire, killing 44, 1921.

25--First manned balloon flight in England, James Tytler, 1784.

26--Charles A. Lindbergh dies, 1974.

27--The **first radio message from an airplane** is transmitted by J.A.D. McCurdy at Sheepshead Bay, NY, 1910.
The **world's first jet**, the Heinkel He-178, 1939.

28--International Air Traffic Assn. (IATA), 1919.

29--McDonnell Douglas DC-10, 1970.
Beech StarShip, 1983.

30--Glenn Curtiss wins the first Bennett Cup, Reims, France, 1909

31--Grumman F8F Bearcat, 1944.

31-Sept. 1--First transpacific flight is made in a Curtiss, 1925.

SEPTEMBER

1--The letter N is adopted as the international designation of U.S. civil aircraft, 1929.
    First aerial refueling of a jet bomber is accomplished, 1953.
2--**Grumman F4F Wildcat**, 1937. **Hawker Tempest**, 1942.
    Japanese surrender ends World War II, 1945.
3--Maj. James Doolittle set a new 294 mph world speed record in winning the Thompson
    Trophy Race in a Gee Bee R-1, 1932. **Mooney M-20**, Kerrville, TX, 1953.
4--Louise Thaden becomes the first woman to win the Bendix Trophy, Beech 17, 1936.
    Lockheed JetStar, 1957.
5--**Roscoe Turner** wins the Thompson Trophy at 283.42 mph, 1938.
6--**Michel Detroyat**, flying a Caudron, wins the 1936 Thompson Trophy Race at a record
    speed of 264.26 mph.
7--**First pilot fatality**--Eugene Lefebvre in France, 1909.
8--**First midair collision**--at Wiener-Neustady, Austria, 1910.
9--**Curtiss R2C-1** racer, 1923.
10--**Cessna 182**, Wichita, KS, 1955.
11--**First transcontinental mail flight**--San Francisco, CA, 1920.
12--**Taylor E-2 "Cub"**, Lock Haven, PA, 1930; **Cessna 150**, Wichita, KS, 1957.
13--Supermarine S.6B wins possession of the **Schneider Trophy** at 340.108 mph, 1931.
14--Igor Sikorsky test flies the **first American helicopter** at Stratford, CT, 1939.
15--**Cessna Citation 500**, Wichita, KS, 1969.
16--Canadian Aviation Corps, 1914.
17--Lt. Thomas Sefridge is the **first airplane
    fatality**, killed during a flight with Orville
    Wright, 1908.
18--U.S. Air Force established, 1947.
19--First balloon ascent with humans,
    Versailles, 1783.

September 15, 1969

20--Wilbur Wright accomplishes the first
    complete circle flight in an aircraft, Simms Station, OH, 1904.
21--**Boeing B-29**, Seattle, WA, 1942.
22--Col. David Schilling makes the **first nonstop transatlantic jet flight**, 1950.
23--Roland Garros is the first person to fly across the Mediterranean (470 miles), 1913.
24--James Doolittle accomplishes the **first instrument takeoff and landing**, 1929.
25--McDonnell Aircraft Corp., St. Louis, MO, 1939.
26--**de Havilland Vampire**, 1943.
27--Geoffrey deHavilland dies in the crash of a D.H.108 tailless research aircraft, 1946.
28--**First retractable landing gear** is used on Dayton-Wright racer at Gordon Bennett
    Cup race, 1920.
    Two Douglas World Cruisers complete the **first flight around the world**, Seattle,
    WA, 1924.
29--A Deperdussin sets a world speed record at 126.666 mph, 1913.
    Supermarine S.6B sets seaplane record of 407.5 mph, 1931.

September cont.

30--**First around-the-world helicopter flight**--Ross Perot, Jr and J.W. Coburn average 35.4 mph in Bell 206-L, Ft. Worth, TX, 1982.

OCTOBER

1--Robert Stanley flies the **first American jet**, the Bell XP-59A, at Rogers Dry Lake, CA, 1942.

2--The rocket-powered **Messerschmitt Me 163A** sets a world speed record of 623.85 mph, 1941.

3--Capt. William Knight flies the North American X-15A to unofficial world speed record **Mach 6.70** (4,534 mph) at Edwards AFB, CA, 1967.

4--USSR launches **Sputnik** into earth orbit, 1957.

4-5--Hugh Herndon and Clyde Pangborn fly from Sabishiro, Japan to Wenatchee, WA in 41:13 hours, 1931.

5--A French pilot scores the **first aerial victory** when he shoots down a German Aviatik, Brimont, France, 1914.

6--John Macready and O.G. Kelly stay aloft in their Fokker T-2 for a record 35 hours, 18 minutes, 30 seconds, San Diego, CA, 1922.

7--The **Learjet 23**, Wichita, KS, 1963.

8--**Curtiss P-40**, 1938.

9--J.H. Banning and T.C. Allen make the **first transcontinental flight by Negro pilots**, 1932.

10--**Travel Air "Mystery Ship"** beats military aircraft in winning the National Air Races at 194.9 mph, 1929.

October 7, 1963

11--**Westland Whirlwind**, 1938.

12--**Cessna T-37**, Wichita, KS, 1954.

13--Cyrus Bettis wins the last Pulitzer Trophy Race in a Curtiss RC3-1 Army Racer at 248.99 mph, 1925.

14--Capt. Charles Yeager makes the **first supersonic flight** in a Bell X-1 at Mach 1.015 (670 mph), Muroc Army Air Field, CA, 1947.

15--**LaGuardia Field** is dedicated in Flushing, NY, 1939.

16--**First aircraft flown in England** by American S.F. Cody, 1908.
First aircraft downed in the Battle of Britain by RAF pilots, 1939.

17--Lt. V.C. Griffin makes the first takeoff from the *U.S.S Langley*, America's first operational carrier, 1922.

18--Charles Lindbergh receives Order of the German Eagle, with star from Hermann Goering, 1938.

19--**Pan American Airways**, 1927.

20--**Bell XH-40** (UH-1 "Huey") Fort Worth, TX, 1956.
Bell Aircraft founder Larry Bell dies, Niagara Falls, NY, 1956

21--**Northrop YB-49** flying jet bomber, 1947.

22--An Italian Bleriot makes a reconnaissance flight from Tripoli to Azizia, becoming the **first airplane used in war**, 1911. Republic F 105, 1955.

October cont.

23--A Macchi MC.72 racing seaplane sets a world record speed at 440.68 mph, 1934.

24--The de Havilland D.H.88 Comet *Grosvenor House* wins the England-Australia MacRobertson race at an average speed of 158.9 mph, 1934.

25--**First coast-to-coast air service**--TWA, 1930.

26--**North American P-51** prototype, 1940.

27--Curtiss XF9C-1 successfully hooks up with airship USS *Los Angeles*, 1930.

28--The court martial trial of air advocate Col. William "Billy" Mitchell begins in Washington, 1925.

29--Alfred Leblanc wins Gordon Bennett Trophy at 68.2 mph in a Bleriot, Belmont Park, NY, 1910

30--**Japan Air Lines** formed, 1928.

31--Type Certificate is awarded to the **Piper J-3 Cub**, 1937.

NOVEMBER

1--Ninety-Nines is formed in Valley Stream, NY, 1929.

2--Hughes H-4 **"Spruce Goose"**, Long Beach, CA, 1947.

3--USSR launches *Sputnik II*, carrying a dog, 1957.

4--**Beech Model 17-R "Staggerwing"**, Wichita, KS, 1932.

5--Bert Acosta wins the Pulitzer Trophy at 176.7 mph in Navy Curtiss CR-1, Omaha, NE, 1921.

6--**Hawker Hurricane**, 1935.

7--**Piper PA-30 Twin Comanche**, 1962; **Falcon 50**, 1976.

8--A MiG-15 is downed by an F-80C in Korea in the **first jet dogfight**, 1950.

9--C.A. Butler arrives in Australia, after a flight from England in Comper Swift light plane, 1931.

10--**Sikorsky S-55**, 1949. The US Air Force reveals the existence of top-secret F-117 "stealth" fighter, 1988

11--**Convair B-58 Hustler**, Ft. Worth, TX, 1956.

12--Maj. Adrian Drew flies a McDonnell F-101A to official world speed record of 1,207.6 mph, 1957.

13--Paul Cornu makes the **first vertical flight**,--at five feet for 20 seconds, Normandy, France, 1907.

14--Lt. Eugene Ely makes the **first flight from a ship**, off the *U.S.S. Birmingham*, 1910.

15--Autogyro developer Juan de La Cierva patents a system of rotor blades with flapping hinges, 1922.

16--**Saab 90 Scandia**, 1946.

17--Dulles Airport is dedicated by President John F. Kennedy, 1962.

18--Boeing XP-9, Dayton, OH, 1930.

19--**Heinkel He 177**, 1939.

November 14, 1910

November cont.

20--Scott Crossfield flies a Douglas D-558-II to **1,328 mph**, 1953.

21--Clyde Cessna, founder of Cessna Aircraft Co., dies, 1954.

22--Lt. Col. R.B. Robinson flies a McDonnell F4H-1F to a record speed of 1,605 mph, Edwards AFB, CA, 1961.

23--Chance Vought XF5U-1 "Flying Flapjack," 1942

24--**D.B. Cooper** bails out of Northwest Airlines 727 with $200,000 ransom, Oregon, 1971.

25--**de Havilland D.H. 98 Mosquito**, 1940.

26--Beech delivers the **last Model 18** after a record 32 continuous years of production, 1969.

27--**First air-to-air rocket**, White Sands, NM, 1951.

28--**First skywriter**--Roscoe Turner, New York City, 1922.

28-29--Bernt Balchen pilots Adm. Richard Byrd in first flight over the **South Pole**, 1929.

30--**First American airplane manufacturing company**--Glenn Curtiss, 1907.

    Maj. Dwight D. Eisenhower is issued a Private Pilot's license at Ft. Lewis, WA, 1939.

DECEMBER

1--Prototype AG-1, the precursor of the **Piper Pawnee**, at Texas A&M, 1950.

2--**Beech Bonanza**, Wichita, KS, 1945.

3--Lafayette Escadrille formed, 1917.

4--Glenn L. Martin dies in Baltimore, MD, 1955.

5--A UC-64 carrying **Maj. Glenn Miller** disappears over the English Channel, 1944.

    Six US Navy planes are the first to disappear mysteriously in the **Bermuda Triangle**, 1945.

December 2, 1945

6--The wooden-winged Heinkel He 162 jet, designed and built in 69 days, makes its first flight, 1944.

7--Surprise Japanese attack on Pearl Harbor, HI signals U.S. entry to World War II.

8--**Bell Model 206**, Ft. Worth, TX, 1962.

9--**Bell X-1** first powered flight, Chalmers "Slick" Goodlin, Muroc, CA, 1946.

    Cessna 208 Caravan, Wichita, KS, 1982.

10--Calbraith P. "Cal" Rodgers arrives in Long Beach, CA after completing the **first transcontinental flight** in 84 days, 1911.

    First man-powered helicopter flight, 6.8 seconds, San Luis Obispo, CA, 1989.

11--Four-engine Sikorsky *Ilya Muromets* crashes on first flight, St. Petersburg, 1913.

12--The first all-metal Junkers flies, 1915.

13--**Piper Aircraft Co.** purchased by M. Stuart Millar, 1986.

14--**Cessna 305** (L-19) flies, 1949.

15--**King Radio Corp.** is formed in Olathe, KS, 1959.

16--UAL DC-8 and TWA Super Constellation collide over New York City, killing 134, 1960.

17--Wilbur and Orville Wright achieve the **first powered, sustained flight** at Kitty Hawk, NC, 1903.

    ☐ **Col. Billy Mitchell** is found guilty of discrediting Army aviation, 1925.

December 17 cont.

    ☐ Boeing, Pratt & Whitney and Chance Vought merge to form **United Aircraft and Transport Corp.**, 1928.

    ☐ **Douglas DC-3** first flight, 1935.

    ☐ The USAAF 509th Composite Group is formed under Col. Paul Tibbets to practice dropping atomic bombs, 1944.

    ☐**Boeing B-47** flies, 1947.

    ☐ The **Wright Flyer** is returned to the U.S. from Britain, where it had resided in a museum for twenty years., 1948.

    ☐ Hughes NOTAR (no rotor) helicopter flies, 1981

18--All-plastic **Windecker Eagle** is certificated, Midland, TX, 1969.

19--Harold Pitcarin's autogiro makes its first successful flight, 1928.

20--Beech 17-R negative-stagger biplane receives Approved Type Certificate, 1932.

21--**General Dynamics F-111**, Ft. Worth, TX, 1964.

22--**Lockheed SR-71** first flight, 1964.

23--"Voyager" lands after seting **unrefueled, nonstop around-the-world record** of 24,986 miles in 216 hours, 4 minutes, 1986.

24--Arado Ar 234B-2s bomb Belgium in the **first operation by jet bombers**, 1944.

25--Raymond Delmotte sets a landplane speed record of 314.33 mph in Caudron 460, Chartres, France, 1934.

26--I.V. Fedorov becomes the **first Russian pilot to break the sound barrier**, 1948.

27--Japanese *Hosho* is first purpose-built aircraft carrier, 1922.

28--National Air Transport Assn. (NATA) is established, 1940.

29--The first Grumman aircraft, the XFF-1, Valley Stream, NY, 1931. **Consolidated B-24 Liberator**, 1939.

December 14-23, 1986

30--First Russian production jet, MiG-15, first flight, 1947. Boeing 747 certificated, 1969.

31--Cessna Aircraft Co. founded, Wichita, KS, 1927. First flight of **Cessna 120**, 1945.

XII. AVIATION CALENDAR

AVIATION CALENDAR

Wed., July 30-Tue., August 5--45th EAA International Fly-In, Oshkosh, WI.

August 6-10--**CBAA Convention and Trade Show/Air Show**, Abbotsford, B.C., Canada

August 14-17--**Taiwan Aerospace Technology Exhibition**, Taipei World Trade Center, Taipei, Taiwan.

August 19-24--**Mosaero '97**, Zhukovsky Airport, Moscow, Russia.

September 6-7--**Mid-Eastern EAA Fly-in**, Marion, OH

September 21-25--**NBAA events, all in Dallas, TX:**

    September 21--NBAA Human Factors in Aircraft Maintenance Workshop

    September 21-22--NBAA Flight Operations Manual Workshop

    September 21-22--NBAA Human Factors Workshop

    September 21-22--NBAA 6th Annual Tax Conference

    September 22--NBAA Inspection Authorization Renewal Certification Review

    September 23-25--**50th NBAA Convention**, Dallas Convention Center

September 23-26--**Inter Airport '97 Frankfurt**, Frankfurt Main Airport, Germany.

September 23-25--**Marine Corps League**, Quantico, VA

September 25-28--**Marine Corps Aviation Association** annual Reunion and Symposium, Sheraton New Bern, Cherry Point, NC

September 25--**NBAA Maintenance Manual** Workshop, Dallas, TX

September 26--**NBAA Small Flight Department Management Workshop**, Dallas, TX

September 26-October 3--**Air Traffic Control Association** 42nd Annual International Technical Program and Exhibits, Sheraton Washington Hotel, Washington, DC

September 30-Oct. 3--**Helitech '97**, Redhlill Aerodrome (near Gatwick), Surrey, UK

October 5-9--**Association of Air Medical Services**, Cincinnati, OH

October 8-12--**Aviation Expo/China '97**, China Intl. Exhibition Center, Beijing, China

October 9-12--**Copperstate EAA Fly-in**, Mesa, AZ

October 10-12--**Southeast EAA Fly-in**, Evergreen, AL

October 10-12--**East Coast EAA Fly-in**, Wilmington, DE

October 13-15--**Association of the U.S. Army**, Washington, DC

October 17-19--**Southwest EAA Fly-in**, Kerrville, TX

October 23-26--**AOPA Expo '97**, Marriott's Orlando World Center, Orlando, FL

November 15-16--**Oregon Air Fair**, Portland Convention Center

November 16-20--**Dubai '97**, Dubai International Airport, Dubai, U.A.E.

December 2-7--**International Maritime & Aerospace Exhibition**, Langkawi, Malaysia

December 10-11--**ATC '97 Asia Pacific Air Traffic Control Exhibition & Conference**, Bali, Indonesia

January 28-30--9th **NBAA Schedulers & Dispatchers Conference**, Anaheim, CA

February 15-17--**Heli-Expo '98**, Anaheim Convention Center, Anaheim, CA

February 24-March 1--**Asian Aerospace '98**, Changi Exhibition Centre, Singapore

March 12-14--**Women in Aviation International Conference**, Denver, CO

April 22-26--**Aviation Africa '98**, Johannesburg, South Africa

May 12-15--**TRANSTEC**, the International Transport Exhibition of Russia, St. Petersburg

May 13-15--**Combination NATA-PAMA** (National Air TransportationAssn.-Professional Aviation Maintenance Assn.) **Annual Meeting**, Kansas City, MO

1998 cont.

May 18-20--**Regional Airline Association Convention**, Minneapolis, MN
May 18-25--**ILA '98**, German Aerospace Industries Assocition, Schonefeld Airport,
    Berlin-Brandenburg, Germany
June 30-July 2--**American Helicopter Society Forum & Exhibition**, Washington, DC
July 15-19--**U.S. Air and Trade Show**, Dayton International Airport, Dayton, OH
July 29-August 4--**46th EAA International Fly-In**, Oshkosh, WI
September 7-13--**Farnborough International '98**, Farnborough, UK
September 24-27--**Marine Corps Aviation Association** annual Reunion and Symposium,
    Worthington Hotel, Fort Worth, TX
October 19-21--**51st NBAA Convention**, Las Vegas, NV
October 23-25--**AOPA Expo '98**, Palm Springs, CA

1999

July 29-August 4--**47th EAA International Fly-In**, Oshkosh, WI.
October 12-14--**52nd NBAA Convention**, Atlanta, GA
December 30--**Pilots beyond age 60** will not be permitted in FAR Part 121 operations
    that were formerly Part 135.

2000

January 1--All aircraft with an MTOW of 75,000 lb. or more that operate to or from US
    airports must meet FAR Part 36, **Stage 3** noise levels.
July 26-August 2--**48th EAA International Fly-In**, Oshkosh, WI.
October 10-12--**53rd NBAA Convention**, New Orleans, LA

2001

October 23-25--**54th NBAA Convention**, Dallas, TX

Information sources

Aerospace: Wichita Perspective, J. Zimmerman, Wichita, KS, 1966
Aircraft Bluebook Price Digest, Overland Park, KS
Aircraft Piston Engines, W. Smith, McGraw-Hill, NY, 1981
Aircraft of WWII, Bill Gunston, Octopus Books Ltd., London, 1980
Aircraft Owners & Pilots Assn., Frederick, MD
Airpac, Inc., Edmond, OK
All the U.S. Air Force Airplanes, Waters, Hippocrene Books, NY, 1982
Aviation & Aerospace Almanac, McGraw-Hill, New York, 1997
American Aircraft of WWII, David Mondey, Temple Press, Feltham, Middlesex, 1982
Aviation International News, Midland Park, NJ
Aviation Week & Space Technology, McGraw-Hill, NY
British Aircraft of WWII, David Mondey, Temple Press, Feltham, Middlesex, 1982
Business & Commercial Aviation, Rye Brook, NY
Cessna Aircraft Co., Wichita, KS
Complete Illustrated Encyclopedia of the World's Aircraft, Chartwell, Secacus, NJ, 1978
Chronicle of Aviation, Chronicle Communications, Ltd., UK, 1992
Confederate Air Force, Midland, TX
Experimental Aircraft Assn., Oshkosh, WI
Federal Aviation Administration, Washington, DC
Flying Magazine, Greenwich, CT
General Aviation Manufacturers Assn., Washington, DC
General Aviation News & Flyer, Tacoma, WA
German War Birds, Kenneth Munson, New Orchard Editions, Poole, Dorset, 1986
The Golden Age of Air Racing, Schmid & Weaver, EAA, Oshkosh, WI, 1991
Information Please Almanac, Houghton Mifflin Co., Boston
Jane's Fighting Aircraft of World War II, Military Press, NY, 1987
Mooney Aircraft Co., Kerrville, TX
National Aeronautic Assn., Washington, DC
The National Air and Space Museum, Bryan & Abrams, Inc., NY, 1979
National Aviation Hall of Fame, Dayton, OH
National Business Aircraft Assn., Washington, DC
1001 Flying Facts & Firsts, Joe Christy, TAB, Blue Ridge Summit, PA 1989
Piston Aero Engines, Bill Gunston, Patrick Stephens, Ltd, 1993
Piper Aircraft, Peperell & Smith, Air-Britain Ltd., Kent, 1987
Piper Aircraft Co., Vero Beach, FL
Quest for Performance, L.K. Loftin, Jr., NASA, 1985
Raytheon Aircraft, Wichita, KS
Reno Air Racing Assn., Reno, NV
Smithsonian Book of Flight, Walter Boyne, Orion Books, NY, 1987
U.S. Bombers, Lloyd Jones, Aero Books, Fallbrook, CA, 1974
U.S. Civil Aircraft, Juptner, Aero Publishers, Fallbrook, CA, 1971
U.S. Department of Transportation, Washington, DC
U.S. Warbirds, Bill Gunston, New Orchard Editions, Dorset, 1985
Vintage Airplane Catalogue, Flightline Press, Baltimore, 1989
Whittle, John Golley, Smithsonian Press, 1987
The World Almanac and Book of Facts, Pharos Books, NY
World Aviation Directory, McGraw-Hill, NY
World Encyclopedia of Civil Aircraft, Enzo Angelucci, Crown Publishers, NY, 1978

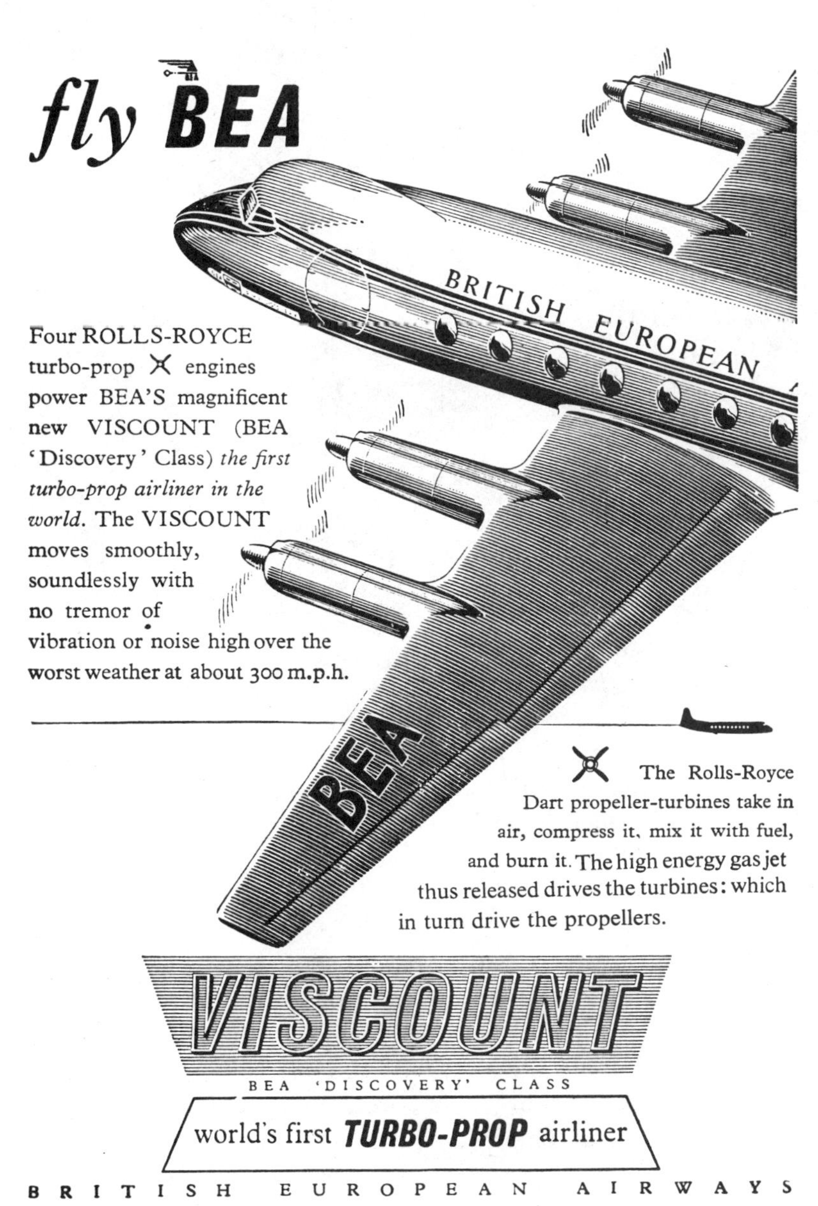
fly BEA
BRITISH EUROPEAN
Four ROLLS-ROYCE turbo-prop engines power BEA'S magnificent new VISCOUNT (BEA 'Discovery' Class) the first turbo-prop airliner in the world. The VISCOUNT moves smoothly, soundlessly with no tremor of vibration or noise high over the worst weather at about 300 m.p.h.
BEA
The Rolls-Royce Dart propeller-turbines take in air, compress it, mix it with fuel, and burn it. The high energy gas jet thus released drives the turbines: which in turn drive the propellers.
VISCOUNT
BEA 'DISCOVERY' CLASS
world's first TURBO-PROP airliner
BRITISH EUROPEAN AIRWAYS

X-1 sets world speed and altitude marks

B-29 mother ship prepares for takeoff from Muroc test facility with the record-breaking Bell X-1 nestled in its bomb bay.

MUROC DRY LAKE, CALIF. --The experimental Bell X-1 smashed the world speed and altitude records this spring as Capt. Charles Yeager attained a speed of 957 mph (Mach 1.45) in the bullet-shaped rocket aircraft and flew to an altitude of 64,000 ft.

The speed record was set on March 26 in the same aircraft in which Yeager became the first person to fly faster than the speed of sound when he reached 700 mph (Mach 1.06) last October.

On a May 26 flight, Yeager broke the world altitude record of 54,046 ft. which had been set by an Italian Caproni before the war.

The aircraft (serial 46-062) is nearly 30 ft. long, has a wingspan of 28 ft., and is powered by a single four-chamber rocket engine designed and built by Reaction Motors which develops 6,000 lb. of thrust. Two separate stainless steel fuel tanks hold 311 gallons of liquid oxygen and 293 gallons of diluted ethyl alcohol which provide a total of five minutes duration at full throttle. The X-1 is taken to its operating altitude by a B-29 which holds the rocket craft in a sling under the fuselage.

Massive Airlift is lifeline for Berlin

BERLIN, WEST GERMANY--Operation Vittles, what might prove to be the world's most ambitious airborne operation, began here June 26 on the orders of Gen. Lucius Clay, military commander of the US Zone of Berlin. For two days, the two million people living in the Western Sector of the city have been effectively under siege after Soviet authorities severed all road, rail and canal links to West Germany.

The cause of the action started with an argument about currency reform among the occupying powers in Berlin, but Western allies characterize the action as a forced attempt to abandon this pocket of democracy behind the "Iron Curtain." Because the three Western zones of the city need thousands of tons of food and fuel, so the USAF is airlifting supplies from Frankfurt and Wiesbaden, and the RAF is flying out of Wunsdorf and Bückeburg.

Transatlantic records

An Air France Constellation makes the first non-stop Paris-New York flight in 16 hours, 1 minute on April 28.

July 20, Sixteen U.S. Air Force Lockheed Shooting Stars completed the first west-east crossing of the Atlantic by jet aircraft, flying from Selfridge Field, Mich. to Stornaway. Scotland in 9 hours, 20 minutes.

Orville Wright dies

DAYTON, OHIO--Orville Wright, who in 1903 was the first man to achieve powered, sustained flight, died here Jan. 30 after a heart attack. He was 76. His older brother, Wilbur, died in 1912, and Orville had sold his interests in Wright Aircraft in 1915 and retired.

Flyer returns to US

WASHINGTON, DC--On the 45th anniversary of the first flight at Kitty Hawk, the historic Wright "Flyer" was returned to the US. Because of an ongoing dispute between the Wright Brothers and Smithsonian secretary Samuel P. Langley whether their airplane or Langley's *Aerodrome* was the first to make a successful flight, the *Flyer* had--at Orville Wright's request--been housed since 1928 in London's Science Museum.

Low-cost 4-place Cessna debuts

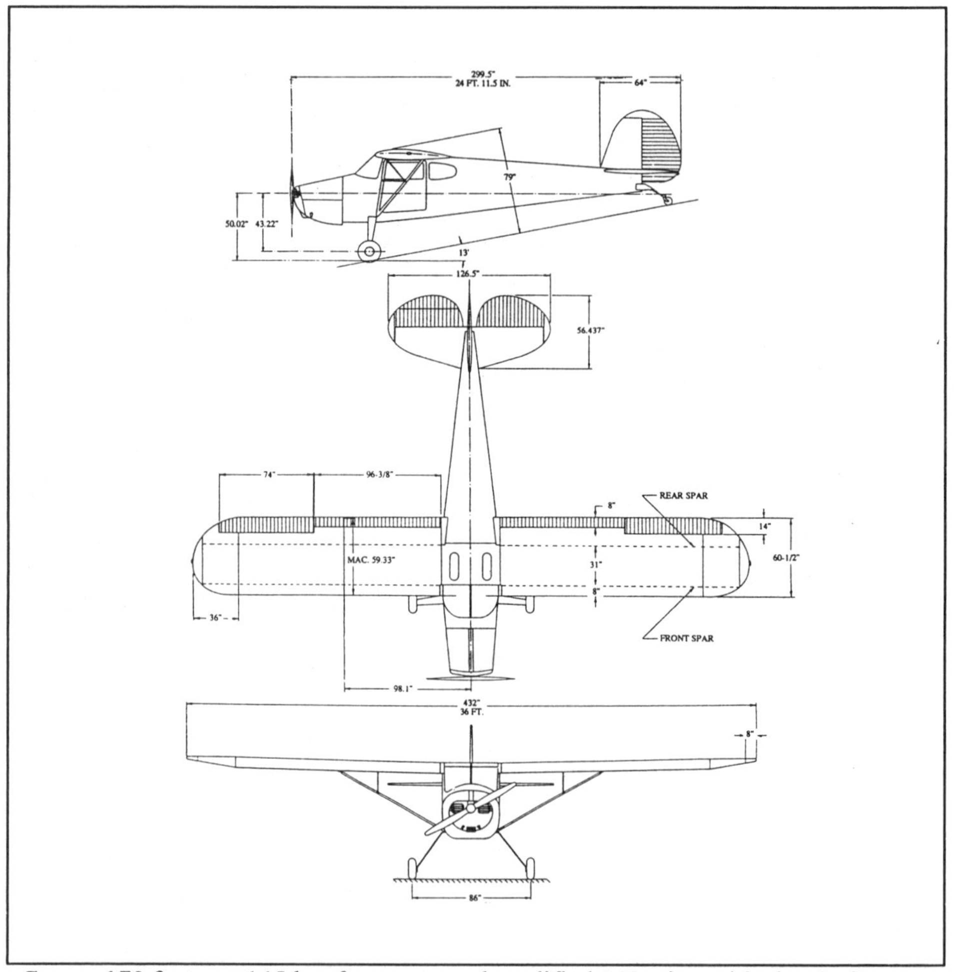

Cessna 170 features 145 hp, four seats and modified 140 wing with simple flaps

WICHITA, KAN.--Cessna Aircraft Co., buoyed by the postwar sales success of their two-place 120 and 140 models, unveiled their new 140 mph four-place Model 170 at a base price of just over $5,400.

With an all-metal fuselage and V-braced, fabric-covered wings, the 170 has a similar look to its predecessors, but its 30-in. longer fuselage and 4-in. wider cabin provides "ample" room for four adults and 120 lbs. of luggage, according to the company.

The 145 hp six-cylinder Continental engine produces cruise speeds well in excess of 120 mph and give the 2,200 lb. aircraft a take off roll of 584 ft. and landing roll of 711 ft., 690 fpm rate of climb, and 15,000-ft. service ceiling.

Jet engine inventor finally gets his due

LONDON--On recommendation of the Royal Commission on Awards to Inventors, Air Commodore Sir Frank Whittle, 41, will be given £100,000 (about $403,000 US) for his role in the invention of the jet engine.

Ironically, Whittle, who began experimenting with the gas turbine concept while serving as an RAF Flight Officer in 1928, had been compelled to do most of his research without the approval, compensation, or even interest of the British government. In spite of the obstacles, the Whittle turbo-jet became operational in 1937, the first British jet flew in May 1941, and the Gloster Meteor became the first Allied operational jet at the close of World War II.

Whittle's health forced him to retire from the RAF in April of this year after 25 years' service.

In addition to the monetary award, Whittle was made a Knight Commander of the Order of the British Empire in July, and was subsequently knighted by King George VI.

Twelve U.S. General Aviation manufacturers delivered a total of 7,037 single-engine aircraft in 1948, worth $32.4 million. The unit total was down 55% and dollar value was 45% less than the previous year.

New auto design utilizes aero engine

CHICAGO, ILL.--The first of Preston Tucker's radical new rear-engined automobiles rolled from a converted B-29 engine plant here in April, powered by a modified Franklin 6-335 aircraft engine which had originally been developed for Bell Helicopter. In addition to the unique powerplant, the Tucker features disc brakes, popout safety windshield, an "uncrushable" passenger compartment, and is slated to sell for about $2,500.

NY Idlewild opens

NEW YORK--What was a marsh 15 miles from Times Square three years ago was dedicated July 31 as New York International Airport, the largest in the US.

Named for the former Idlewild Beach golf course which it displaced, its six runways began operations in July.

Johnson and Mustang top Thompson Trophy

CLEVELAND, OHIO.--Anson Johnson, flying a nearly stock P-51 Mustang, outlasted defending champion Cook Cleland to take first place money in the 1948 Thompson Trophy.

Allison test pilot Charles Brown sat on the pole with a qualifying speed of 418.3 mph in his P-39 *Cobra II*, and turned the fastest lap in the race at 413 mph before a faltering engine slowed him down late in the race and allowed the hard-charging Johnson to catch him.

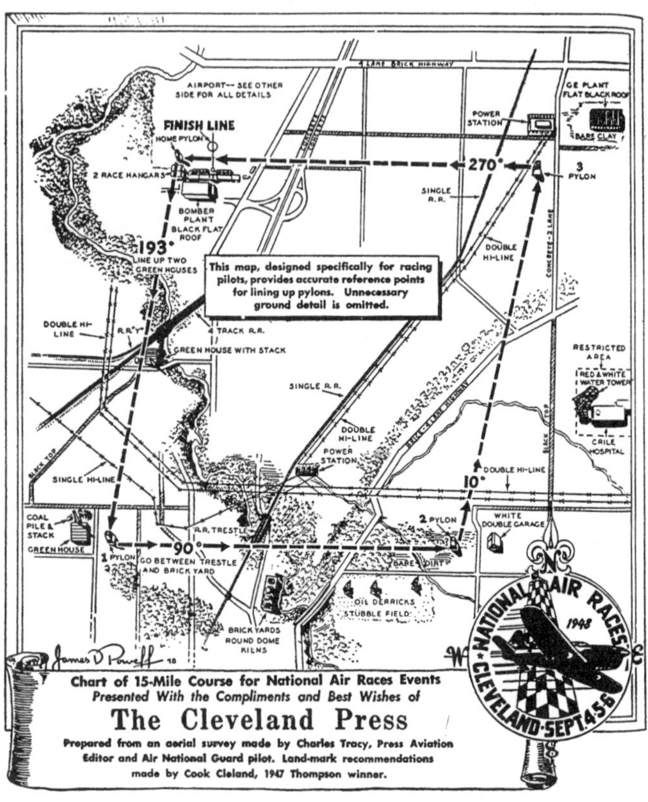

Finally, on lap 19, Brown went out with a vapor lock. The favored team of 1947 winner Cleland and Dick Becker, both aboard 4,000 horsepower F2G-1s, both experienced severe engine misfires which blew the engine cowls off their Corsairs early in the race.

Johnson's average speed of 383.77 mph was nearly 13 mph slower than Cleland's 1947 record, but he was able to best runner-up Bruce Raymond in the race by nearly a full lap and 20 mph average. Third-place Wilson Newhall was the only other airplane remaining of the ten who started the 20-lap, 300-mile event.

Invention may replace radio tubes

Three American physicists--John Bardeen, Walter H. Brattain, and William Shockley--have developed a device for controlling flow of electric current which is being called a point-contact transistor. It is hoped that the invention's simplicity and small size may be useful in reducing size and weight of electronic equipment by eliminating vacuum tubes.

Boeing hopes its YL-15 Scout is what the Army wants as its next liaison aircraft

Boeing's Scout slated for Army

WICHITA, KAN.--The giant plant which produced thousands of B-17 and B-29 bombers during WWII here is gearing down to hopefully produce a 2,111-lb., 104 mph puddle jumper for an upcoming U.S. Army aircraft competition.

The L-15 "Scout" features a two-place gondola fuselage with rear-facing, full-swivel observer's seat, and can be disassembled in 40 minutes and towed behind a Jeep at 45 mph, or can be shipped by air or truck.

With 145 hp Lycoming O-290 power, the Scout can operate off a football field, and spend up to 10 hours in the air.

Other companies vying for the Army contract include Wichita neighbor Cessna Aircraft with its XL-19 "Bird Dog" design, and Consolidated Vultee's folding-wing six-seat XL-13.

Getting into the Senate by helicopter

Soon after Congressman Lyndon Johnson (D-Tex.) attended a demonstration of the new Bell helicopter, one of his aides suggested using the unique aircraft during his campaign to become a U.S. Senator.

Bell president Larry Bell agreed, and furnished Johnson with a loudspeaker-equipped Model 47.

Despite the 100+ degree heat of the Texas summer, Johnson and Bell pilot Joe Mashman flew to as many as 30 towns and cities a day during the campaign. "LBJ," as he is called by friends, trailed early, but came from behind to defeat his opponent by a slim margin of just 87 votes.

Air safety questioned in wake of crashes

The burgeoning business of air travel may be exacting its toll in human life, say some critics. With an abundance of ready pilots and new or converted military transport aircraft debuting almost monthly, the issue of safety seems to be of secondary concern.

1948 turned out to be one of the deadliest so far in commercial aviation history, with 279 persons losing their lives in airliner crashes. Most of the accidents occurred at or near busy airports, and pilot associations throughout the world called for an immediate overhaul of antiquated airways, navigation, and traffic control systems.

Jan. 29--A British South American Airways Tudor 4 disappears on a flight from the Azores to Bermuda with 34 aboard...*March 2*--nineteen die when a Sabena DC-3 (the last one built) crashes at Brussels...*April 5*--A Soviet fighter collides with a BEA Viking near Berlin, killing fifteen...*April 15*--A Pan Am Constellation crashes in Ireland, killing thirty...*June 17*--Nineteen passengers die in the crash of a United DC-6 at Mount Carmel, Penn....*July 4*--an RAF York collides with a SAS DC-6 in England, killing 39 in Britain's worst peacetime disaster...*Aug.1*--a Latécoère 631 flying boat disappears on a flight to West Africa with 52 aboard...*Aug. 29*--Thirty-seven die in the crash of a Northwest Martin 2-0-2 at Winona, Minn...*Oct. 21*--A KLM Constellation crashes near Prestwick, Scotland, killing thirty-four.

Convair testing improved B-36

FT. WORTH, TEXAS--A new version of Convair's giant six-engine B-36 intercontinental bomber took to the skies July 8, powered by six Pratt & Whitney R-4360-41 Wasp Major powerplants uprated to 3,500 horsepower each.

The engines furnish the 227,000-lb. B-36B with a top speed of 381 mph and service ceiling of 42,000 feet. Its range is 10,000 miles.

While the company already has assurance of orders for B-36Bs, it has been working on two additional variants for several years. The "C" version employs six special Wasp VDT (Variable Discharge Turbine) engines capable of producing 4,300 hp with water injection, and the "D" will likely be a model with J-47 turbojets used in conjunction with the piston engines.

Airline offers "coach" fare

In response to non-scheduled operators offering cut-rate deals, Capital Airlines began the first scheduled "coach" service November 4 on service between New York and Chicago, offering a one-third reduction of standard fare.

Capital justified its position by offering the service in off-peak periods,by reducing inflight services and adding high-density seating.

US services set up joint transport unit

WASHINGTON, DC--The air-lift components of the U.S. Air Force and U.S. Navy were combined February 4 with the creation of the Military Air Transport Service(MATS), under the command of USAF Lt. Gen. Laurence Kuter.

The new service will integrate the Air Transport Command and the smaller Navy Air Transport Service into an organization which will fulfill the air transportation demands of all U.S. military services..

Quad-jet fighter may be Curtiss' last

MUROC DRY LAKE, CAL.--The XP-87 Blackhawk, the giant four-jet night fighter that may prove to be the last airplane built by the ailing Curtiss company, made its first flight here March 1.

Curtiss has found itself without the management or design expertise to compete in the new field of high-speed jets, and even though the XP-87 is powered by four 3,000-lb. thrust Westinghouse J34 engines, the airplane apparently lacks necessary speed or maneuverability

1948 Prices (f.o.b. factory)

Aircraft

Aeronca 7AC Champ......................$2395	Hudson Super Six 4-dr.....................$1749
Beech Bonanza V35.......................$9445	Kaiser 4-dr.......................................$2244
Cessna 190....................................$12750	DeSoto Deluxe 4-dr.........................$1825
Luscombe 8A.................................$3195	Dodge Deluxe 4-dr...........................$1718
Navion...$7750	Ford Deluxe 4-dr..............................$1346
Piper PA-11 "90"............................$2595	Lincoln 4-dr......................................$2554
Stinson 108-2.................................$6289	Mercury 4-dr....................................$1660
Swift GC-1B...................................$3995	Nash 4-dr..$1587

Automobiles

	Oldsmobile 66 4-dr..........................$1334
Buick Special 4-dr..........................$1673	Packard Eight 4-dr............................$2249
Cadillac 61 4-dr..............................$2833	Plymouth Deluxe 4-dr.......................$1441
Chevrolet Stylemaster 4-dr..............$1371	Pontiac Torpedo 4-dr........................$1641
Chrysler Royal 4-dr.........................$1545	Studebaker Champion 4-dr...............$1635

Jet prototypes fly

The new Lockheed TF-80C two-seat trainer prototype of the F-80 Shooting Star made its first flight in March. It is powered by a 4,600-lb. General Electric J33-A-35.

In August, the Northrop XF-89 Scorpion took to the air for the first time. Its twin Allison J35-A-35 turbojets are mounted side-by-side in the fuselage beneath the cockpit. Designed as an all-weather interceptor, it has a gross weight of 42,241 lb., a maximum speed of 636 mph, and an initial rate of climb of 8,360 fpm.

Performance evaluation of the XF-89 resulted in the cancellation of 88 Curtiss XP-87 aircraft ordered earlier.

Engine builders crank out new postwar models

With the postwar emergence of new light aircraft designs, major U.S. engine manufacturers have switched their priorities to peacetime projects.

Franklin, who ignored other customers to provide sixes for the overly-optimistic Repbulic Seabee market, has found itself having to survive on helicopter business while Lycoming and Continental developed new designs.

With three fours from 65 to 125 hp and a six at 225 hp, Lycoming expanded this year with a 320 hp gear-drive flat eight, and Continental

opted to upgrade its C-140 six to the C-145 and add a C-90 four.

No new radials have appeared since the war, and the turbine is expected to dominate future military and commercial markets.

Mach 1 achieved in fighter

MUROC, CAL.--George Welch, the North American Aviation test pilot who won fame during World War II as the first American pilot to engage the enemy at Pearl Harbor, achieved another milestone in aviation in April. Flying the prototype XP-86 Sabre jet fighter in a shallow dive over the California desert, Welch slipped past Mach 1, making the first supersonic flight in an aircraft designed for combat.

USAF changes designations

Since the establishment of the United States Air Force as an independent branch of the Armed Forces (Sept. 18, 1947), several measures have been instituted to separate its personnel and inventory from Army and Navy. In the latest, a June 11 regulation has redesignated function or capability classification of Air Force aircraft:

Type	Old	New
Attack	A	B or F
Amphibious	AO	A
Bomber	B	B
Fighter	P	F
Glider	CG	G
Helicopter	R	H
Reconnaissance	F	R
Research	S	X

British break sound barrier

Sept. 6--De Havilland test pilot John Derry has become the first Briton to break the sonic barrier. Diving vertically from 45,000 ft., he lost control of his DH 108, but managed to recover by 23,500 ft., and in the process indicated Mach 1.04 on the airspeed indicator. Geoffrey De Havilland, chief test pilot and eldest son of the company's founder, died in a 1946 attempt at sonic speed in the same model.

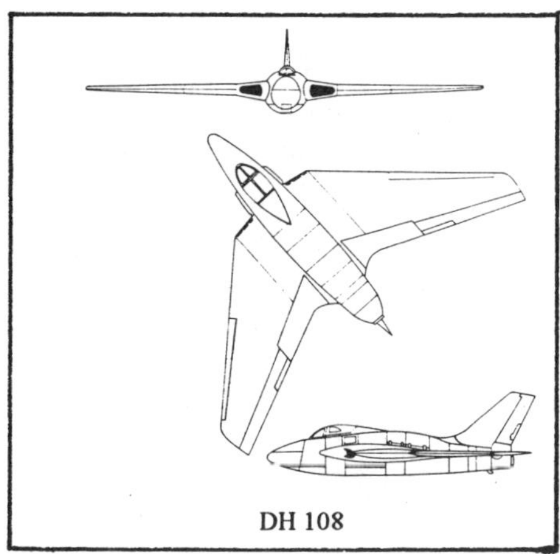

DH 108

An ultra-light airplane

Ken Coward, a California aeronautical engineer, has designed and built what is surely the world's smallest airplane.

Of all-metal construction, the Wee Bee is scaled down so far that the pilot is outdoors--lying prone atop the fuselage. The wing span is 18 feet and the empty weight is 210 pounds. Power is supplied by a 20 hp two-stroke engine from a surplus unmanned target drone.

When they say "Burn off" fog, they really mean it!

BLACKBUSHE AIRPORT, ENGLAND--A newly installed fog dispersal system (FIDO) was activated here for the first time in late November. It comprised a series of gas pipes feeding burners that, when ignited, literally burn away the damp air.

Activated a 4:12 p.m., it enabled a Vickers Viking charter flight to depart for Gibraltar enroute to Accra with urgently needed bank notes for the government.

It is said that FIDO improved visibility from 30 yd. to 600-800 yd.

First purpose-designed turboprop airliner flown

WISLEY, SURREY--The Vickers-Armstrong Type 630 Viscount, the first transport designed specifically for turboprop engines, made its maiden flight here on July 16.

Powered by four Rolls-Royce Dart engines, the 40-passenger transport will cruise at 302 mph at 20,000 ft. The prototype aircraft will be operating on the London-Paris route of BOAC.

In the news...

Postwar inflation keeps prices rising...Congress approves $5.3 billion for Marshall Plan aid to war-torn nations in Europe through the Foreign Assistance Act...the U.S. recognizes new state of Israel on May 14 and admits 205,000 war refugees from Europe.

President Truman orders peace-time draft and desegregation of the country's Armed Forces...The House Unamerican Activities Committee charges former State Department official Alger Hiss with passing secret documents to Whittaker Chambers for transmission to the Soviet Union.

The Kinsey Report *Sexuality in the Human Male* is published...President Truman wins an upset reelection over Republican candidate Thomas E. Dewey of New York despite Progressive and Dixiecrat walkouts from the Democratic Party.

Mohandas K. Gandhi, 78, is assassinated in New Delhi, India...Communists take over the government in Czechoslovakia and form a People's Republic...Afrikaaners come to power in South Africa and pledge a policy of "apartheid"...Korea is divided into two nations, North Korea and South Korea.

The World Council of Churches is formed...a 200-inch reflecting telescope is built at Mount Palomar, Calif...the long-playing record is invented by American Peter Goldmark...German auto, aircraft, engine and tank designer Dr. Ferdinand Porsche builds his first production automobile, the Model 356...heir to the British throne, Prince Charles, is born to Queen Elizabeth II.

Laurence Olivier and *Hamlet* win best actor and best picture Oscars...BeBe Shopp, Miss Minnesota, is named Miss America.

In sports, Bob Mathias wins the Decathlon at the London Olympic Games......Kentucky defeats Baylor for the NCAA Basketball Championship...SMU's Doak Walker is named recipient of the Heisman Trophy, and undefeated Michigan wins both the Rose Bowl and the National Football Championship...The Philadelphia Eagles defeat the Chicago Cardinals 7-0 for the National Football League championship...pro golfer Ben Hogan tops the money winner list with earnings of $36,812 for the year, but Mauri Rose pockets $42,800 for his first place at the Indianapolis 500 with an average speed of 119.814 mph.